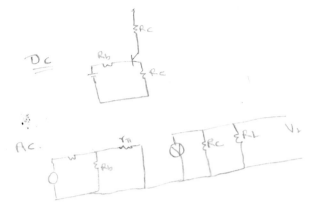

3-31-76

A Sourcebook
of Modern
Transistor Circuits

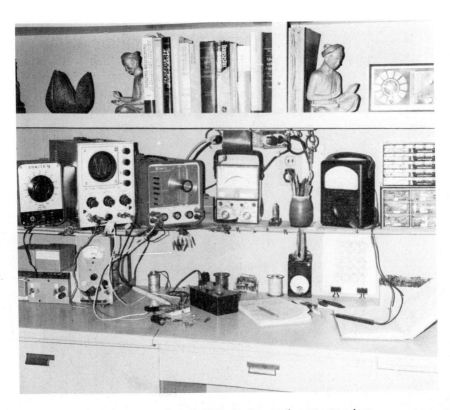

A home electronics laboratory for the experimenter or engineer—the author's. From left to right: a variable-voltage transformer, oscilloscope, oscillator with attenuators, FET volt-ohmyst, ac power jacks, D'Arsonval volt-ohmmeter, and small parts cabinets. Below: power supplies, breadboard circuits, β-tester, reactance chart, and notebook.

A Sourcebook of Modern Transistor Circuits

LAURENCE G. COWLES

Retired,
Formerly Senior Electronic Design Engineer
The Superior Oil Company
Houston, Texas

PRENTICE-HALL, INC. Englewood Cliffs, New Jersey

Library of Congress Cataloging in Publication Data

COWLES, LAURENCE G
   Sourcebook of modern transistor circuits.

   Bibliography: p.
   Includes index.
   1. Transistor circuits. I. Title.
TK7871.9.C688   621.3815'3'0422   75-30748
ISBN 0-13-823419-1

© 1976 by Prentice-Hall, Inc.
Englewood Cliffs, N.J.

All rights reserved. No part of this book
may be reproduced in any form, by mimeograph
or any other means, without permission in
writing from the publisher.

10 9 8 7 6 5 4 3 2 1

Printed in the United States of America

Prentice-Hall International, Inc., *London*
Prentice-Hall of Australia Pty. Limited, *Sidney*
Prentice-Hall of Canada, Ltd., *Toronto*
Prentice-Hall of India Private Limited, *New Delhi*
Prentice-Hall of Japan, Inc., *Tokyo*
Prentice-Hall of Southeast Asia Pte. Ltd., *Singapore*

When a mathematician engaged in investigating physical actions and results has arrived at his conclusions, may not they be expressed in common language as fully, clearly, and definitely as in mathematical formulae? If so, would it not be a great boon to such as I to express them so?—translating them out of their hieroglyphics, that we also might work upon them by experiment.

> Michael Faraday in a letter to James Clerk Maxwell,
> Nov. 13, 1857
>
> *The Selected Correspondence of Michael Faraday*,
> L. Pearce Williams, Cambridge University Press, 1971

Contents

PREFACE   xvii

1

PRACTICAL TRANSISTOR CIRCUIT THEORY   1

    1.1   Iterated Circuits
    1.2   Symbols
    1.3   Feedback
    1.4   The Miller Effect
    1.5   Transistors
    1.6   The transistor gain-impedance relation
    1.7   Ohm's law and dc current-voltage relations
    1.8   Collector power dissipation
    1.9   Transistor power ratings and heat sinks
    1.10   The decibel, or dB
    1.11   The Transistor Current Gain β

2

SINGLE-STAGE AMPLIFIERS   16

    2.1   Single-stage amplifiers
    2.2   Applications using a single transistor

Contents

- 2.3 Threshold-sensitive circuits
- 2.4 Collector feedback amplifiers
- 2.5 The emitter feedback amplifier
- 2.6 A typical emitter feedback stage
- 2.7 High-gain single-stage amplifiers
- 2.8 CE stage with Q-point adjustment for large signal applications
- 2.9 Combined collection and emitter feedback
- 2.10 High gain CE stage with gain control
- 2.11 Universal single-stage amplifier
- 2.12 Oscilloscope preamplifier
- 2.13 Common-collector amplifiers
- 2.14 High input impedance CC stage
- 2.15 CC stage for driving a low-impedance load
- 2.16 CC-relay amplifiers
- 2.17 Complementary emitter followers

3

TWO-STAGE TRANSISTOR AMPLIFIERS   38

- 3.1 Two-stage capacitor-coupled amplifier
- 3.2 Two-stage direct-coupled amplifier with feedback
- 3.3 Direct-coupled amplifiers with series feedback
- 3.4 High-input impedance CE-CE amplifiers with feedback
- 3.5 High-gain, general-purpose, two-stage amplifier
- 3.6 General-purpose line amplifier (audio)
- 3.7 High-gain audio-frequency amplifiers
- 3.8 High-gain emitter-feedback amplifiers
- 3.9 Two-stage combined feedback amplifier
- 3.10 Emitter-coupled two-stage amplifier
- 3.11 CE-CE high-gain pair amplifier
- 3.12 CE-CB cascode pair amplifier
- 3.13 The Darlington CC-CC amplifier
- 3.14 Darlington pair for power loads
- 3.15 CE collector feedback pairs
- 3.16 Phase inverters
- 3.17 Low-impedance phase inverter

4

FIELD-EFFECT TRANSISTOR AMPLIFIERS 59

4.1 RC-coupled FET stages
4.2 Single-stage RC-coupled FET amplifiers
4.3 High-gain FET stages
4.4 FET-transistor amplifiers
4.5 Oscilloscope preamplifier
4.6 High-impedance remote pickup
4.7 Low-impedance remote pickup
4.8 Low-power dc preamplifier
4.9 Two and three-stage FET transistor amplifiers
4.10 FET-transistor pair amplifier for low-impedance loads
4.11 Temperature compensated FET transistor amplifier
4.12 Broadband direct-coupled amplifier
4.13 FET with bootstrapped input
4.14 FET transistor dc amplifier
4.15 FET millivoltmeter

5

MOS FET AMPLIFIERS 77

5.1 Biasing single-stage MOS amplifiers
5.2 Two-stage resistance-coupled MOS-FET amplifiers
5.3 A wide-band MOS amplifier
5.4 High-gain three-stage MOS-FET amplifiers
5.5 High-gain low-noise low-drift dc amplifiers
5.6 MOS FET electrometer or picoammeter
5.7 MOS FET high-frequency amplifiers
5.8 MOS FET's in RF and UHF amplifiers

6

POWER SUPPLIES AND REGULATORS 87

6.1 Ac power supplies
6.2 Half-wave rectifiers

- 6.3 Bridge rectifiers
- 6.4 Voltage doublers
- 6.5 Regulated voltage doublers
- 6.6 Regulated power supply for integrated circuits
- 6.7 Typical low-power 30-V dc power supply
- 6.8 Line-operated power supply for a 2-W amplifier
- 6.9 Zener-diode regulators
- 6.10 Regulator for low-power 9-V equipment with a 12-V battery
- 6.11 Transistors as Zener diodes
- 6.12 Low-impedance regulator
- 6.13 Shunt transistor regulators
- 6.14 Series transistor regulators
- 6.15 Low-power voltage regulator
- 6.16 One-ampere voltage regulators
- 6.17 Current limiters
- 6.18 Fifteen-ampere short-circuit-protected voltage regulator
- 6.19 Capacitance multipliers
- 6.20 Replacing filter capacitors with Zener diodes
- 6.21 Battery power supplies

7

LOW-NOISE AMPLIFIERS AND PREAMPLIFIERS 112

- 7.1 Noise in circuit components
- 7.2 Low-noise audio amplifiers
- 7.3 Low-noise collector feedback amplifier
- 7.4 Low-impedance, low-noise amplifiers
- 7.5 JFET high-impedance low-noise amplifiers
- 7.6 Nonblocking amplifier
- 7.7 Low-noise op-amp preamplifiers
- 7.8 Phonograph preamplifier
- 7.9 Transistor phonograph preamplifier

8

AUDIO POWER AMPLIFIERS 125

- 8.1 Low-power class-A amplifiers
- 8.2 Pocket-radio audio amplifiers

Contents　xi

8.3　Low-power line-to-line class-A amplifiers
8.4　Single-side class-A 2-W amplifier
8.5　Line-operated 1-W audio amplifiers
8.6　Transformer-coupled 4-W amplifiers
8.7　Power pairs
8.8　Push-pull 2-W amplifiers
8.9　Class-B 2-W amplifier and driver stage
8.10　Single-ended class-B 10-W amplifier
8.11　Complementary-symmetry amplifiers
8.12　Complementary-symmetry op-amp power amplifiers
8.13　Low-power complementary-symmetry amplifiers
8.14　A 20-W quasi-complementary-symmetry amplifier

9

LINEAR INTEGRATED CIRCUITS　144

9.1　Basic IC circuits
9.2　Signal addition and subtraction
9.3　Instrument amplifier
9.4　IC integrators
9.5　Analog computers
9.6　IC amplifier with a single power supply
9.7　IC amplifier for low-impedance loads
9.8　RC coupled IC amplifier
9.9　Bridge amplifiers
9.10　High-accuracy bridge amplifier
9.11　Resistance-indicating amplifier
9.12　IC feedback switch

10

FILTERS　160

10.1　Resistance-capacitance filters
10.2　Amplifier interstage coupling capacitors
10.3　Active filters
10.4　Miller-feedback filters
10.5　Practical low-pass filter
10.6　An IC low-pass filter
10.7　Active filters (higher order)
10.8　Filter types and nomenclature

xii Contents

    10.9   Active high-pass filters
    10.10 Active low-pass filters
    10.11 CE stage filters
    10.12 Butterworth filters
    10.13 Butterworth 3-pole filters with equal capacitors
    10.14 Chebyshev filters
    10.15 Bessel filters
    10.16 Op amps in active filters
    10.17 Active filters using ICs
    10.18 High-pass filters
    10.19 Band-pass filters
    10.20 Band-pass filters using op amps
    10.21 Band-reject filters
    10.22 Active band-reject filters

11

TUNED AMPLIFIERS   195

    11.1   Tuned circuits and the Q-factor
    11.2   Tuned audio-frequency amplifiers
    11.3   Twin-T tuned amplifiers
    11.4   Narrow-band high-Q tuned amplifiers
    11.5   Low-impedance tuned CB-CE amplifiers
    11.6   Tuned-amplifier gain limitations
    11.7   Tuned IF amplifiers
    11.8   Tuned amplifier example
    11.9   UHF tuned amplifiers
    11.10 JFET tuned RF amplifiers
    11.11 JFET UHF amplifiers

12

VIDEO AMPLIFIERS   209

    12.1 Video CB amplifier
    12.2 Video CC amplifier
    12.3 Video line-to-line amplifiers
    12.4 Video iterated stage
    12.5 Single-stage CRT video driver
    12.6 Direct-coupled high-frequency amplifiers

Contents   xiii

12.7 High-impedance video amplifiers
12.8 CE-CE medium-frequency amplifier
12.9 High-voltage video amplifiers
12.10 Integrated-circuit video amplifiers
12.11 MOS-FET video amplifier
12.12 Compensating techniques

13

DIODE CIRCUITS AND APPLICATIONS   **225**

13.1 Diode switches
13.2 Diode switches in dc power supplies
13.3 Spark and noise suppressors
13.4 Diode meter protection
13.5 Preventing transistor breakdown
13.6 Diode switches in low-level circuits
13.7 Diode steering circuits
13.8 Diode modulators and choppers
13.9 Diode AM modulator
13.10 Diode detectors
13.11 UHF diode voltmeter
13.12 Diode mixers
13.13 Varactor diode tuning
13.14 Diode clipping circuits
13.15 Base clipper (noise suppressor)
13.16 Low-level limiter and noise suppressor
13.17 Ignition noise suppressors
13.18 Diode clamp
13.19 Diode gate

14

OSCILLATORS AND INVERTERS   **250**

14.1 Sinusiodal oscillators
14.2 Resonant circuit oscillators
14.3 Tuned emitter and tuned collector oscillators
14.4 Tuned-circuit power oscillators
14.5 Crystal oscillators
14.6 Phase-shift oscillators
14.7 Stepped low-frequency phase-shift oscillator

xiv   Contents

14.8   Twin-T R-C oscillators
14.9   Wien bridge oscillator
14.10 Nonsinusoidal oscillators
14.11 Multivibrators
14.12 Ringing converter or flyback oscillator
14.13 Blocking oscillators
14.14 Variable-frequency pulse oscillator
14.15 Inverters
14.16 Converters
14.17 High-voltage converters
14.18 UHF oscillators
14.19 Parasitic oscillations and feedback instability

15

SPECIAL-PURPOSE CIRCUITS—AF AND AGC 280

15.1   An inexpensive hearing aid or intercom
15.2   Tone controls
15.3   Feedback tone control
15.4   Low-distortion AF mixer
15.5   A 7-W high-fidelity amplifier
15.6   Power amplifier with IC driver
15.7   Audio crossover network
15.8   Automatic gain-control amplifiers
15.9   Gain-control circuits
15.10 Audio AGC amplifiers
15.11 Instrument AGC amplifiers
15.12 Radio-frequency AGC amplifiers
15.13 IC amplifiers with AGC

16

TRANSISTOR SWITCHING CIRCUITS 302

16.1   Transistor switches and relays
16.2   Switching amplifiers
16.3   All-purpose latch-up switch
16.4   Transistor SCR equivalent
16.5   Transistor breakdown pair
16.6   Temperature- or light-operated switch

Contents xv

16.7   A burglar and fire alarm
16.8   Flashlight-operated TV sound switch
16.9   Switching inverters
16.10 Flip-flops
16.11 Multivibrators
16.12 Monostable multivibrators
16.13 Schmitt triggers
16.14 Transistor and FET gates
16.15 MOS-FET choppers

17

LABORATORY INSTRUMENTS AND METHODS   321

17.1   Laboratory instruments
17.2   Laboratory techniques
17.3   Volt-ohmmeters
17.4   Signal attenuators
17.5   Breadboards
17.6   Transistor testing
17.7   Transistor current-gain meters
17.8   Laboratory power supplies
17.9   Auxilliary power supply
17.10 Voltage regulators and current limiters
17.11 Inexpensive shop meter
17.12 High-frequency RF voltmeters
17.13 Linear AC voltmeter
17.14 Laboratory voltage standards

ANNOTATED BIBLIOGRAPHY   343

APPENDIX   345

Transistors
Junction field-effect transistors
Insulated-gate field-effect transistors
Miscellaneous semiconductor devices
Zener diodes—400 mW and 1W
Signal and rectifier diodes
Semiconductor interchangeability
Composition resistor values

Standard resistance multiples
RMA resistor color code
Color code examples
Decibel formulas and tables
A practical decibel table
Frequency classifications
Thermal resistances
Maximum recomended temperatures
Reactance-Frequency Chart

INDEX   **353**

Preface

The existing number and variety of semiconductor circuits have created the need for an up-to-date handbook of practical transistor and integrated circuits. This book, a product of my own needs for reliable circuits, offers both experimenters and designers a graded series of selected and carefully designed transistor circuits. The circuits, shown with component values and readily available semiconductors, may be used either for design or for study to obtain a working knowledge of practical circuits without an encumbering theory and mathematics. Circuits of equipment that may be purchased in manufactured or in kit form are not included and not needed in a sourcebook.

I have tried to describe mainly those amplifier, diode, and switching circuits that are used by experienced designers. Each circuit is a complete package that may be used alone or combined with others to achieve known performance characteristics. To make the book useful as a design manual I have included charts and tabulated data to help a designer change a circuit in order that it may serve many different purposes. An annotated bibliography is included for persons needing supplementary information. The references are selected carefully to supply working circuits and to serve practical needs without requiring resources beyond the reach of a small home library.

Laurence G. Cowles

List of Symbols

C_I	Input capacitance
C_M	Miller effect capacitance
C_N	Neutralizing capacitance
C_{OB}	Collector-to-base capacitance
e_I	AC input voltage
e_O	AC output voltage
e_p	Peak voltage or peak-to-peak voltage
e_S	AC signal voltage or generator voltage
f_β	Beta cutoff frequency
f_c	Cutoff frequency (half power, or 3 db)
f_h	High frequency cutoff
f_l	Low frequency cutoff
f_T	Current gain-bandwidth product
G_i	Current gain
G_v	Voltage gain
G'_v	Voltage gain with feedback
i_b	AC base current
I_B	DC base current
i_c	AC collector current
I_C	DC collector current
I_{DSS}	Zero bias drain current
i_e	AC emitter current

List of Symbols

I_E	DC emitter current
I_I	AC input current
I_L	AC load current
i_o	AC output current
P_O	AC output power
R_A	Bias resistor (usually adjustable)
R_B	Base resistor
R_C	Collector resistor
R_E	Emitter resistor
R_f	Feedback resistor
R_I	Input resistance
R_L	Load resistance
R_S	Generator or source resistor
S	Usually R_B/R_E or R_f/R_L; approximately the dc current gain
V_B	DC base voltage (to ground)
V_{BB}	DC base supply voltage
V_C	DC collector voltage (to ground)
V_{CC}	DC collector supply voltage
V_D	DC drain voltage
V_{DD}	DC drain supply voltage
V_E	DC emitter voltage (to ground)
V_{GS}	DC gate-to-source voltage
V_P	FET pinchoff voltage
V_R	DC regulated voltage
V_Z	DC zener diode voltage
α	CB short-circuit current gain, i.e., $-h_{fb}$; approximately 1
β	CE short-circuit current gain, h_{fe}; approximately 50
ω	Frequency in radians per second ($2\pi f$)
Ω	Ohms

Abbreviations

A	Ampere
B	Base
C	Collector
CB	Common base
CC	Common collector
CE	Common emitter
D	Drain
dB	Decibel (see Appendix)
dBm	Decibel referred to 1 mW
E	Emitter
G	Gate
GHz	Gigahertz
Hz	Hertz
IC	Integrated circuit
kHz	Kilohertz
kΩ	Kilohm

mA	Milliampere
mH	Millihenry
mV	Millivolt
pF	Picofarad
p-p	Peak-to-peak
Q	Factor of merit
Q-point	Quiescent point
S	Source (FET)
V	Volts
μA	Microampere
μF	Microfarad
μH	Microhenry
Ω	Ohms

A Sourcebook
of Modern
Transistor Circuits

1 Practical Transistor Circuit Theory

Transistor circuits offer the experimenter many challenging opportunities to construct useful and interesting electronic devices. With a collection of circuits one can build audio and public address systems, communication and ham radio equipment, alarms, metal finders, model controls, electronic measurement and test instruments, and many other devices.

This book brings designers and experimenters a collection of compatible circuits that may be combined to produce their own electronic designs. The design process is simplified because the circuits and their performance characteristics are given without the encumbering theory of textbooks. Occasionally, the reader may wish to calculate the power rating required for a resistor, the voltage gain of a stage, or the cutoff frequencies of an amplifier. These calculations are easy enough, but the circuits in this book can be understood and used without struggling with theoretical relations and complicated equations.

More than likely the reader has already examined some of the circuits in this book and is eager to see them operate. Chapter 17 offers suggestions for constructing experimental circuits, describes easily built test equipment, and shows how to assemble a small but adequate electronic laboratory. In this chapter we define and explain terms like *decibel*, *cutoff frequency*, β (*beta*), and *feedback*. With only the simplest mathematical relations we review **Ohm's law**, **dc power** calculations, and the ***transistor gain-impedance relation***, which is useful in describing transistor amplifiers. The reader may skip most of the first chapter and return later when better prepared to use the introductory material, but one should at least understand why so many circuits are presented as *iterated circuits* with equal source and load impedances.

2 Practical Transistor Circuit Theory

1.1 ITERATED CIRCUITS

Many circuits in this book are iterated circuits that have identical source and load impedances, so they may be connected one to another or in a continuing series of identical units. As illustrated in Fig. 1.1, most of the circuits have a

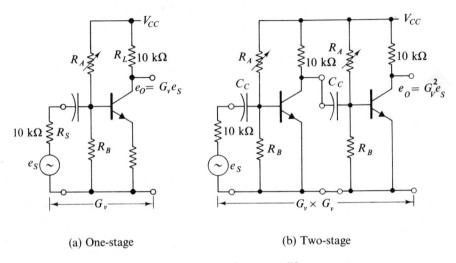

(a) One-stage   (b) Two-stage

Figure 1.1. Iterated stage amplifiers

10-kΩ source impedance in series with a signal voltage e_S, and a 10-kΩ collector load resistor with an open-circuit output voltage e_o. If several iterated stages that are designed for the same impedance level are coupled in series and the voltage gain of each stage is known, the voltage gain of 2 stages is the product of the iterated stage gains, and the voltage gain of n identical stages in series is the gain of 1 stage raised to the nth power. The advantage to the experimenter of working with iterated stages is that one well-designed stage may be coupled to another because the impedances are known to be correct for an optimum energy transfer. Thus, you may combine many of the circuits in this book without making extensive changes. Except when a circuit is intended to couple a low-impedance source to a high-impedance load, and vice versa, the circuits are designed with a 10-kΩ collector load resistor and a 10-kΩ signal-source impedance.

The use of 10 kΩ as the source and load resistor for the examples of iterated circuits is a somewhat arbitrary choice that for the most part comes from the need to use small signal circuits designed to consume relatively low power and to have resistor values that may be made either larger or smaller by a factor of 10. In this way the experimenter may change the impedance level of most circuits either up or down, knowing that the new circuit should operate without either overloading the semiconductors or reducing the current levels enough to impair the performance characteristics markedly.

Iterated stages with the 10-kΩ impedance level are approximately the optimum choice for audio applications. For applications at higher frequencies the impedance level should be reduced to 1 kΩ, or even to as low as 100 Ω. The impedance level of the 10-kΩ circuits in this book may be changed, provided all resistors are reduced (or increased) by a common factor. If the impedances are reduced because the operating frequency range is higher, it may not be necessary to increase the capacitor values, and in some applications the capacitors may be reduced by a common factor.

When the operating frequency range is not changed, the impedance values shown in the circuits may often be increased or decreased by a factor of 10 without a noticeable effect on the circuit performance, provided the capacitors are increased when the resistances are decreased, and vice versa. However, because the upper-cutoff frequency of most circuits depends on the relatively small collector-to-base capacitance of the transistor and on the Miller effect, an increase of the resistor values usually produces a corresponding decrease of the upper-cutoff frequency, while decreasing the resistors of a circuit may increase the upper-cutoff frequency. On the other hand, the impedance values used in power circuits cannot usually be changed easily from the values given in a circuit.

When a signal source has an internal impedance that is several times larger than 10 kΩ, a Darlington impedance step-down stage may be used as an input stage to couple the high-impedance source to a 10-kΩ iterated stage. Similarly, a Darlington stage may be used to couple a 10-kΩ iterated stage to a low-impedance load. A low-impedance source is most efficiently coupled to a higher-impedance input by means of a transformer. Similarly, power is conserved by coupling an amplifier to a low-impedance load by means of a step-down transformer. In power applications the output transformer usually does not match impedances but is designed to present an impedance that makes the output stage deliver a maximum amount of undistorted output power.

In many circuits a 2-to-1 reduction of the source or load impedance does not greatly affect the circuit performance except to increase or reduce the gain, as the case may be. The fact that most circuit examples have significant feedback usually makes possible a small change of any single component without a marked change of the gain or frequency characteristics. However, if the element changed reduces the feedback, a circuit may not perform as well with changes of the supply voltage or the ambient temperature. Except for small adjustments, the experimenter is urged to examine carefully the effect of changes he may make in circuits.

1.2 SYMBOLS

The descriptions in this book are simplified by use of a uniform set of symbols to designate the components and performance characteristics of circuits. For convenient reference, a list of the symbols and their names is shown on page xix in the front of the book. The most frequently used symbols are shown in Fig. 1.2, which is the circuit of an emitter feedback stage. The

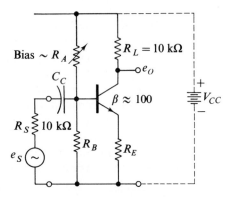

Figure 1.2. Circuit symbols

amplifier has a collector load resistor R_L, an emitter resistor R_E, and a base resistor R_B. The transistors generally shown are *npn* silicon transistors because they are somewhat more readily available than *pnp* transistors. In most of the low-frequency circuits almost any general-purpose, or audio-frequency, transistor may be substituted for the transistor shown, provided the transistor current gain β is at least 50, and preferably over 100. While *npn* transistors and a positive collector supply voltage V_{CC} are usually indicated, the experimenter may substitute complementary *pnp* transistors and use a negative supply in most circuits. A few of the circuits use FETs or MOS FETs that are not readily available in complementary types.

Most circuits in the book may be made operative by adjusting a single bias resistor, designated by R_A with an arrow through the resistor. The value given for the bias resistor is a value the experimenter may expect to find satisfactory. However, the bias required in a transistor stage usually varies with the supply voltage and other factors, so changing the bias resistor may produce a better quiescent operating point (*Q*-point). The optimum collector voltage V_C is usually indicated on a circuit diagram. In linear amplifiers the best collector *Q*-point for the output stage has the collector voltage one-half of the supply voltage, unless otherwise indicated.

The capacitors shown in the circuits are sized to give the amplifiers a flat, or constant, frequency response throughout the audio-frequency range, 30 Hz to 20 kHz. The capacitors shown in the figures may be expected to put the -3 dB low-frequency cutoff at the frequency $f_l = 30$ Hz, and the high-frequency cutoff f_h is in excess of 20 kHz, unless otherwise indicated. The experimenter may lower the low-frequency cutoff by increasing the capacitors, and the high-frequency cutoff may be lowered by adding a shunt capacitor across one of the collector load resistors. The high-frequency cutoff is usually increased only by decreasing the resistors in a circuit.

The signal source is generally designated as a constant voltage e_S in series with an equivalent 10-kΩ source impedance R_S. The iterated voltage gain of an amplifier without feedback is indicated by G_v, the ratio of the ac output voltage e_O to the ac source voltage e_S. If the amplifier has feedback, the voltage gain e_O/e_S is indicated by G'_v. The prime symbol is used to call attention to the fact that the value given exists when the amplifier has feedback. If the voltage gain is given with the feedback removed or bypassed, the prime is omitted. Occasionally the voltage gain is measured from the input base to the output collector, and the voltage ratio is shown as e_O/e_I with the input voltage e_I shown in the circuit diagram.

1.3 FEEDBACK (REFS. 1, 2, 9)

A transistor is such a variable circuit element that a satisfactory amplifier cannot be designed without some form of feedback. Feedback reduces the effect of temperature and of a different transistor on the amplifier Q-points and on the overall voltage gain. With feedback the amplifier performance is determined mainly by the feedback resistors and is relatively independent of transistor characteristics, provided the transistors are in new condition.

Feedback exists when the amplifier input signal is obtained by comparing the output signal with the input signal. With negative feedback the input to the amplifying section of the amplifier is the difference between the input and output signals, and the amplified signal tends to make the output more like the desired input signal. Feedback reduces the gain of an amplifier and more stages are required, but the advantages far outweigh the cost of a few additional stages. The effectiveness of feedback in improving an amplifier is approximately proportional to the gain reduction, and the cost of the improved performance is the reduced gain. A high quality of performance dictates low stage gain and makes the design expensive. However, a satisfactory compromise is usually secured by reducing the stage gain to one-fourth the transistor current gain β, and the stage is said to have **significant** feedback.

Most of the feedback circuits shown in this book should be stable as long as the output leads of the amplifier or stages are well separated or shielded in high-gain circuits. If an amplifier is occasionally found to be oscillating, a small capacitor, perhaps 100 to 500 pF, connected from a collector to ground may be found adequate to prevent oscillation. In stubborn cases the collector supply may need to be bypassed by am 0.1n μF capacitor connected near an input stage.

Feedback may be applied in series with the input signal or in shunt with the signal. Series feedback tends to increase the input impedance, whereas shunt feedback reduces the input impedance. A transistor stage with an unbypassed emitter resistor has series feedback, and the input impedance of the stage is approximately the resistance of the base resistor R_B. We may refer to R_B as the *input impedance*. If the bias resistor is less than 10 times the base resistor, the input impedance is approximately the resistance equivalent of R_A in parallel with R_B. At high frequencies the input of an amplifier has a capacitance component that shunts the input and reduces the gain. The input capacitance depends on the collector-to-base feedback capacitance and is discussed separately as the *Miller effect*.

Perhaps the most important use of feedback in transistor amplifiers is in making the current gain of a stage relatively constant instead of changing with the transistor current gain β. When the current gain of a stage is stabilized by feedback, we call the current gain S, as shown in Fig. 1.3. Thus, the current gain of a stage with significant feedback is found by recognizing the feedback resistors and calculating the resistance ratio S. The current gain of a stage with series feedback is approximately the ratio of the base resistor R_B divided by the emitter resistor R_E, or $S = R_B/R_E$. The S-factor of a stage with shunt feedback is the ratio of the collector-to-base resistor R_f divided by the load resistor R_L, or $S = R_f/R_L$. The input impedance of a stage with shunt feedback is R_f/G'_v. With series feedback the transistor input impedance is high and the stage input impedance is R_B.

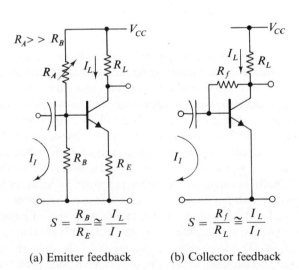

(a) Emitter feedback   (b) Collector feedback

Figure 1.3. S-factor (current gain) definitions

1.4 THE MILLER EFFECT (REFS. 1, 9)

The high-frequency cutoff of an amplifier is often produced by the collector-to-base (or gate-to-drain) feedback capacitance in the first stage, as illustrated by Fig. 1.4. Because the collector ac voltage is G_v times the signal input volt-

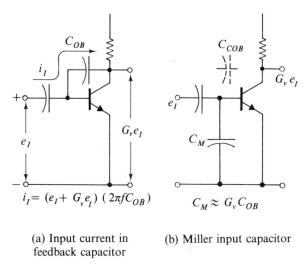

(a) Input current in feedback capacitor

(b) Miller input capacitor

Figure 1.4. Miller effect input capacitance

age, the signal input current through the feedback capacitance is increased as if the stage input capacitance is G_v times the feedback capacitance. Thus, the apparent increase of the input capacitance by the amount $G_v(C_f)$ is called the **Miller input capacitance**. With a stage gain of 10 or more the input capacitance is mainly the Miller feedback capacitance and the effect of the base-to-emitter, or the gate-to-source, input capacitance may be neglected.

Because a relatively small feedback capacitance is made G_v times more effective in reducing the high-frequency response of an amplifier, the experimenter should try to minimize the feedback capacitance, especially in high-impedance, high-gain stages. When the high-frequency cutoff of an amplifier is limited by the Miller effect, the cutoff frequency f_h may be increased by reducing the input or source impedance, by reducing the feedback capacitance, and by reducing the voltage gain. The feedback capacitance C_{OB} of a transistor is usually 1 pF for RF types and 10 pF for low-power audio types. Thus, with a voltage gain of 30 the input capacitance may exceed 300 pF, which makes the shunt input impedance less than 25 kΩ at 25 kHz. With a 10-kΩ source impedance the high-frequency cutoff is just above the audio range. With a source impedance greater than 10 kΩ, an amplifier may not be satisfactory for high-fidelity audio applications. The gate-to-drain capacitance of an audio-type FET is approximately 2 pF, and a voltage gain of 15 may be expected. Thus, the equivalent signal-generator impedance seen by the gate should not exceed 100 kΩ in audio-frequency applications of FETs.

1.5 TRANSISTORS

The transistors shown in most circuits in this book are general-purpose silicon *npn* audio types. In a few examples an RF or a power transistor is indicated. For most purposes the reader may substitute any transistor at hand or more readily available as long as the substitution has the characteristics of a typical audio transistor. The transistor should have a current gain β that is between 50 and 200, a collector-emitter breakdown voltage BV_{CEO} of 30 to 50 V, and a power dissipation rating of 200 mW or more. In circuits showing a 2N1711 the metal case TO-5 device is indicated when a collector power rating of at least 500 mW is needed. In most circuits any good transistor may be made to operate satisfactorily by changing only the bias resistor.

Circuits for RF applications need RF transistors with a gain-bandwidth f_T rating comparable with that of the device shown in a circuit. The f_T rating of the plastic audio and the 2N1711 transistor is 50 MHz or better. Substitutions for power transistors are generally more difficult because the demands on power transistors are exacting and many devices are inferior, even though more expensive. The power transistors shown in the circuits are carefully selected and should be readily available. If a substitution must be made, the transistor should have a similar voltage breakdown rating and have a comparable minimum β at a specified collector current. Use of a transistor with a higher current rating than that of the original device may result in an inferior performance. The power dissipation rating of a substitution device should be equivalent or a little better.

The more readily available and inexpensive FETs usually have inferior characteristics that make them poor substitutes for a good quality device. The FETs used in a resistance-coupled audio amplifier should have a g_m of 1000 μmho with less than 1 mA drain current. A device with 500 μmho and less than 1 mA is better than a device with 1000 μmho and 2 mA or higher drain current. For RF and switching applications a high g_m is required, and a drain current exceeding 2 mA is acceptable. A low gate-to-drain capacitance and a high power dissipation rating may be required of a substitution device in RF circuits.

The silicon diodes are generally not specified by a 1N number. If a diode type is not indicated, almost any 1 A, 200 V, silicon diode may be used. The diodes used as RF detectors and in similar high-frequency applications should be signal-type diodes intended for high-frequency service. A rectifier-type diode has a long charge-storage time and too high a shunt capacitance to be used in place of a signal-type diode. Inexpensive hot-carrier diodes may be used to replace almost any high-frequency diode shown in circuits published before 1970. Hot-carrier diodes usually give improved performance and greater reliability.

1.6 THE TRANSISTOR GAIN-IMPEDANCE RELATION (REF. 1)

The gains and impedances of a transistor are closely related, and their relationship can be expressed in a simple formula. Once understood, this formula makes it possible for one to see at a glance just how a given amplifier can be expected to operate. The formula applies to almost any transistor amplifier, including amplifiers with feedback, so that it is of great value and utility. We will call this formula the *transistor gain-impedance relation* (TG-IR) and use the word *transistor* as a reminder that the relation is not useful for circuits that include vacuum tubes or field-effect devices.

The derivation of the transistor gain-impedance relation is simple. The amplifier represented in Fig. 1.5 has a pair of input and a pair of output

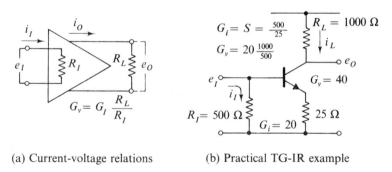

(a) Current-voltage relations   (b) Practical TG-IR example

Figure 1.5. Transistor gain-impedance relation (TG-IR)

terminals, respectively. The ac signal input voltage is designated by e_I and the ac signal output voltage by e_O. We represent the input impedance of the amplifier by the resistor R_I and the output load resistor by R_L. Also, we designate the input current by i_I and the output current by i_O. By application of Ohm's law we know that the input voltage is equal to the product of the input current and the input resistance:

$$e_I = i_I R_I \qquad (1.1)$$

The output voltage is likewise the product of the output current and the load resistance:

$$e_o = i_o R_L \tag{1.2}$$

Dividing Eq. (1.2) by Eq. (1.1), we have:

$$\frac{e_o}{e_I} = \frac{i_o R_L}{i_I R_I} \tag{1.3}$$

Now, we recognize the voltage ratio e_o/e_I as the voltage gain of the amplifier and the current ratio i_o/i_I as the current gain of the amplifier. Calling the voltage gain G_v and the current gain G_i, we can write Eq. (1.3) in the form:

$$G_v = G_i \frac{R_L}{R_I} \tag{1.4}$$

Equation (1.4) is the transistor gain-impedance relation, which we call the TG-IR for short.

This relation tells us at once that if we wish to know the voltage gain of a transistor stage and we know the current gain, we must also know both the load resistance and the input impedance of the stage. Usually we can identify the load impedance R_L. The current gain can be determined by inspecting the circuit. The input impedance of a transistor or of an amplifier can be easily determined by applying a few simple rules. Any of the standard gain formulas require that one be able to identify and find these same quantities. The TG-IR requires nothing new and unifies the understanding of many transistor circuits.

As an example of the use of the TG-IR, let us consider an amplifier comprised of a series of *identical* R-C coupled or direct-coupled stages, as shown in Fig. 1.6(a). The input impedance of each stage is the load impedance of the previous stage. Hence $R_I = R_L$, R_L/R_I in the TG-IR is unity, and the stage voltage gain is numerically equal to the stage current gain.

In practical amplifiers it is usually necessary to limit the current gain to about 20 per stage by local feedback or by overall feedback. Hence, one does not expect the voltage gain of transistor amplifiers to exceed about 20 per stage. As the TG-IR shows, voltage gain in excess of the tolerable current gain must be obtained by using an R_L/R_I ratio of more than 1. If the impedance ratio is less than 1, as is often the case, then one must expect voltage gains correspondingly lower than 20.

As a second example consider an emitter follower which has a voltage gain of approximately 1, as shown in Fig. 1.6(b). If the emitter follower is designed to make the input impedance independent of the transistor characteristics, β in particular, the feedback resistors must hold the base-to-emitter current gain to no more than 50. The TG-IR shows that if $G_v = 1$ and G_i

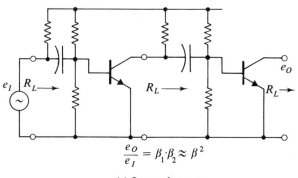

(a) Iterated stages

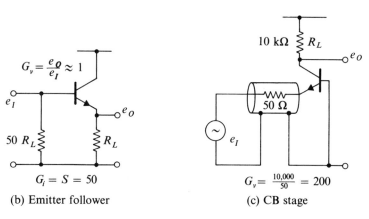

(b) Emitter follower (c) CB stage

Figure 1.6. Transistor gain-impedance examples

$= 50$, then $R_I = 50 R_L$. Thus, the input impedance of an emitter follower with significant feedback cannot be more than 50 times the load impedance. Where higher ratios of input-to-output impedance are required, we must use additional common-collector (CC) stages or step-down transformers.

For a third example consider a common-base (CB) transistor amplifier, Fig. 1.6(C). The current gain of a CB transistor is 1, and the TG-IR shows that a voltage gain of more than 1 is only obtained by making R_L greater than the input impedance R_I. If we wish to couple a 50-Ω line to a 10-kΩ iterated stage, we can use a CB stage with $R_I = 50\ \Omega$ and $R_L = 10{,}000\ \Omega$, and the TG-IR shows that the CB input stage should give a voltage gain of approximately $10{,}000/50 = 200$.

Amplifiers with FETs cannot be analyzed with the aid of the TG-IR. The voltage gain of FET stages must be evaluated separately from that of transistor stages because FETs do not have a characteristic current gain. The voltage gain of each FET stage is usually between 5 and 10 in a common-source amplifier; it is 1 if the stage is a common-drain stage with significant source feedback. The voltage gain of FET amplifiers with feedback may be estimated from the values of the feedback resistors in exactly the same way as for transistor amplifiers.

1.7 OHM'S LAW AND DC CURRENT-VOLTAGE RELATIONS (REF. 3)

Everyone knows Ohm's law! However, it is not always clear how Ohm's law may be used in three different forms. As an example, referring to Fig. 1.7, we

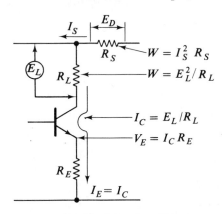

Figure 1.7. Ohm's law in a CE stage

may measure the voltage drop across resistor R_L with a voltmeter, and, using the resistance marked on the resistor, we calculate the current from $I_C = E_L/R_L$. If we wish to know what the emitter voltage is, we use the known collector current I_C and the emitter resistor R_E and calculate $V_E = I_C R_E$. We have assumed in using I_C that the base current does not make I_E significantly larger than I_C.

The third form of Ohm's law is illustrated by the problem of finding the resistor that may be used in the power supply filter when the current is known and we select a voltage drop by which the supply may be reduced. In this problem we calculate the resistor with the formula $R_S = E_D/I_S$, where E_D is the voltage drop allowed across R_S when the current is I_S. These three examples of Ohm's law illustrate the calculation of a current I, a voltage V, and a resistance R. We have used Ohm's law in each of the forms, $I = E/R$, $E = IR$, and $R = E/I$.

Today's transistors have high current gains, and for all practical purposes we may assume the collector and emitter currents are equal. Thus, the voltage drop in the collector or emitter resistor may be used to find the collector current. The base-emitter voltage may be used as an approximate indication of correct or incorrect biasing. Small signal transistors are always designed to operate with a base current that makes the base-emitter voltage (V_{BE}) 0.25 V for germanium and 0.6 V for silicon transistors. Silicon power transistors are driven with V_{BE} as low as 0.5 V at cutoff to as high as 1.5 V when full ON. Germanium power transistors may be driven from 0.1 V to 1.0 V.

Feedback usually makes the base current and voltage of relatively minor importance. The best indicators of proper transistor operation are the values of the collector voltage and current. In resistance-coupled circuits we use the collector Q-point voltage, and in transformer-coupled circuits and radio circuits we use the voltage drop across the emitter resistor as the best measure of the collector current.

1.8 COLLECTOR POWER DISSIPATION (REF. 2)

The power formula $W = E \times I$ may be used in each of three forms, as with Ohm's law. The product form is simple enough when the voltage and current through a resistor or a transistor are known, but in most cases we wish to know the power dissipated when either E or I is known. Using Fig. 1.7 as an example again, we know the voltage drop across R_L and the value of R_L. By combining Ohm's law and the power formula we calculate the power, using the formula $W = E_L^2/R_L$. In a similar way we may wish to know the power rating required for the series resistor R_S when we know I_S. For this we calculate the power dissipated in R_S using the formula $W = I_S^2 R_S$. Occasionally we are given the power rating of a device and wish to know the maximum voltage that can be allowed without overheating the device. In this type of problem we use the power formula in the form $E = W/I$ or $E = \sqrt{WR_L}$. Similarly, the permissible current may be calculated from $I = W/E$ or $I = \sqrt{W/R_S}$.

In an RC stage the transistor power input is maximum when the Q-point voltage is one-half the supply voltage, and the transistor should be rated to dissipate $W = V_{CC}^2/R_L$. In a power stage the transistors may have to dissipate 2 to 4 times the maximum power output, or the collector input power $V_{CC}I_C$ if greater.

1.9 TRANSISTOR POWER RATINGS AND HEAT SINKS (REF. 3)

Most low-power transistors safely dissipate 200 mW at room temperatures, and 100 mW collector dissipation is reasonably conservative for transistors in a warm location near a transformer or a heat sink. Transistors in a resistance-capacitor-coupled amplifier with a 1-kΩ load resistor may be operated with a 28-V collector supply without exceeding a 200-mW rating.

For greater reliability and higher ambient temperatures the load resistor should be at least 2 kΩ or the supply should not exceed 10-to-20 V. The power dissipated by a collector resistor may be as high as $V_{cc}I_c$ watts.

If the fingers can be held in continuous contact with a transistor, one may safely assume the transistor is operating at a conservatively cool temperature. A power transistor is probably operating near the maximum temperature rating when a drop of water on the case remains as long as ten seconds before evaporating. A case temperature near the boiling point of water is too high for a conservative and reliable design unless the ambient temperature is as high as can be expected under all normal operating conditions.

Transistors are cooled by attaching a heat sink. The low-power transistors have a high thermal resistance between the junction and the case, so there is little advantage in attaching a large heat sink. A clip-on heat sink with an area 3 times the area of the transistor case provides all the heat sink that may be used effectively. Power transistors have a low thermal resistance between the collector and case and are sometimes attached to a large heat sink. Unless a blower is used to move air, there is not much advantage in using an elaborate heat sink with deep radiating fins. Often a 10-to-30 square centimeter flat sheet of 3 mm thick aluminum provides enough heat sink to protect each power transistor of a power amplifier. Power transistors are sometimes operated class B or as switches and may be used with a relatively small heat sink.

1.10 THE DECIBEL, OR DB

The decibel is a convenient logarithmic measure of the ratio of two numbers, although, strictly speaking, the dB is a measure of power ratios only. For example, suppose a 3-stage amplifier has a voltage gain of 4 in each stage. We can multiply $4 \times 4 \times 4 = 64$ and say that the overall voltage gain is 64 times. On the other hand, once we know that a voltage ratio of 4 is 12 dB, we can simply add 12 dB three times and say that the overall gain is 36 dB. The dB units give the best measure of gains and circuit performance and have the advantages of addition over multiplication. If we need the voltage gain indicated by 36 dB, we may use a table and find the gain is 64. However, we soon learn that 30 dB is a little more than 30, 32 to be exact, and 6 dB is a gain of 2, so we know the overall gain is $32 \times 2 = 64$.

Once a few of the commonly used dB values are learned, it is relatively easy to work with the dB sums instead of gain products. For example, we know a 3-stage amplifier with a gain of 8 in each stage has 18 dB per stage for a total of 54 dB. We know also that 54 dB is 6 dB less than 60 dB, which makes the total voltage ratio equal to $1000/2 = 500$. However, after becoming familiar with the dB units, we do not often convert dB units back to ratios. Because the ratio of two numbers is more meaningful than their difference, the use of logarithmic scales makes frequency response curves symmetrical, with slopes indicating the number of stages that produce cutoff. Gain curves plotted on a linear scale have a lopsided shape that is meaningless and misleading.

Although dB units are commonly used to measure voltage or current ratios, the dB is defined as a measure of power ratios. A gain measured in dB has only a limited meaning unless the input and output impedances are given or the gain is specified as a power gain. In an iterated stage where the input and load impedances are equal, the stage gain may be given in dB units without specifying the impedances. If the gain is 30 dB, for example, a table shows that the current and voltage gains are 31.6, and the power gain is 1000, the square of 31.6. Thus, when the voltage gain e_o/e_s is given in dB for an iterated stage, the current and power gains are the same in dB units.

However, in a CC stage where the voltage gain is 1 and the current gain is 50, the voltage gain is 0 dB, the current gain is 34 dB and the power gain is 17 dB, or a power ratio of 50. In a CB stage where the current gain is 1, 0 dB, and the voltage gain is 200, or 46 dB, the power gain is 23 dB, or a power ratio of 200. Thus, when the impedances change from input to output, we must use the dB values carefully. A practical dB table is given in the Appendix.

1.11 THE TRANSISTOR CURRENT GAIN β (BETA)

The current gain of a transistor is an important figure of merit that is always given in the manufacturer's data sheets. When the transistor is used with the emitter as the common terminal, the base is the input, and the collector current, the output, is 20 to 200 times the base current. The ratio of the collector current to the base current is the CE current gain called $β$. When the base is the common terminal, the CB current gain is $α$, alpha, which is approximately 1.

The current gain $β$ varies with the operating Q-point, with the temperature, and within a given transistor type may vary from 15 to 200 or more. Because $β$ is so variable, practical transistor circuits must be made relatively independent of $β$ by feedback. To ensure significant feedback we should know that $β$ is at least 50, but it is not necessary to know the exact value of the current gain. The CE current gain may be measured with an ohmmeter using a method described in Chapter 17.

2 Single-Stage Amplifiers

This chapter describes some of the many forms of single-stage amplifiers that are the basic building blocks of linear circuits. Most of the amplifier examples are designed as iterated stages that may be connected in series with an efficient transfer of signal power. The use of single-stage amplifiers with discrete components gives the designer a flexibility in his choice of interstage coupling circuits, a wide choice of semiconductor characteristics, high output power, and the simplicity of a single-side power supply. An important and often overlooked advantage of discrete circuits is that higher gain may be obtained more easily above 1 kHz than with ICs.

Transistors have three terminals; two terminals are required for the input and two for the output. Therefore, it is necessary to make at least one terminal common to both input and output. Amplifiers are designated as **CE**, **CC**, or **CB**, depending on which terminal is common.

The CE stage offers both current and voltage gain and is therefore the most generally adaptable. In an iterated stage we can obtain voltage and current gains of 30 to 50, which is a power gain of 900 to 2500, 30 to 34 dB.

The CC stage is used to couple a high impedance source to an impedance that is a factor of 30 to 50 times lower. The CC stage transfers the voltage at the high-impedance input to the low-impedance output without a voltage change. Thus, the CC stage has a voltage gain of 1, a current gain of 30 to 50, and a power gain of 30 to 50, 15 to 17 dB. The CC stage has a wide-band frequency response and is easily direct-coupled to the collector output of almost any transistor amplifier. When connected to a prior-stage collector, the CC stage does not require bias resistors and is useful for driving low-power, low-impedance loads.

The CB stage has a current gain of 1, and voltage gain is obtained only by making the output load impedance greater than the input impedance. Thus, the voltage gain is 100 when the load impedance is 100 times the input impedance, and the power gain is 100, 20 dB. The CB stage is used mainly in high-frequency applications and for coupling low-impedance signal sources to higher impedance loads. In low-frequency applications the single-stage amplifiers in order of importance are the CE, CC, and CB. This chapter is mainly concerned with the iterated CE stage.

2.1 SINGLE-STAGE AMPLIFIERS

A transistor is not often used by itself, and a practical amplifier usually has two or three resistors and perhaps one or two capacitors. The resistors are needed to establish dc bias currents and voltages so that the transistor will amplify without distortion, and the capacitors are used to isolate the transistor stage from associated circuits that might disturb the desired operating point, the Q-point.

About the simplest form of a transistor stage is the CE amplifier shown in Fig. 2.1. The stage is shown with a 10-kΩ collector resistor connected to a

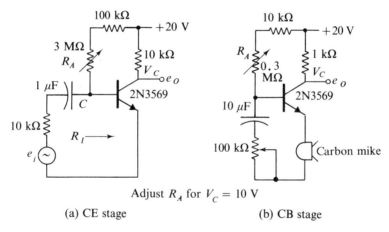

Figure 2.1. Single-stage amplifiers without feedback

+20 V supply, and the bias is supplied by connecting the variable resistor R_A to the collector supply. A fixed 100-kΩ resistor is used in series with the bias resistor to limit the maximum bias current to about 0.2 mA. The bias in this stage is adjusted to make the collector voltage about 10 V. When so adjusted, the amplifier will operate as long as the room temperature does not depart more than 5 or 10°C (9 or 18°F) from the initial temperature.

The simplicity of the CE stage amplifier gives it several disadvantages. Besides the temperature sensitivity, the bias adjustment and the amplifier gain depend on the transistor β, and the input impedance is relatively low and varies with β. If β is 100, the input impedance βh is 2600 Ω. If the stage is driven by a 10-kΩ source, the transistor input impedance reduces the input signal by a factor of 5, and the iterated gain is approximately β. With a slightly more complicated circuit we can turn this signal loss to an advantage by reducing the gain with feedback instead of shunting the source. On the other hand, the stage may be used for a temporary application at room temperatures. A low-impedance microphone connected to the base input results in a voltage gain of 250, 48 dB, and a low-impedance or a carbon microphone may be connected in series with the emitter for a 40 dB gain. When the signal source is connected in the emitter lead, the emitter current supplies the required microphone dc current, and the input capacitor should be connected to ground. If connected to ground through a 100-kΩ potentiometer, the potentiometer may be used as a gain control.

Reducing all the resistors in the CE stage circuit by a factor of 10 decreases the input impedance by a factor of 10 and increases the emitter current to 10 mA. Most carbon microphones require about 10 mA, and the low-impedance circuit makes a better microphone amplifier. Reducing the input impedance of the CE stage to 260 Ω reduces the gain only about 3 dB.

2.2 APPLICATIONS USING A SINGLE TRANSISTOR

The circuit described in Sec. 2.1 may be used to measure the current gain of a transistor. With a 500-Ω collector resistor and a 10-V or larger collector supply we adjust the bias resistor to make the collector voltage one-half the supply voltage. The transistor current gain is the value of the bias resistor in kilohms. Thus, if the bias resistor is 100 kΩ, the current gain β is 100. We may explain the reason β is 100 by observing that doubling the collector resistor makes the collector voltage-drop equal to V_{CC}. The voltage-drops across R_A and R_L are then approximately equal, and the collector current is larger than the base current by the ratio $R_A/R_L = 100$ k$\Omega/1$ k$\Omega = 100$. The reason for not using a 1-kΩ collector resistor and adjusting for zero collector voltage is that the current gain should be measured with the collector voltage greater than zero. Except to measure β this simple circuit is impractical and should be used only rarely. A better method of measuring β by using an ohmmeter and a single resistor is decribed in Sec. 17.6.

A single transistor may be used to increase the sensitivity of a current-indicating meter. The circuit used to increase the sensitivity of a 50-μA meter to 0.5 μA full scale is shown in Fig. 2.2. This is not to say that the 0.5 μA meter can be accurately calibrated, but it may be calibrated approximately and can be very useful in special purpose applications. When used as shown in Fig. 2.2(a), the meter indication is linearly proportional to the input current and is β times more sensitive. Silicon transistors with a high value of β should be used and in some cases may have to be selected to obtain current gain at the low-current end of the meter scale. The reason that the meter cannot be

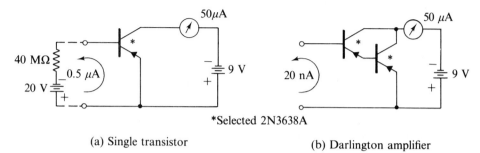

*Selected 2N3638A

(a) Single transistor   (b) Darlington amplifier

Figure 2.2. Dc meter amplifiers

calibrated easily at such low currents is that the base-emitter voltage drop varies with the current and is 0.5 V to 1 V or more. If the current source is from a high voltage source or is independent of the load, a calibration may be obtained by using a 10-V battery and a 20-MΩ resistor, provided the transistor voltage drop may be neglected or estimated.

For still higher current sensitivity two transistors may be used as a Darlington pair, as shown in Fig. 2.2(b). With the pair current gains exceeding 3000 are obtained easily, and with a 50-μA meter full-scale sensitivities exceeding 20 nA may be obtained.

2.3 THRESHOLD-SENSITIVE CIRCUITS

Another application for a transistor without a bias circuit is for sensing a small change of voltage. A threshold detector may be built using the characteristic that a 60 mV increase of the base-emitter voltage of a transistor causes a 10-to-1 change in the collector current. The threshold detector illustrated in Fig. 2.3(a) may be used to operate a sensitive relay when the voltage applied

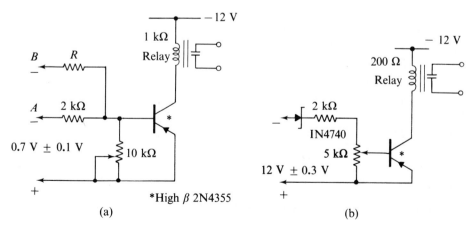

*High β 2N4355

(a)   (b)

Figure 2.3. Threshold detectors

to terminal A changes from 0.7 to 0.8 V. The 2-kΩ resistor is used only to protect the transistor from an accidental overvoltage. Some change of the threshold voltage may be obtained by adjusting the potentiometer, or the voltage may be applied to the base through a 2-kΩ series resistor without the adjustment.

For sensing a 10 per cent change of a higher voltage the input is connected to a series resistor R of approximately 3 kΩ per volt, as at B in Fig. 2.3(a). With a 39-kΩ resistor the potentiometer may be adjusted to sense the relatively small voltage changes of a 12-V battery or of an instrument battery. For sensing still smaller changes of a high-voltage source a Zener diode may be connected in series with the 2-kΩ resistor, as shown in Fig. 2.3(b). With a 1-kΩ relay the threshold sensitivity is decreased very little by inserting the Zener diode, and the decreased sensitivity indicated in Fig. 2.3(b) is caused mainly by the use of a less sensitive 200-Ω relay. While the base-emitter voltage of a transistor decreases 2 mV per °C, the threshold detector may usually be adjusted to operate without a temperature correction over a temperature range of ± 10°C.

The circuit in Fig. 2.4 shows how a less sensitive 50 to 100 Ω, 12-V relay may be operated by connecting a second transistor as a switch and current amplifier. Circuit (a) turns the relay on with a voltage rise, and circuit (b) turns the relay off with a voltage rise. There are numerous applications for threshold detectors in relay circuits because most relays turn off only when the voltage is reduced to less than 20 per cent of the on voltage. The threshold sensing transistor turns a relay off when the voltage is reduced to approximately 95 per cent of the turn-on voltage. Relays manufactured to

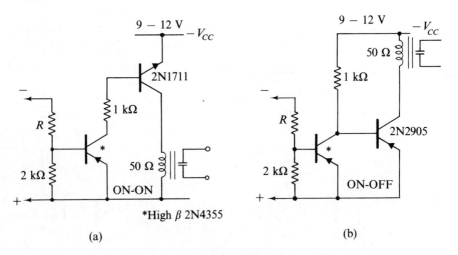

Figure 2.4. Threshold-sensitive relay amplifiers

duplicate this performance are generally much more expensive than an inexpensive relay and a transistor.

In power supplies transistors are used as threshold detectors to sense and prevent an overload condition when the input current exceeds a given value or the load voltage is lower than the input voltage by a value that implies an overload. Sensing circuits of this type are illustrated in the regulated power supply circuits.

2.4 COLLECTOR FEEDBACK AMPLIFIERS

The collector feedback stage is the simplest amplifier that possesses the advantages of feedback in making the amplifier less dependent on the transistor and the operating temperature. The collector feedback stage is often used because an amplifier may be constructed with a transistor and only two resistors. An amplifier in this form is shown in Fig. 2.5. As a practical rule the bias resistor should be between 10 and 50 times the collector resistor, with the higher values used with high β transistors. If the bias resistor is adjusted to make the collector voltage about one-fourth the supply voltage, the stage has significant feedback.

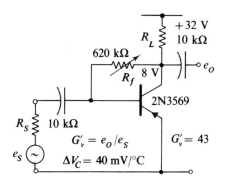

Figure 2.5. Collector feedback stage with significant feedback

The collector feedback stage has the disadvantage that the bias resistor should be selected for a particular transistor, and significant feedback is obtained only by reducing the collector Q-point voltage to a relatively small fraction of the available supply. The low value of collector voltage generally limits application of this circuit to low signal applications and to circuits offering at least a 20-V supply voltage. The collector feedback does have the advantage that, if the transistor β doubles, the Q-point voltage is only halved, whereas without feedback the collector current is doubled and may reduce the Q-point voltage to cutoff, 0 V.

The collector feedback stage has two other important disadvantages. The input impedance is generally about 1/100th the load resistance and the input impedance varies with the load, so that there is an interaction between stages that may cause instability, oscillation, or unexpected problems. For these reasons the collector feedback stage cannot usually be used in a multistage amplifier and is not recommended for operation with low supply voltages.

An interesting variation of the collector feedback stage is shown in Fig. 2.6. This stage is operated on a +90-V supply, and the bias resistor is only 2.5 times the collector resistor. Thus, the collector Q-point voltage is only 2 V, but there is so much dc feedback that the stage is operative over a $\pm 30°C$ temperature range. The feedback stage is shown as operating between a capacitor-coupled 1-kΩ source and load with an iterated voltage gain of 38, 32 dB. With the load disconnected the ac feedback increases and the voltage gain approximately doubles. Thus, the ac gain is relatively independent of the load, and the effective internal impedance of the stage is about 2 kΩ.

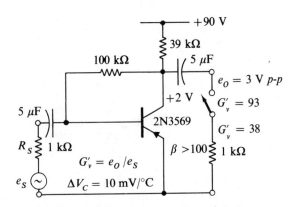

Figure 2.6. Collector feedback stage with 90-V supply

2.5 THE EMITTER FEEDBACK AMPLIFIER (REFS. 1, 2)

The most popular and most commonly understood form of resistance-coupled transistor amplifier is the emitter feedback stage illustrated in Fig. 2.7. This amplifier was developed in the early days of the transistor, when transistors were relatively temperature-sensitive and variable, one from another. The emitter feedback stage may be the better choice for some applications, although in others it presents serious limitations.

As described in the introductory chapter, the best way to begin a design is to select the collector resistor by choosing a high value (100 kΩ) for a low-current, low-level design, or by using a low value (100 Ω to 1000 Ω) for a power-stage driver. The stage illustrated in Fig. 2.7 uses an intermediate

Sec. 2.5   The Emitter Feedback Amplifier

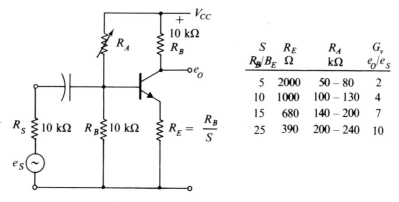

| S | R_E | R_A | G_v |
R_B/B_E	Ω	$k\Omega$	e_o/e_S
5	2000	50 – 80	2
10	1000	100 – 130	4
15	680	140 – 200	7
25	390	200 – 240	10

Figure 2.7. CE stage design examples

resistor value of 10 kΩ. Once the load resistor is selected, the base resistor R_B should be equal to the load resistor, the emitter resistor R_E should be one-tenth the load resistor, and the bias resistor should be 10 to 15 times the load resistor. With these values the stage input impedance is approximately R_B, the voltage gain V is $R_L/R_E = 10$, and the current gain S is $R_B/R_E = 10$. When driven by a 10-kΩ source R_S, the iterated voltage gain is only 4.

Although commonly used, the emitter feedback stage has serious disadvantages, particularly in circuits using supply voltages below 20 V. The limitations of the emitter-feedback stage and their causes are described in the references. For present purposes we need only say that the voltage gain of a stage R_L/R_E should not exceed the magnitude of the supply voltage, the supply voltage must be reasonably fixed, and the bias resistor may have to be carefully selected or specified with a 5 per cent tolerance. The curves shown in Figs. 2.8, 2.9, 2.10, and 2.11 are included to assist the designer in selecting a bias resistor and to show how the collector Q-point voltage may vary with a change of the resistor or the supply voltage. If the stage is to be used with large signals, the collector voltage should be close to one-half the supply voltage. As shown in Fig. 2.9 for $S = 10$, a bias resistor that is about 8 times the base resistor R_B makes the collector voltage 5 V independent of the supply. If the Q-point is independent of the supply and the base is driven by a source with a high ac impedance, the output will be relatively free of power supply ripple. However, when R_A is only 8 times the base resistor, the Q-point collector voltage is too low for satisfactory operation with large signals.

The emitter-feedback stage is a common choice for large signal applications, such as a power-stage driver. For these purposes the stage should have an S-factor of 5 to 10, and the collector supply should be 20 to 30 V. A supply voltage less than 12 V cannot be recommended for a power stage unless the S-factor is 5 or less. Higher S-factors should be avoided because they cause a 2-to-3 times greater Q-point shift with temperature, and the Q-point is

24 Single-Stage Amplifiers Ch. 2

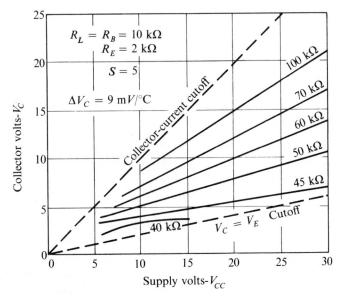

Figure 2.8. Collector characteristics with $S = V = 5$

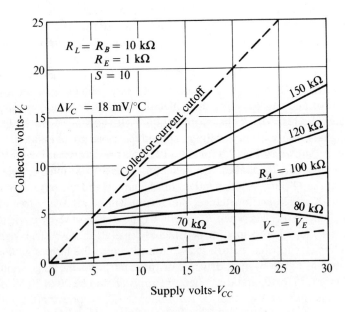

Figure 2.9. Collector characteristics with $S = V = 10$

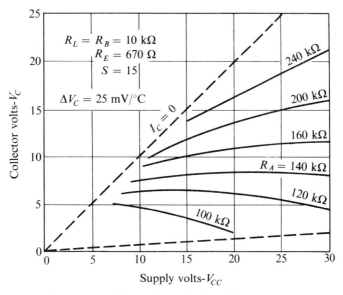

Figure 2.10. Collector characteristics with $S = V = 15$

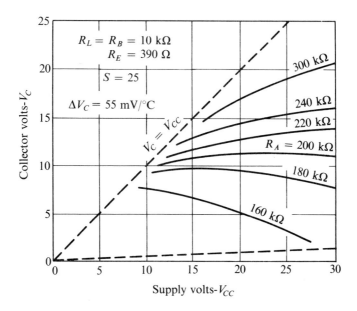

Figure 2.11. Collector characteristics with $S = V = 25$

undesirably sensitive to a change of the bias resistor or the supply voltage. An undistorted, large signal may be obtained only with a relatively fixed Q-point that is well away from collector cutoff.

The table in Fig. 2.7 may be used as a guide in selecting component values for an emitter feedback design, and Figs. 2.8, 2.9, 2.10, and 2.11 may be used to select a satisfactory Q-point and bias resistor. The bias resistor curves show that a bias resistor that is satisfactory for a 30-V collector supply may not be satisfactory for a 20-V supply, because the Q-point is too close to cutoff for even a moderately large signal. An advantage of a stage with $S = 5$ or 10 is that a bias resistor may be found that makes the Q-point follow changes of the supply voltage. This advantage is particularly important at the lower end of the bias curves, where the maximum undistorted signal amplitude may be only 1 or 2 V.

2.6 A TYPICAL EMITTER FEEDBACK STAGE

The component values shown in Fig. 2.12 are typical of those used in an emitter feedback stage. The amplifier has a 10-kΩ source and load impedances, $S = 10$, $V = 10$, and the input impedance is a little less that 10 kΩ. The overall voltage gain of the iterated stage is only $e_o/e_s = 4$, and with a 20-V supply the maximum output voltage is 6 V rms, or 17 V p-p.

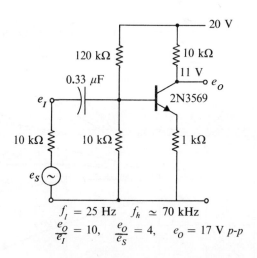

Figure 2.12. CE stage with $S = 10$

The high-frequency cutoff of the stage shown in Fig. 2.12 is at 70 kHz and is determined by the collector-to-base feedback capacitance and the voltage gain. Lowering all the resistors in the stage by a factor of 10 increases the high cutoff to about 0.5 Mhz, and increasing the resistors by a factor of 10 lowers the high cutoff to 7 kHz. As these cutoff frequencies show, the impedance level selected for a circuit may be determined by the required upper frequency response. Moreover, a stage with higher S and V factors has proportionally lower cutoff frequencies. The low-frequency cutoff of the stage is determined by the coupling capacitor, the series impedance of the source, and the base resistor. For the stage illustrated the low-frequency cutoff is the frequency at which the reactance of the 0.33 μF capacitor equals 20 kΩ, which is 25 Hz.

In low-impedance designs the heat dissipated by the transistor may be an important limiting factor. The transistor must dissipate the maximum power when the collector voltage is one-half the supply voltage, and this power is $V_{CC}^2/4\,R_L$. With $V_{CC} = 20$ V and $R_L = 100$, 1 watt must be dissipated, which is about 3 times the maximum permissible power for a plastic 2N3569, or about 20 per cent more than is permissible for a TO-5 2N1711. A 500-Ω load resistor is perhaps the lowest value that should be used with the 2N3569. The 2N1711 may be used with a small heat sink, or the collector resistor should be at least 300 Ω.

2.7 HIGH-GAIN SINGLE-STAGE AMPLIFIERS

Probably the most practical high-gain single-stage amplifier is a collector-feedback stage in the form illustrated by Fig. 2.13. This CE stage has collector feedback to stabilize the Q-point, but the feedback is removed for ac signals by the capacitor, which is connected to the center point of the feedback resistor. An iterated stage with the collector feedback bypassed has a voltage gain

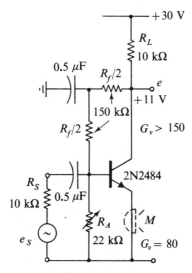

Figure 2.13. Collector feedback stage, bypassed and biased

e_o/e_s, which is approximately equal to the transistor β. In the example shown, the voltage gain is greater than 150, 44 dB, with equal source and load impedances.

The input impedance of a CE stage without ac feedback is always low compared with the load resistor and is about 2 kΩ for the stage illustrated. Thus, the signal at the base is about a factor of 5 lower than the generator signal e_s, and with a low-impedance source we may expect an even higher voltage gain. With a 100-Ω dynamic microphone replacing the 10-kΩ source, the voltage gain is 600, 56 dB.

The stage shown in Fig. 2.13 is ideally suited for use as an amplifier for a carbon microphone. The emitter provides 2-to-3 mA dc to activate a microphone connected in series. This current may be increased by reducing all resistors by the same factor. With the microphone in the emitter the stage operates as a CB amplifier with collector feedback to control the Q-point. For a maximum gain the base should be bypassed to ground, as shown, and the capacitor at the midpoint of the feedback resistor is no longer required. With a carbon microphone the voltage gain is 30 to 100, 30 dB to 40 dB, in the circuit shown, but reducing the resistors to increase the microphone current reduces the voltage gain. The size of the base bypass capacitor may be determined by trial, and in some applications may be omitted. The required capacitor does not need to be increased when the resistors are decreased because the gain and the feedback are reduced.

With a 10-Ω source the CB stage has a voltage gain of 600 and offers an exceedingly simple circuit with a feedback-stabilized Q-point. However, with a source impedance of less than 1 per cent of the collector resistor the amplifier tends to distort input signals exceeding a few mV ac. The distortion may be avoided by connecting a resistor in series with the source or by using the amplifier in the CE configuration with the signal connected to the base.

2.8 CE STAGE WITH Q-POINT ADJUSTMENT FOR LARGE SIGNAL APPLICATIONS

Experiment shows that a bias resistor R_A connected as shown in Fig. 2.13 generally raises the Q-point enough to permit large signals without otherwise affecting the circuit relations previously described. Unless the S-factor is quite low, values of R_A that are from 2 to 10 times the value of R_L will raise the collector Q-point voltage to as high as one-half the collector supply voltage. (Sometimes an amplifier will tolerate a higher output signal if the resistor is connected across the base-emitter junction.)

The use of a bias resistor has distinct advantages with low supply voltages where otherwise it is impossible to use significant feedback. The bias resistor makes possible the use of an S-factor of 30, as shown in Fig. 2.13, while keeping a Q-point voltage that permits higher signal levels and a wider ambient temperature range than is possible with the stage shown in Fig. 2.5. A 12-V supply can be used with a 6-V Q-point when S is reduced to 8 and $R_A = 10$ kΩ. Thus, whether or not R_f is bypassed, the addition of R_A considerably improves the collector-feedback stage and makes the amplifier a more attractive choice in applications using a low supply voltage. A disadvantage of the bias resistor is that it tends to increase the temperature sensitivity of the collector Q-point. When using the resistor, the temperature shifts of a collector-feedback stage and an emitter-feedback stage are almost the same if both have the same S-factor and the same base resistor. The collector-feedback stage has the advantage of twice the voltage gain, and increasing the bias resistor reduces the temperature shift.

2.9 COMBINED COLLECTION AND EMITTER FEEDBACK (REF. 2)

The input impedance of a CE stage may be increased by emitter feedback, and, except with high source impedances, the voltage gain is made independent of the transistor β. Thus, a combination of dc collector feedback and ac emitter feedback may make the best all-purpose, single-stage amplifier configuration. A circuit with the combined collector and emitter feedback is shown in Fig. 2.14. This amplifier has a voltage gain of 40, instead of 150 as in Fig. 2.13, because the emitter resistor provides significant feedback. In the construction of this type of CE stage the emitter resistor is omitted and the Q-point is adjusted by selecting the collector feedback resistors and the base

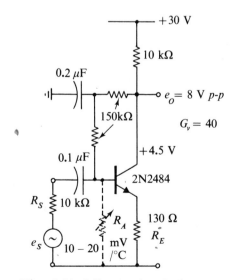

Figure 2.14. Collector-emitter feedback stage

bias resistor, if used. Finally, an emitter resistor is added that reduces the voltage gain by a factor of 4. The emitter resistor should be selected with the intended source impedance in use; otherwise, the ac feedback changes when the signal source is changed.

With significant feedback a stage with the combined collector and emitter feedback has an input impedance that is about 3 times the source impedance, and the input coupling capacitor may be a factor of 4 smaller than without the ac feedback.

The capacitors shown in Fig. 2.14 are sized to make the low-frequency cutoff fall at 30 Hz to 40 Hz, and the high-frequency cutoff is at 50 kHz and depends on the collector-to-base feedback capacitance. The feedback bypass capacitor C_f is larger than the input capacitor because the feedback signal is 40 times the base signal. If the feedback resistors are reduced (or increased), the bypass capacitor should be increased (or decreased). Generally it is better to find the correct capacitor values for a different circuit by experimental trial when the circuit is first assembled.

2.10 HIGH-GAIN CE STAGE WITH GAIN CONTROL

The amplifier shown in Fig. 2.15 is a simplified form of a combined collector and emitter feedback stage. Because the S-factor is $10^6/10^4 = 100$, and β is 150, there is relatively little collector feedback, and the emitter feedback increases the input impedance from approximately 2 kΩ to at least 10 kΩ with only a 2-to-1 gain reduction. Although this circuit does not have significant feedback, the amplifier has the advantages of unusual simplicity, few components, and better performance characteristics than a stage with emitter-feedback biasing. The stage may be used with supply voltages of from 5 V to 50 V, is relatively insensitive to temperature, and permits peak-to-peak output

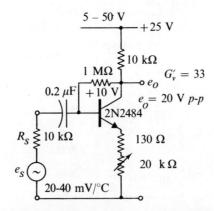

Figure 2.15. Simplified collector-emitter feedback stage

signals exceeding one-half the supply voltage. A further advantage is that the gain may be reduced 40 dB by increasing the emitter resistor. This increase tends to raise the Q-point voltage and permits larger output signals than can be obtained with a collector-feedback gain control. With a 20-kΩ variable resistor, as shown, the voltage gain is decreased 40 dB.

The simplified collector and emitter feedback stage has the disadvantage that the Q-point and the performance characteristics change with the transistor β, so adjustments are required at assembly. However, with both collector and emitter feedback the effects of load changes do not change the input impedance as much as with collector feedback alone, so the stage may be used in a multistage amplifier where a stage with collector feedback cannot be used.

2.11 UNIVERSAL SINGLE-STAGE AMPLIFIER

An amplifier for room-temperature applications with sufficient flexibility to be called a universal amplifier is shown in Fig. 2.16. With the capacitors in use and a 1-μF input capacitor the stage has a voltage gain $e_o/e_s = 100$ and a frequency response from 50 Hz to 5 kHz. With the capacitors removed and $C_I = 0.25$ μF, the gain is 10 with a frequency response from 20 Hz to 35 kHz. A particularly useful characteristic of the amplifier is that with a fixed bias resistor $R_A = 47$ kΩ, the voltage gain is 20 dB with a 5-V supply, and it increases to only 22 dB with a 40-V supply. Similarly, the peak-to-peak output voltage is approximately eight-tenths the supply voltage for the same range of supply voltage. By adjusting the bias resistor a slightly higher output voltage may be obtained with a fixed supply voltage. With the bias resistor removed the voltage gain is 1 to 2 dB higher, and the undistorted peak output signal is reduced by a factor of approximately 3.

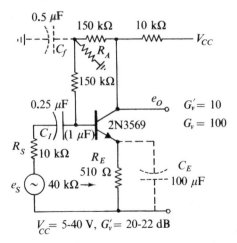

Figure 2.16. Universal single-stage CE amplifier

With the resistors shown in Fig. 2.16 there is significant feedback, and the input impedance is 40 kΩ. The addition of a capacitor across R_A does not increase the gain unless the emitter capacitor is used also. However, a gain of 30, 30 dB, may be obtained by using the capacitor C_f with the emitter resistor reduced to 180 Ω. In summary, the stage is recommended particularly for use with the capacitors removed, except for C_I, and with significant ac feedback the stage may generally be used in multistage audio-frequency applications. With the feedback removed by using all 3 capacitors the iterated gain is increased to 100, 40 dB, and the input impedance is reduced to 3 kΩ. This amplifier is not recommended in a multistage amplifier with the ac feedback bypassed.

Because the amplifier has a somewhat universal character, Table 2-1 gives, for 6 values of the supply voltage, the optimum bias resistor, the iterated voltage gain, and the peak-to-peak output voltage.

Table 2-1 Universal Amplifier Design Data—Fig. 2.16

V_{CC}	V_C	R_A(kΩ)	G_v (dB)	e_O (p-p)
30	17	33	22	27
25	14	33	22	23
20	11.5	33	22	18
15	7.5	47	22	12
10	5.5	47	21	7
5	2.5	100	20	4

2.12 OSCILLOSCOPE PREAMPLIFIER

The oscilloscope amplifier, Fig. 2.17, is interesting because the voltage gain may be changed from 20 dB to 0 dB by reversing the collector supply voltage. Reversing the supply voltage forward-biases the collector diode and reverse-biases the emitter diode, and the transistor acts as a switch to connect the input to the output. With the component values shown in Fig. 2.17 the voltage gain is 20 dB between the 3-dB cutoff frequencies of 20 Hz and 350 kHz. With the supply voltage reversed the upper cutoff frequency increases to 500 kHz because there is no Miller effect when the voltage gain is 1, 0 dB.

The amplifier input impedance is about 20 kΩ up to 60 kHz, the broadband noise is less than 5 μV peak-to-peak, and maximum output voltage is 2.5 V rms. The amplifier is useful as an oscilloscope preamplifier for ac measurements in low-impedance circuits. The circuit may be assembled on a GR-type plug for easy insertion and removal, and the ability to reduce the gain by reversing the supply voltage makes the amplifier useful in many applications.

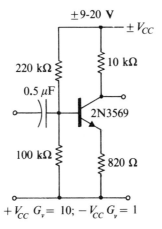

Figure 2.17. Oscilloscope preamplifier

The input impedance of the amplifier may be increased by increasing all the resistors by a factor of 5, but the input capacitance reduces the upper cutoff frequency to 8 kHz when the source impedance is 100 kΩ. However, the high-impedance amplifier is useful for audio-frequency measurements requiring an input impedance of 100 kΩ, but the amplifier shows a 10-dB loss with the supply voltage reversed. Both amplifiers operate satisfactorily on an 8-V to 30-V supply voltage with an adjustment of the bias resistor.

2.13 COMMON-COLLECTOR AMPLIFIERS

The CC amplifier, or emitter follower, is a broad-band amplifier used to couple high-impedance circuits into low-impedance loads. The amplifier transfers the input voltage to the output load without requiring voltage step-down, as in a transformer. Thus, the CC stage provides an impedance step-down and a current gain that makes the ac emitter voltage follow the base signal with a loss of only 1 or 2 per cent.

The current gain of a CC transistor alone is β, which means that a CC stage should have significant feedback to limit the dc current gain. Without S-factor control, the Q-point of a CC stage varies with β and temperature, and both the input impedance and the ac load current vary similarly with β and temperature.

A CC amplifier is made to be independent of β by using a low S-factor circuit design similar to the example shown in Fig. 2.18. For operation within a limited temperature range the S-factor of the CC stage may be designed to be as high as one-half the minimum β that is expected over the temperature range. Hence, with transistors that have current gains exceeding 100, a CC stage may have an S-factor as high as 50. Because the load is in the emitter

34 Single-Stage Amplifiers Ch. 2

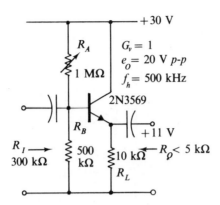

Figure 2.18. CC step-down stage with $R_I = 300\ \text{K}\Omega$

circuit, the dc emitter voltage is usually high enough to make the V_{BE} voltage change with temperature relatively unimportant.

For large-signal outputs we may make the bias resistor R_A about equal to the base resistor R_B so that the emitter Q-point voltage is a little less than one-half the collector supply voltage. This Q-point allows a maximum signal output. With the dc emitter voltage at the midpoint the power dissipation is the same in the transistor as in the resistor and is easily calculated. The S-factor of the CC stage is the parallel value of R_A and R_B divided by R_L, and the input impedance is then SR_L.

The principal application of the CC amplifier is in coupling high-impedance circuits into low-impedance loads. The amplifier transfers the input voltage to the load without requiring voltage step-down, as in a transformer. The practical form of the CC amplifier has a power gain and a current gain of about 10 and is used to replace transformers in many circuits.

2.14 HIGH INPUT IMPEDANCE CC STAGE

The CC stage illustrated in Fig. 2.18 is designed to present a high input impedance and couple the input signal to a following stage with a 10-kΩ input impedance. The stage transfers the input signal to the emitter load with a negligible voltage change and a 15-dB power gain. The input impedance equals the resistance R_B in parallel with the bias resistor R_A, and, depending on the supply voltage and the bias used, the impedance is approximately one-half to two-thirds the resistance R_B.

A low-power CC stage on a 10-V supply may be biased with $R_A = R_B$, while with a high-voltage supply the power dissipation may be limited by making R_A 2 or 3 times R_B. For large signal outputs with a low-voltage supply R_A may equal R_B, as in Fig. 2.19. When a maximum power output is required with a low-voltage supply, the bias resistor may be less than one-half R_B. A

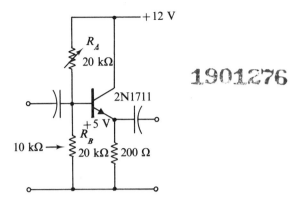

Figure 2.19. CC stage with $R_L = 200\ \Omega$

transistor operated near collector voltage breakdown is sometimes protected by connecting a resistor $R_C = R_E$ in series with the collector.

The high-frequency cutoff of the stage illustrated in Fig. 2.18, using almost any low-power transistor, is above 100 kHz. For audio applications the resistors may be a factor of 10 larger if higher input and load impedances are desired. Increasing the resistors by a factor of 5 reduces the high-frequency cutoff to approximately 20 kHz.

2.15 CC STAGE FOR DRIVING A LOW-IMPEDANCE LOAD

The CC stage shown in Fig. 2.19 is designed to couple a 10-kΩ iterated-stage amplifier into a low-impedance load. To obtain a maximum power output the resistors R_A and R_B should be adjusted to permit a maximum output voltage. Usually, the optimum value for R_B is 25 to 50 per cent larger than R_A. With a 20-V to 30-V supply voltage R_A and R_B may be made of equal value, or the CC stage may be direct-coupled to a preceding stage.

The power that must be dissipated by the transistor may be an important problem in the application of a CC stage used to drive a low-impedance load. The maximum power dissipated by the transistor equals the dc power in the load when the voltage across the load is one-half the supply voltage. Thus, for the example shown the maximum power dissipated is $(\tfrac{1}{2}V_{CC})^2/R_L = (6)^2/200$, or 0.18 W. In room temperature applications a silicon transistor should operate satisfactorily over the long term if the maximum power is limited to one-half or less, the 25°C ambient temperature rating.

If the transistor used in Fig. 2.19 has a 0.5-W, 25°C rating and we wish to ensure a maximum 0.25-W power dissipation, $(12/2)^2/R_L = 0.25$; solving for R_L we obtain $R_L = 144\ \Omega$. However, if $V_{CC} = 30$ V, we find $R_L = 900\ \Omega$. The examples show that for loads requiring a maximum output voltage and with supply voltages between 12 V and 30 V, a load resistor as low as 1 kΩ may be used if the transistor is capable of dissipating 0.5 W or more. Transistors rated to dissipate 50 mW at room temperatures may be used with loads exceeding 1 kΩ and low supply voltages or 10 kΩ loads with a 30-V supply. For occasional applications a power transistor may be connected as a CC stage to drive a low-impedance load.

2.16 CC-RELAY AMPLIFIERS

An example of a single CC stage used as a 0.5-W nonlinear amplifier is shown in Fig. 2.20. The amplifier is shown with a photo-conductive cell that may be used with a low-power transistor to operate a 12-V relay and also with a magnetic switch to turn off the relay. The photo cell and the magnetic switch may be interchanged, or either may be used alone, as the application requires. The CC stage provides a current gain of nearly 50, and the relay requiring 60 mA is controlled with a switching current of 3 mA or less.

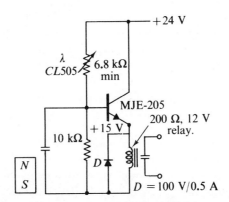

Figure 2.20. CC relay amplifier

When a switch is used to connect the turn-on bias, a series 6.8 kΩ resistor is required to limit the bias current. When a light cell is used, a resistor may be required if the cell is operated with a high light intensity. If a switch or a low-resistance light cell is connected across the base resistor to turn off the relay, a 6.8-kΩ bias resistor is used to supply the turn-on bias.

The relay amplifier may be used with a supply voltage equal approximately to the rated relay voltage by using the switch to connect the base to the supply without a series resistor. The current-limiting bias resistor is needed only when the supply voltage is too high to be used as the relay voltage. However, when the switch connects the base to the collector supply, the relay voltage is approximately 0.6 V less than the supply voltage. Hence, a 12-V relay may require a slightly reduced spring tension if the supply voltage is only 12 V. The diode across the relay may be almost any low-voltage diode that is

2.17 COMPLEMENTARY EMITTER FOLLOWERS

A complementary emitter follower uses a pair of *npn* and *pnp* transistors that are connected in parallel and made to conduct on alternate half cycles, as in a class-B amplifier. The complementary emitter follower is particularly useful for driving capacitor loads or capacitor-coupled loads where the charged capacitor may reverse-bias the base-emitter junction of a simple emitter follower. With the complementary emitter follower the upper transistor charges the capacitor, and the lower transistor provides a low-impedance circuit for discharging the capacitor. The load is driven on both half cycles by a low impedance, and the amplifier may carry a greater load because each transistor operates only 50 per cent of the time.

The complementary emitter follower shown in Fig. 2.21(a) is used mainly for driving capacitor loads where the capacitance C is less than 10 μF. The transistors are germanium transistors that do not require a diode to reduce cross-over distortion. If the stage is used with a supply voltage exceeding 12 V, the transistors may be protected from second breakdown by connecting a 50-Ω, or larger, resistor in series with both collectors.

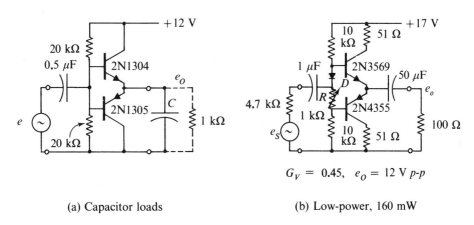

(a) Capacitor loads   (b) Low-power, 160 mW

Figure 2.21. Complementary emitter followers

The complementary emitter follower in Fig. 2.21(b) uses silicon transistors and needs the diode D and series resistor R to prevent cross-over distortion. Sometimes the adjustable resistor may be replaced by a second diode, in which case both a germanium and silicon diode should be tried while observing the distortion. Because the stage does not have overall feedback, there is usually a small residual distortion at low signal levels.

3 Two-Stage Transistor Amplifiers

Transistors offer important advantages of circuit simplicity and of more effective use of feedback when they are used as direct-coupled pairs or as 2-stage amplifiers with overall feedback. For many applications a 2-stage amplifier has as much voltage gain as is needed in a single package. With the transistors direct-coupled in pairs fewer components are needed, and the transistor gain is more fully utilized.

The amplifiers shown in this chapter are designed to operate over a wide range of supply voltage, with considerable flexibility in the amount of feedback employed. Most of the 2-stage amplifiers use complementary transistors in a CE-CE configuration that uses the feedback to limit the overall gain and to reduce the internal output impedance. The CE-CE pair, using like transistors, both *npn* or *pnp*, permits easy separation of the dc and ac feedback in order to maintain high ac gain, with the Q-point stabilized by dc feedback. The Darlington pair is useful for wide-band impedance transforming applications, while the emitter-coupled pair and the cascode may be used in wide-band applications when voltage gain is desired without limiting the input impedance by Miller-effect feedback.

The circuits of this chapter include high-gain, wide-band, and high-impedance feedback amplifiers, and cascode and emitter-coupled stages for high-frequency applications. Phase inverters and Darlington-pair amplifiers are described which may be used to couple high impedances to low-impedance loads.

3.1 TWO-STAGE CAPACITOR-COUPLED AMPLIFIER

Like vacuum tube amplifiers, early transistor amplifiers used a coupling capacitor between each stage or transformer coupling. By about the year 1960 the development of the planar transistors made possible the direct coupling of transistors in pairs, which eliminated a capacitor and one or two resistors. With fewer components the overall gain is increased and feedback is more easily applied. Moreover, with feedback the circuit performance is less dependent on the transistor characteristics and may be predicted by examining the feedback resistors.

The 2-stage capacitor-coupled amplifier in Fig. 3.1 illustrates the earlier method of constructing RC-coupled amplifiers. Each stage has 13 dB gain when driven by a 10-kΩ source, and a fixed bias resistor is satisfactory for supply voltages from 20 to 30 V. For the RC-coupled 2-stage amplifier the gain is 20, 26 dB, and the frequency response is from 50 Hz to 50 kHz.

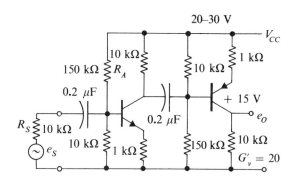

Figure 3.1. RC-coupled 2-stage amplifier

3.2 TWO-STAGE DIRECT-COUPLED AMPLIFIER WITH FEEDBACK (REF. 1)

The RC-coupled amplifier is converted to a direct-coupled pair by removing the interstage bias resistors and the coupling capacitor and by readjusting the bias resistor as shown in Fig. 3.2. With these changes the amplifier has a higher gain, 33 dB, and the frequency response is from 35 Hz to 40 kHz between 3-dB cutoffs. However, the amplifier is relatively sensitive to a change of the supply voltage, and the bias must be changed if the supply voltage changes more than ± 1 V.

The performance of the amplifier may be greatly improved by removing the local feedback in the first stage and using the increased gain to provide overall feedback, as shown in Fig. 3.3. In the improved form the amplifier has a gain of 32, 30 dB, a peak output voltage of 10 V rms, and a frequency

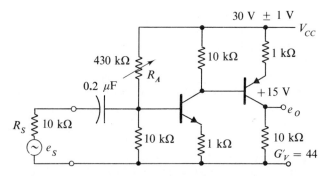

Figure 3.2. Direct-coupled 2-stage CE-CE amplifier

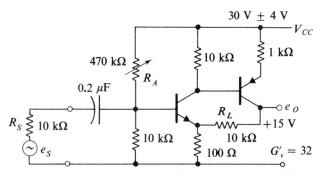

Figure 3.3. Direct-coupled CE-CE amplifier with feedback

response from 35 Hz to 160 kHz. With overall feedback the bias resistor should be within ± 10 per cent of the optimum value, and the performance is satisfactory for a supply voltage change of 8 V. Thus, the direct-coupled amplifier is simpler than one with RC coupling and has a higher gain and performance characteristics adequate for almost any application. For limited applications requiring more gain both emitter resistors may be lowered by a factor of 3. The gain is increased to 100, 40 dB, with a frequency response from 50 Hz to 120 kHz, and the low-frequency response may be lowered to 20 Hz by making the input capacitor 0.5 μF.

With a 12-V supply voltage the feedback amplifier using the circuit shown in Fig. 3.3 has only the advantage of simplicity because the gain and the sensitivity to a supply voltage change are about the same as for the capacitor-coupled amplifier. Significant feedback may be obtained by reducing the feedback resistor R_L and the gain, but the amplifier remains impractically sensitive to a voltage change for the reasons explained in Sec. 2.5. With the 30-V supply the CE stages are less sensitive to a supply voltage change and the loop gain is higher, so that the amplifier may have a higher gain with significant feedback, or more feedback may be used to improve the Q-point stability.

3.3 DIRECT-COUPLED AMPLIFIERS WITH SERIES FEEDBACK

Important advantages of the direct-coupled pair amplifiers are that feedback may be easily applied and the gain and impedance characteristics of the amplifier become almost independent of the transistors as long as the transistors remain undamaged by accidental abuse. A further advantage of using transistor pairs with feedback is that high-β transistors generally produce better performance characteristics with feedback, whereas selected low-β transistors are often required in circuits lacking significant feedback. For example, an integrated circuit with more than 10 transistors is often used for a gain of 10 or less, which can be obtained with a single transistor. The IC with feedback has many unselected transistors, but the circuit performance with feedback is accurately determined by the feedback circuit. The single stage with a selected transistor has relatively inferior characteristics.

The feedback amplifier shown in Fig. 3.4 is similar to the amplifier described in the preceding section except that the second-stage resistors are lowered to reduce the output impedance by a factor of 20. The voltage gain is reduced by increasing the feedback in order to maintain a 10-kΩ input impedance. The voltage gain of the amplifier is 20 from base to collector, which is approximately the ratio of the collector resistor to the emitter resistor, and the voltage gain from the source to the load is 10. The advantages gained by the feedback are that the input resistance is increased to 10 kΩ, the internal output impedance is reduced from 510 to 100 Ω, and the voltage gain

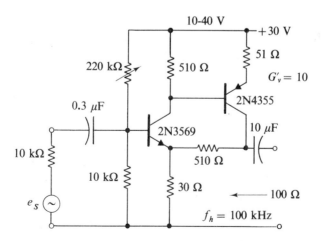

Figure 3.4. Wide-band CE-CE amplifier with feedback

is a factor of 10 times less sensitive to a change in the supply voltage. However, these advantages are not all available simultaneously, since, for example, a low-impedance load reduces the ac feedback current and reduces the ac input impedance. Thus, using feedback to obtain several advantages simultaneously usually requires an amplifier with 40 dB to 60 dB gain without feedback and a gain reduction by a factor of 10 to 100, which is referred to as *20 dB to 40 dB feedback.*

For many applications the amplifier shown in Fig. 3.4 makes a very satisfactory single-sided amplifier. The feedback amplifier has a voltage gain of 10 with a frequency response from 30 Hz to 100 kHz between cutoffs. The input impedance is 10 kΩ, and the amplifier may be used with a bias adjustment on supply voltages from 10 V to 40 V.

3.4 HIGH-INPUT IMPEDANCE CE-CE AMPLIFIERS WITH FEEDBACK

Two forms of the direct-coupled series feedback amplifier that are designed for high-input-impedance applications are shown in Figs. 3.5 and 3.6. The first amplifier (Fig. 3.5) has a voltage gain of 100, 40 dB, an input impedance exceeding 200 kΩ, and, with a low-impedance source, the high-frequency cutoff is at 10 kHz. When the bias supply is regulated by the 6-V Zener diode, the amplifier operates satisfactorily on supply voltages from 15 V to 30 V. Observe that the 1-kΩ emitter resistor is needed to protect the input stage transistor from collector current overload with high-input signals.

In the second amplifier (Fig. 3.6) the input impedance is increased by increasing the feedback and reducing the gain to 10, 20 dB. The 1-kΩ emitter

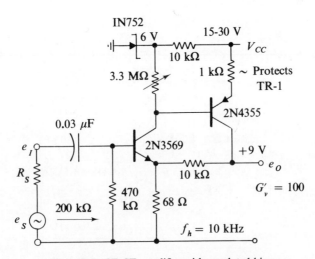

Figure 3.5. CE-CE amplifier with regulated bias

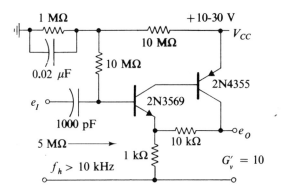

Figure 3.6. CE-CE high-input-impedance amplifier

resistor of the first stage eliminates the need for a current-limiting resistor in the second stage, and the increased feedback makes it possible to operate the amplifier on supply voltages from 10 V to 30 V, with the bias supplied through the 10-to-1 voltage divider. Both the high-impedance amplifiers exhibit an unusual insensitivity to a change of the supply voltage. The success of the high-gain amplifier comes from the regulated bias and of the low-gain amplifier from the feedback.

3.5 HIGH-GAIN, GENERAL-PURPOSE, TWO-STAGE AMPLIFIER

The amplifier shown in Fig. 3.7 may be used to obtain more than 60 dB gain (iterated) with supply voltages from 5 V to 40 V. The characteristics of

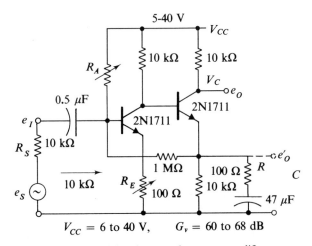

Figure 3.7. High-gain general-purpose amplifier

the amplifier are remarkably independent of the supply voltage, providing the bias resistor is adjusted to maximize the peak-to-peak output signal with the given supply voltage. With a 20-V supply the voltage gain is 66 dB, the peak-to-peak output is 12 V, and the frequency response between 3 dB points is 30 Hz to 20 kHz. When the bias resistor is adjusted to give the maximum peak-to-peak output voltage, the collector voltage is usually 0.7 V_{CC}, and the input impedance is a constant 10 kΩ. With a low supply voltage the frequency response extends from 20 Hz to 13 kHz, and with a high supply voltage it is from 30 Hz to 35 kHz.

The gain of the amplifier may be adjusted by changing the resistor R that is used in series with the emitter capacitor and, with both R and C removed, the voltage gain is 30 dB to either side of the output transistor, the input impedance is 10 kΩ, and the frequency response is 15 Hz to 22 kHz with supply voltages from 8 V to 40 V. A 5-kΩ variable resistor R_E may be used for gain control, but the bias resistor should be reduced to 2 MΩ with a 30-V supply, and the maximum output signal is reduced to 8 V, peak-to-peak. With the minimum gain, 40 dB, the upper cutoff frequency is reduced to 10 kHz, and the input impedance is 100 kΩ.

With the amplifier constructed as shown in Fig. 3.7, the relatively high gain may be exchanged for a higher input impedance merely by connecting a resistor in series with the input. This exchange degrades the signal-to-noise ratio a little over values obtainable with an amplifier designed for a minimum noise, but the noise is not significantly higher if the series resistor is less than 100 kΩ. A more satisfactory way of increasing the input impedance is, of course, to construct the amplifier with all resistors increased by a factor of 10 or more. As compared with an OP amp, the 2-stage amplifier requires about twice as many components, but the gain-bandwidth, 40 MHz, is higher than that of most OP amps, and, with only 2 stages, the high-gain and wide-frequency response is obtained without requiring the shielding, compensating adjustments, and care otherwise needed to obtain an equivalent performance.

Table 3-1 shows, for 5 different supply voltages, the optimum bias resistor and the most important characteristics of the amplifier shown in Fig. 3.7. For intermediate voltage values the characteristics may be easily interpolated.

Table 3-1 High-Gain Amplifier Design Data—Fig. 3.7

V_{CC} (Volts)	V_C (Volts)	R_A (M-ohms)	e_O (p-p V)	G_v (dB)	f_l (Hz)	f_h (kHz)
40	30	15	26	68	25	40
30	23	15	18	67	25	35
20	14	8	12	66	30	20
12	8	5	6	63	25	15
6	4	3	2.5	60	10	13

3.6 GENERAL-PURPOSE LINE AMPLIFIER (AUDIO)

A remarkably versatile amplifier for coupling a moderately high-impedance source to a low-impedance line is illustrated in Fig. 3.8. The output stage Q-point remains close to the optimum $V_{CC}/2$ for supply voltages from 8 V to 30 V once the feedback resistor is adjusted. The gain may be varied from 0 dB to 46 dB without changing the Q-point, and the product of the gain and input impedance varies only from 100 kΩ to 300 kΩ. Moreover, with a source impedance equal to the input impedance the frequency response extends to at least 30 kHz for all gain and V_{CC} values. With a source impedance lower than the input impedance the frequency response may extend to 300 kHz and make necessary an interstage capacitor C to prevent high-frequency oscillations.

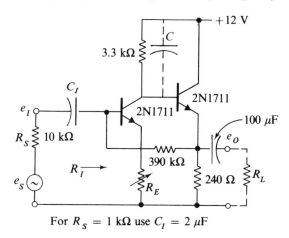

For $R_S = 1$ kΩ use $C_I = 2$ μF

Figure 3.8. General-purpose line amplifier (audio)

The amplifier may be coupled to loads or a line with an impedance as low as 50 Ω, with the 50 Ω load reducing the gain only 2 dB. The maximum peak-to-peak output voltage is 10 V with a 500 Ω load, 6 V with 200 Ω, and 3 V with 50 Ω. Table 3-2 shows the voltage gain, input impedance, and input

Table 3-2 Line Amplifier Design Data—Fig. 3.8

G_v (dB)	$G_v = e_O/e_I$	R_E (ohms)	C_I (μF)	R_I (k ohms)
46	200	0	2	1.2
30	30	68	0.3	10
20	10	300	0.1	24
10	3	1000	0.05	50
0	1	3000	0.03	100

capacitor needed to assure a 3-dB cutoff at 30 Hz. For many audio applications this general purpose amplifier may be adjusted by selecting only one resistor, R_E, to give a required gain or input impedance. The dc feedback is able to maintain the optimum Q-point for a relatively wide temperature change.

The amplifier in Fig. 3.8 has approximately 60 dB power gain, and the maximum power output is approximately equal to $V_{CC}^2/6000$ with only 8 per cent efficiency. The transistor should be capable of dissipating a power that is at least $V_{CC}^2/1000$ W, which limits the supply voltage to 22 V when a 0.5-W transistor is used, or 10 V with a 0.1-W device. The 2N1711 is rated for 0.8 W and may be operated with a 20-V to 28-V supply voltage, depending on the reliability desired.

3.7 HIGH-GAIN AUDIO-FREQUENCY AMPLIFIERS

The audio-frequency amplifier of Fig. 3.9 is useful when an exceptionally high-gain amplifier is required (70 to 80 dB) with enough flexibility available to use supply voltages from 8 to 50 V. The amplifier has a Darlington input pair direct-coupled to a CE output stage and has a control for adjusting the gain. Feedback from the output emitter to the input makes the amplifier performance relatively independent of the temperature and supply voltage.

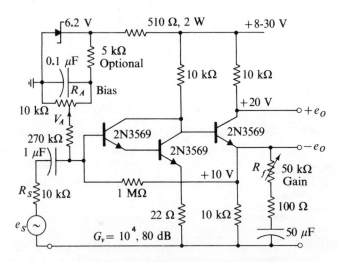

Figure 3.9. High-gain audio-frequency amplifier

The gain of the amplifier is adjusted by changing the feedback resistor R_f. With $R_f = 0$ the feedback is limited by the 100-Ω resistor, and the voltage gain is 80 dB with a 30-V collector supply. The gain measured from a low-impedance source to the output collector is down 3 dB at 20 Hz and 20 kHz, and the input impedance is 10 kΩ over this frequency range. When R_f is 50 kΩ, the gain is reduced to 200, 46 dB, the input impedance is 5 kΩ, and the output may be used as a phase inverter with peak-to-peak output signals of nearly 10 V.

The amplifier operates quite well for supply voltages down to 8 V, provided the bias is adjusted more positively as the collector supply is lowered. Because of the inverse relation between the optimum bias and the supply voltage, the amplifier operates better with a fixed bias voltage than when connected through a resistor to the collector supply. When the supply voltage is 30 V, the bias voltage V_A may be zero. Hence, the bias filter and the regulator may be eliminated. When the supply voltage is as low as 8 V and V_A is 3.5 V, the maximum output voltage is 1 V rms, and the voltage gain is 70 dB. The input impedance is 10 kΩ, and the 3 dB high-frequency cutoff is reduced to 10 kHz. With or without a change of the bias voltage, the voltage gain increases approximately with the supply voltage, and the relatively small change with temperature may be easily compensated. The amplifier has a simplicity, flexibility, and advantages in ac applications that cannot be obtained with most ICs because the gain-bandwidth product is in excess of 200 MHz.

3.8 HIGH-GAIN EMITTER-FEEDBACK AMPLIFIERS

The high-gain amplifier described in Sec. 3.7 may be reduced to a simpler form when 20 dB lower gain is acceptable. The CE-CE amplifier shown in Fig. 3.10 requires only half as many components but offers a voltage gain of 1000, 60 dB, with a 10-kΩ source impedance, and a frequency response from 35 Hz to 30 kHz, the 3 dB cutoffs. The gain is 60 dB with the switch closed

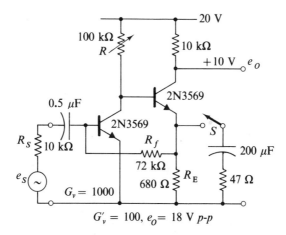

Figure 3.10. High-gain CE-CE emitter-feedback amplifier

and 100, 40 dB, with S open. The input impedance for either position of the switch is between 3 and 5 kΩ. With the emitter fully bypassed by 1000 μF, the overall voltage gain is 5000, 74 dB, and the upper cutoff frequency is reduced to 7 kHz.

With either type of ac feedback the amplifier performs very well, as the supply voltage is varied upward from 17 V, and the Q-point bias is adjusted by changing the first-stage collector resistor R or the second-stage emitter resistor. The high-gain amplifier shown in Fig. 3.10 is entirely satisfactory for room-temperature applications. Amplifiers in this form are designed for wide temperature range applications by reducing the dc voltage gain of the second stage until the Q-point drift is acceptable. With further adjustments the first-stage drift can be made to compensate for the second-stage drift over a part of the temperature range.

3.9 TWO-STAGE COMBINED FEEDBACK AMPLIFIER

The 2-stage amplifier shown in Fig. 3.11 shows how 2 stages designed for use with equal load and source impedances may be coupled to give twice the gain of a single stage. The amplifier is interesting because the overall gain is 64 dB, which is unusually high for an amplifier with only 2 semiconductors, and significant feedback makes possible only a 10 dB change in gain as the supply is reduced from 30 V to 5 V.

The 2 stages are essentially alike except that the output stage is biased for a maximum peak output voltage. With a 20-V supply the second collector

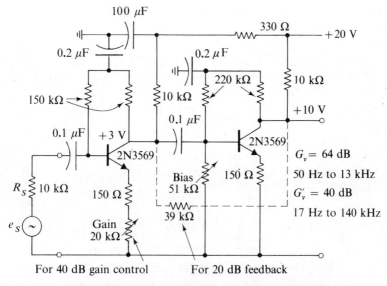

Figure 3.11. Two-stage combined feedback amplifier

voltage is 10 V, and a 17-V peak-to-peak sine wave signal output may be obtained. The frequency response is from 50 Hz to 13 kHz with the upper cutoff frequency produced by the Miller effect that exists in both stages. However, the bandwidth of the amplifier may be increased, especially toward higher frequencies, by connecting a feedback resistor from collector to collector. The overall gain is reduced 20 dB with a 39-kΩ resistor, and the high-frequency cutoff is increased to 140 kHz. The low-frequency cutoff is reduced a factor of 3 to 17 Hz. The amplifier is stable with 23 dB feedback, but 10 dB feedback is sufficient for audio-frequency applications.

The broad-band equivalent input noise is relatively low (8 μV peak-to-peak), the frequency response decreases approximately 40 dB/decade below the low-frequency cutoff, and 20 dB/decade above the high-frequency cutoff.

3.10 EMITTER-COUPLED TWO-STAGE AMPLIFIER

The amplifier shown in Fig. 3.12 has identical CE and CB stages that are direct-coupled at the emitters. The signal is applied at the base of the first stage and returns to ground through the second-stage base. Because of the

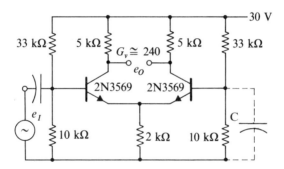

Figure 3.12. Emitter-coupled amplifier

relatively high value of the emitter resistor R_E, the current leaving one emitter enters the other and forces the second stage to follow the first with the same signal amplitude but with reversed polarity. The amplifier responds as if the input signal were applied from base to base, so there is no signal current in the emitter resistor. The voltage gain from the input to either side is 120, 42 dB, and the differential collector-to-collector gain is 240, 48 dB. Analysis of the dc circuit relations shows that the Q-point responds to the collector-supply voltage and the bias, similar to an emitter feedback stage with the S-factor equal to 2.5.

The amplifier has the gain of a CE stage with the emitter bypassed and is used when high gain is required and an emitter bypass capacitor is not permitted, as in an IC. The amplifier has the advantages of a CE stage without the usual phase turnover, or it may be used as a high-gain phase inverter with signals of both polarities available at the collectors. For the foregoing applications the amplifier is limited to the audio-frequency range.

In high-frequency applications the input-stage collector may be connected to the power supply without the resistor, and the amplifier is a form of cascode that gives high gain without the Miller feedback. When the Miller effect is removed, the upper cutoff frequency is increased to about 1 MHz, provided the second-stage base is bypassed to ground. The base capacitor C may be selected by trial to boost the high-frequency cutoff. In many respects the emitter-coupled pair is a remarkably versatile amplifier. With cross-connected feedback the emitter-coupled pairs are used as flip-flops, trigger circuits, and frequency dividers.

3.11 CE-CE HIGH-GAIN PAIR AMPLIFIER

An npn and a pnp transistor may be direct-coupled and used as a single high-β transistor. An amplifier using the pair, as shown in Fig. 3.13, has an input impedance of 0.5 MΩ and a voltage gain of 80. When driven by a high-impedance source (0.5 MΩ), the 3-dB high-frequency cutoff is at 2.5 kHz. The response may be extended to higher frequencies by decreasing either the load impedance R_L or the source impedance R_S. However, the frequency response may be moved to a 2-to-3 times higher frequency by adding a neutralizing capacitor and variable resistor, as indicated in Fig. 3.13. Because neutralization is a form of positive feedback, the amplifier may be unstable or oscillate when the frequency response is increased more than 6 dB in the cutoff region. For use with a lower impedance source the frequency response

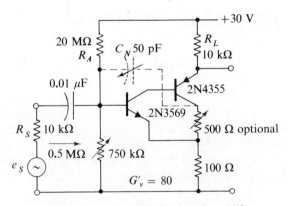

Figure 3.13. CE-CE high-gain pair amplifier

may be made to cover the audio band by decreasing all the resistors by a factor of 10. For room-temperature applications the amplifier has an unusually low component count, a 50:1 impedance step-down, and 38-dB voltage gain.

3.12 CE-CB CASCODE PAIR AMPLIFIER

A pair of transistors used in a configuration known as a *cascode* have a CE stage followed by a direct-coupled CB stage. Because the input impedance of the CB stage is relatively low, there is negligible Miller feedback in the first stage, and the cascode has a constant amplitude-frequency response up to relatively high frequencies. The cascode shown in Fig. 3.14(a) has a voltage gain of 40 dB up to 1 MHz and is designed to couple a 1-kΩ source to a 10-kΩ load. The amplifier in Fig. 3.14(b) is designed for iterative applications and has a voltage gain of 42 dB up to 100 kHz with the emitter capacitor. Without it the voltage gain is 23 dB and the high-frequency cutoff is above 0.5 MHz. The high-frequency cutoff is sometimes moved a little higher by connecting a base-to-ground bypass capacitor.

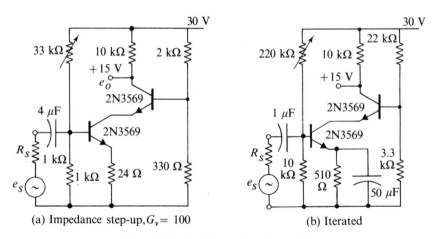

Figure 3.14. CE-CE cascode pair amplifiers

3.13 THE DARLINGTON CC-CC AMPLIFIER

The Darlington compound, shown in Fig. 3.15, provides one of the simplest ways of direct-coupling 2 transistors and is useful as an impedance transformer that has a very wide-band frequency response. The Darlington connection has slightly less than unity voltage gain but gives a very high current gain (β^2), so that the input impedance approaches $R_I = \beta^2 R_L$. The

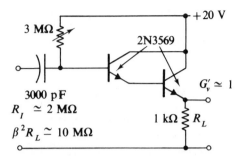

Figure 3.15. Darlington CC pair

input transistor should be selected for its ability to provide a high current gain at one-tenth to one-hundredth the emitter current of the second stage. Often Darlington pairs will use identical transistors with the second stage operated at an emitter current that gives a maximum current gain.

The circuit of a high-gain Darlington pair with S-factor control is shown in Fig. 3.16. The amplifier transfers the input signal at an impedance level of 0.8 MΩ to the load at 1 kΩ. The ac voltage loss in the amplifier is practically negligible, while the dc loss is 1 to 2 V, depending on whether the transistors are low-power devices, as in the figure, or power transistors, as in a power amplifier. The ac power gain in the amplifier is simply the ratio of the input impedance to the load resistance, which is 800, 29 dB. In a power amplifier in which the transistors are operated at high current levels and with lower S-factors, the power gain may be only 10 dB.

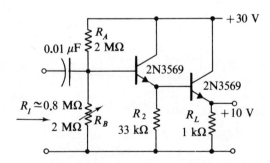

Figure 3.16. CC pair amplifier, 30 dB

A Darlington pair is easily substituted for a single transistor almost any time a higher current gain can be used to advantage. Sometimes three or more stages of emitter followers are found in circuits of a decade ago. However, with today's devices 2 high-β transistors or an FET stage usually provides a better way to obtain high input impedance. Some power transistors are supplied as Darlington pairs as a way of increasing the current and power gains.

These devices are used as if they were a single transistor with a current gain of 3000 to 10,000.

3.14 DARLINGTON PAIR FOR POWER LOADS

The Darlington amplifier is suitable for driving many low-power loads such as relays, solenoids, or electromagnets. For these devices a highly linear input-output relation is not required, and a single power transistor with a low-power driver may be used to drive sizable power loads. A germanium transistor is often used in the output stage for low cost with a silicon transistor for the input stage. The silicon transistor helps to make the amplifier relatively insensitive to temperature changes, but the resistors used to limit the current gain of both stages are perhaps of most importance in making the amplifier reliable and stable over a wide temperature range or suitable for industrial use. A good practical rule is to limit the S-factor to about 20 per stage, or a maximum of 400 overall. Lower S-factors may be needed for high operating temperatures.

The practical form of the Darlington amplifier shown in Fig. 3.17 has an overall S-factor of 200 and an input impedance of about 2 kΩ with a 10-Ω load resistance. By increasing all the resistors by a factor of 100, the load impedance becomes 1000 Ω and the input impedance is about 200 kΩ. For either form of the amplifier the voltage gain is just a little less than 1, which means that the compound is providing a power gain of nearly 200, 23 dB.

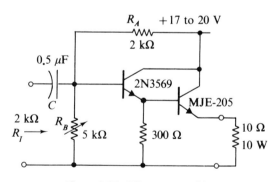

Figure 3.17. CC power amplifier

The circuit in Fig. 3.18 illustrates the use of a CC pair for operating a solenoid or relay with a pulse input signal. With a 12-V solenoid a 14-V square pulse is adequate, provided the time constant RC is 5 to 10 times the pulse time and the time required to operate the solenoid. If it is operated by switching the capacitor, as indicated in Fig. 3.18, the supply voltage must be more than 15 V or a very large capacitor is required. The diode across the solenoid is necessary to protect the output transistor from a high reverse

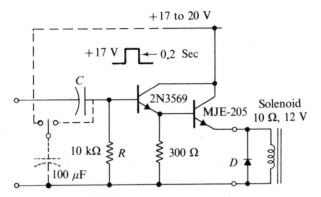

Figure 3.18. CC solenoid pulser

voltage when the solenoid is turned off. However, the diode slows release of the solenoid, and a resistor in series with the diode may sometimes be used to obtain faster release.

A circuit for operating a 12-V dc relay by a low-level ac signal is shown in Fig. 3.19. An 8-V peak-to-peak signal applied from a 10-kΩ source is converted to dc in a voltage-doubling rectifier. The 2-stage amplifier has a gain of approximately 40 dB, and the relay may be operated with input frequencies from 30 Hz to 40 kHz. The operating times are 0.5 s for turn-ON and 0.1 s for turn-OFF. A 1-to-2 V higher supply voltage is required if the transistors are connected as a Darlington pair.

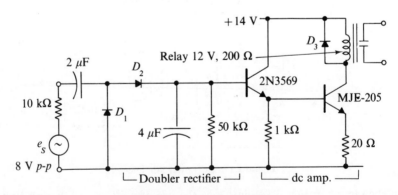

Figure 3.19. Ac–dc relay amplifier

3.15 CE COLLECTOR FEEDBACK PAIRS

The direct-coupled collector feedback amplifier shown in Fig. 3.20 offers rather useful and interesting performance characteristics. The amplifier uses a Darlington pair as a direct-coupled CE amplifier with a 5-kΩ collector load

Sec. 3.15 CE Collector Feedback Pairs

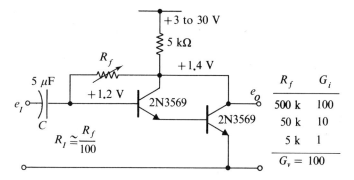

Figure 3.20. Collector feedback CC-CE pair amplifier

resistor. Because there are 2 emitter diodes in series, the input base operates at about 1.2 V above ground. The collector feedback tends to hold the collector Q-point voltage only about 0.2 V above the base voltage. If the collector voltage falls below 1.4 V, the current gain of the first stage falls and the collector voltage rises. If the collector voltage rises above 1.4 V, the bias current and the current gain increase, causing the collector voltage to fall. The collector voltage regulation is so effective that the collector is maintained at about 1.4 V above ground for 10-to-1 collector supply voltage changes and for a considerable change in the ambient temperature. The regulation also makes it possible to change the Darlington current gain as much as 1000 to 1 by varying the feedback resistor without adversely changing the Q-point. This feedback resistor provides one of the simplest ways of changing the current gain of a transistor amplifier. The voltage gain is approximately 100 and is independent of R_f.

The amplifier in Fig. 3.21 is a 3-stage version of the collector feedback pair. In this amplifier advantage is taken of the relatively fixed collector volt-

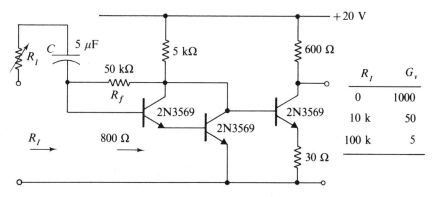

Figure 3.21. Three-stage CC-CE-CE feedback amplifier

age of the first stage in direct-coupling a third stage. The third stage increases the voltage gain by 10, so the overall voltage gain is 1000. The input impedance of the amplifier is 800 Ω, but we may easily add series resistance as an exchange of voltage gain for a higher input impedance. If the series input resistor is made 0.1 MΩ, the voltage gain drops to about 5, 14 dB. If high temperature stability is not required, the feedback resistor can be increased to 500 kΩ and R_I increased to 1 MΩ. The amplifier then has a voltage gain of about 10. The amplifier allows great flexibility in the choice of gains and impedances as long as output signal levels below 1 V rms can be tolerated.

The collector feedback amplifier is easy to use, requires few components, and for room temperature applications can be direct-coupled to a fourth stage. If the Darlington pair shown in Fig. 3.21 is replaced by a single CE stage, the direct-coupled CE-CE amplifier provides a simple replacement for increasing the gain of an existing CE stage without adversely degrading the Q-point stability.

3.16 PHASE INVERTERS

A phase inverter is used to produce a pair of equal but out-of-phase signal sources. Where additional voltage gain is needed, the emitter-coupled stage, shown in Fig. 3.12, has the advantage that phase inversion is obtained with a relatively high gain and without requiring an emitter capacitor. The amplifier tends to be self-balancing, and the signal peaks are symmetrically clipped. However, the emitter-coupled phase inverter has too high an internal output impedance for use as a power-stage driver.

For applications that do not require additional voltage gain a single transistor may be used for phase inversion. The stage shown in Fig. 3.22 is called a *split-load phase inverter* because the load resistor is divided between the emitter and collector. The emitter resistor is usually about one-tenth the base resistor R_B, and the collector resistor is 2 RMA resistor values larger

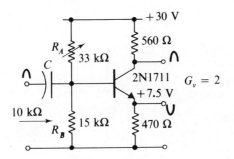

Figure 3.22. Split-load phase inverter

than the emitter resistor. The signals across each of the load resistors equal the input signal, except that the emitter voltage is in phase and the collector voltage is 180° out of phase with the input signal.

The bias resistor is adjusted to make the dc emitter voltage one-fourth the supply voltage. The maximum ac output voltage on each side is 1 V rms for each 8-to-10 V dc of the supply. For a first trial the bias resistor may be made approximately 3 R_B. The voltage gain of the phase inverter is 1 to each side, but from the input to the full load the voltage gain is 2. The effective power gain of the stage shown in Fig. 3.22 is about 40, 16 dB.

A split-load phase inverter is a class-A stage and is not usually suitable as a driver stage for a power amplifier. One of the problems in power applications is that the emitter presents a low-impedance source and the collector side has a higher impedance R_L. With a low-impedance load connected to the phase inverter, the collector signal peaks are clipped and the collector-to-emitter waveform may have considerable second harmonic distortion. The complementary CE-CE feedback pair has a low impedance on both sides and makes a better low-impedance phase inverter for driving power stages.

3.17 LOW-IMPEDANCE PHASE INVERTER

The CE-CC transistor pair shown in Fig. 3.23 is equivalent to a single high-β transistor and may be used as a Darlington pair in many respects. In the circuit illustrated here the pair is used as a phase inverter with the advantage that there is adequate feedback to give both output terminals a low internal output impedance. A phase inverter using a single transistor presents a relatively high impedance on the collector side and does not make a satisfactory low-impedance driver. The phase inverter shown in Fig. 3.23 has an input

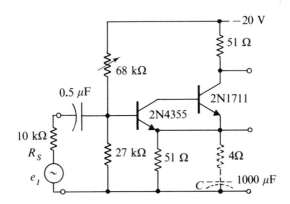

Figure 3.23. Feedback-pair phase inverter

impedance of 20 kΩ and an impedance step-down exceeding 200 with a voltage gain of 0.8 to each side. With the bias adjusted to make the output Q-points -5 V and -15 V, the no-load output signal may be nearly 5 V peak-to-peak, the internal output impedances are less than 1 Ω, and the 2N1711 transistor needs a clip-on heat sink to dissipate 1 W. If different transistors are used, an intermediate base resistor may be needed for the second stage, and trial will determine whether the resistor should be connected to ground or to the collector supply.

The phase inverter may be converted to an amplifier with a voltage gain of 10 by using 4 Ω in series with 1000 μF to bypass the input emitter partially. The amplifier has significant feedback, a power gain of approximately 40 dB, but a power output of only 50 mW to 100 mW.

4 Field-Effect Transistor Amplifiers

The JFET and MOS FET semiconductors give the designer the advantages of high input impedance, simple bias circuits, low noise levels, and a relatively low temperature sensitivity. Field-effect devices provide simple high-impedance circuits but usually produce low-voltage gains when used alone. Hence FETs are often paired with transistors, and, with the high transconductance of transistors, the pairs make feedback an economical way of offsetting the inherent variability of FETs.

The field-effect semiconductors are available in a wide range of characteristics and in n-channel and p-channel types that may be used with either power supply polarity. The junction-type FETs (JFETs) are made for use as amplifiers, switches, or voltage-controlled variable resistors. The switching types usually have a low ON resistance and are OFF until the gate voltage exceeds 6 to 10 V. The RF amplifier types must have a high transconductance and low gate-to-drain capacitance. The low-frequency amplifier types should have a high *figure of merit*, which means a high ratio of transconductance to zero-bias drain current. A high figure of merit implies a pinchoff voltage less than 2 or 3 V, whereas the switching FETs have a pinchoff voltage exceeding 6 V to suppress the possibility of switching by low-energy noise transients.

The insulated gate field-effect devices (MOS FETs) are available with transconductances exceeding 10,000 μmho and low gate-to-drain feedback capacitance. The single-gate and dual-gate FETs with even lower capacitance are used in RF and UHF tuned amplifiers, where they offer several advantages over high-frequency transistors. The insulated gate has an exceedingly high dc resistance that makes these FETs useful in electrometer-type amplifiers where the diode and rectifying characteristics of the JFET must be avoided.

The MOS FETs are available in enhancement types that are normally OFF or in depletion types that are normally ON, like the JFETs. The complementary MOS FETs (CMOS) are extensively used as logic switches and for memory storage in computers and control circuits.

The circuits of low-frequency amplifiers in this chapter usually use FETs with one or two transistors and overall feedback. The circuits are mainly for dc and audio-frequency applications, because circuits for RF and UHF applications are readily available in transistor handbooks and the amateur radio literature. The high-frequency applications are growing rapidly, and the handbook circuits are updated almost annually. However, examples of the most common RF and UHF circuits are included in Chapter 11.

4.1 RC-COUPLED FET STAGES

Field-effect stages may be cascaded to take advantage of the high interstage impedance for filtering or signal-shaping, and resistance coupling is the simplest way of providing both a low-frequency and high-frequency cutoff. However, with a simple 4-component RC-coupling network the narrowest frequency band that can be attained has an equivalent Q of less than 0.5. For recording and voice-frequency applications a narrow-band RC-coupled amplifier has desirable transient characteristics and usually has as narrow a band as can be tolerated.

The RC-coupling circuit shown in Fig. 4.1 is designed to have the narrowest frequency band that can be attained with a 1-kHz peak frequency. As long as the input capacitance of the second stage can be neglected, the peak frequency may be changed without changing the equivalent Q by reducing or increasing both capacitors by the same factor. The peak frequency is at the geometrical mean of the 3-dB low and high cutoff frequencies. At the peak frequency, 1 kHz, the output signal is 6 dB below the drain signal that would exist with the capacitors and R_G removed. The equivalent Q of the coupling circuit with $R_G = 10\, R_L$ is 0.48, and the bandwidth B between the 3-dB

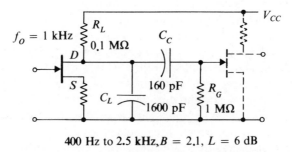

400 Hz to 2.5 kHz, $B = 2.1$, $L = 6$ dB

Figure 4.1. Narrow-band RC coupling, $R_G = 10 R_L$

frequencies is 2.1 times the peak frequency. The cutoff frequencies are 400 Hz and 2.5 kHz.

The peak frequency is unchanged for any values of the capacitors that keep their product unchanged, and for a fixed value of the peak frequency f_0 any change of the capacitors from the values shown in Fig. 4.1 only produces a wider bandwidth. For any values of the resistors and capacitors the peak frequency may be found by the relation

$$f_0 = \frac{1}{2\pi\sqrt{R_L R_G C_L C_C}} \qquad (4.1)$$

Equation (4.1) shows that the center frequency is unchanged as long as the product of the resistors and capacitors has a fixed value. Similarly, the peak frequency is always the mean of the frequencies obtained on a reactance chart when pairing each capacitor with one of the resistors.

With equal values of the drain and gate resistors, as shown in Fig. 4.2, the coupling circuit has a minimum bandwidth that is nearly 3 times the peak frequency. The equivalent Q is 0.35, and the peak-frequency coupling loss is 12 dB. Interchanging the capacitors moves the cutoff frequencies to 250 Hz and 3.8 kHz. $B = 3.6$, and the loss L is 8 dB.

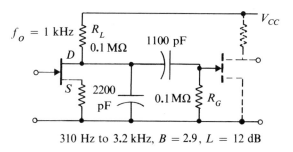

310 Hz to 3.2 kHz, $B = 2.9$, $L = 12$ dB

Figure 4.2. Narrow-band RC coupling, $R_G = R_L$

Interchanging the coupling capacitors of the narrow-band amplifier, as in Fig. 4.3, moves the cutoff frequencies to 100 Hz and 9 kHz, $B = 9$, and the peak-frequency coupling loss is only 1 dB. With a bandwidth 10 times the peak frequency the high and low-cutoff frequencies may be calculated independently, and the frequency response is essentially constant between twice the low cutoff and one-half the high cutoff. The low cutoff is the frequency at which the reactance of the shunt capacitor C_L equals the value of the resistors in parallel. The high cutoff is at the frequency at which the reactance of the coupling capacitor C_C equals the series value of the resistors. These cutoff frequencies may be found with a reactance chart.

The RC-coupling design chart in Fig. 4.4 gives the bandwidth $B = (f_h - f_l)/f_0$ and the peak-frequency loss L for a narrow-band coupling circuit

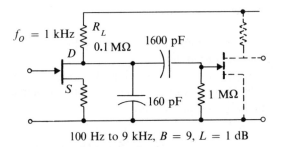

100 Hz to 9 kHz, $B = 9$, $L = 1$ dB

Figure 4.3. Wide-band RC coupling, $R_G = 10 R_L$

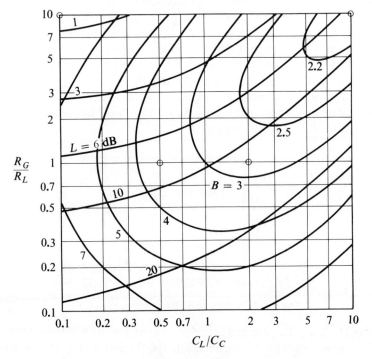

Figure 4.4. Narrow-band RC coupling design chart

with specified ratios of the capacitors and resistors. The circles on the chart refer to the narrow-band circuits in Figs. 4.1 and 4.2; they also refer to the examples with the capacitors interchanged. The chart shows that a narrow band and a low loss require high ratios of the resistors and capacitors. The chart also shows that a given bandwidth may be obtained with a high capacitor ratio that produces a high coupling loss where a lower capacitor ratio may give the same bandwidth with a lower loss. Thus, in a wide-band amplifier a shunt capacitor smaller than the coupling capacitor gives a smaller loss than if the same bandwidth is obtained with a relatively large shunt capacitor.

4.2 SINGLE-STAGE, RC-COUPLED, FET AMPLIFIERS (REF. 1)

Two examples of single-stage FET amplifiers that are designed for RC-coupled applications are shown in Fig. 4.5. The stage in Fig. 4.5(a) is designed for zero-biased operation and has a voltage gain of 10. The stage in Fig. 4.5(b) has significant source feedback and has a gain of 8 that may be increased to 30 by connecting a large capacitor across the source resistor. For either stage the FET may be any general-purpose amplifier type that has a V_P rating less than one-fourth the supply voltage and, preferably, a zero-bias drain current that is less than 1 mA.

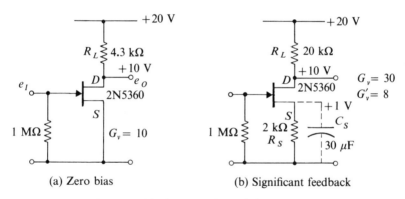

Figure 4.5. Single-stage RC-coupled FET stages

The resistors R_L and R_S are selected by an experimental trial. For zero-bias operation the drain resistor is adjusted to make the dc drain voltage one-half the supply voltage or lower, if the stage is used only with small signals. The stage with significant dc feedback is designed by adjusting the resistors to make the drain voltage one-half the supply voltage while at the same time making the gain with feedback approximately one-fourth the gain with the feedback bypassed. The stage has enough feedback to permit interchanging FETs that may be called typical when there is significant feedback. The capacitor C_S may by connected across the feedback resistor for additional gain and may best be selected by trial, using a signal with the lowest expected frequency.

For many applications the simplicity of zero-bias operation justifies the selection of the FET or an adjustment of the load resistor. The gain with zero bias is approximately the same as with significant feedback. For both stages the peak-to-peak output signal is nearly equal to the supply voltage. However, the stage with feedback has a Q-point that is almost independent of temperature. The stage with zero bias may be operated in ambient temperatures from 0°C to 60°C. The stage with feedback may be operated from -40°C to 100°C, provided the FET is biased to make the Q-point drain current 0.6 mA.

4.3 HIGH-GAIN FET STAGES

The amplifiers shown in Fig. 4.6 illustrate the advantage of using high figure-of-merit FETs, that is, low V_P devices. A high g_m-to-drain current ratio makes it possible to use a high-valued drain resistor and to obtain a high voltage gain. In addition, because the dc drain voltage must be at least as high as V_P, a high figure of merit implies that the drain may be operated at low voltages or from a low supply voltage. The stage using a 2N3086 FET requires a low drain resistor with 6 V on the drain. This stage has a voltage gain of only 2, but with a negative gate bias the drain resistor can be increased until the voltage gain is about 10. The 2N3687 FET stage uses a 10 times higher drain resistor and with zero bias has a voltage gain of 25. A disadvantage of both amplifiers in Fig. 4.6 is that they become inoperative if the drain supply voltage is lowered without lowering the drain resistor. This difficulty can be minimized by using a lower-value drain resistor and accepting a lower gain.

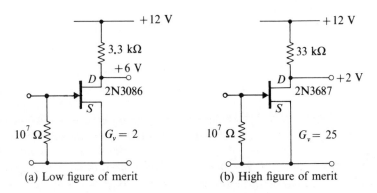

Figure 4.6. JFET RC amplifiers

The amplifiers in Fig. 4.7 show that higher voltage gains may be obtained by using a second FET as a constant current load in place of the drain resistor. When a pair is used as in Fig. 4.7(a), the FETs should be selected so that they are reasonably alike. The small source resistor R_S is used to balance the drain currents and to adjust for a maximum voltage gain. If maximum voltage gain is obtained with the source resistor shorted out, the FETs should be inter-

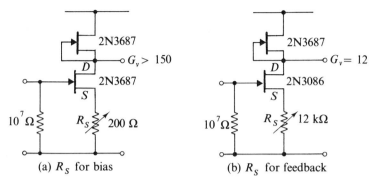

Figure 4.7. JFET amplifier with FET load

changed in order to use the higher I_{DSS} device as the amplifier and to permit drain-current balancing. If the source resistor has to be made so high that the gain is reduced by feedback, a pair of FETs even more alike must be selected.

The amplifier in Fig. 4.7(a) is remarkable for its high voltage gain, high temperature stability, and low noise characteristics. With similar FETs the collector Q-point will show little temperature drift. The voltage gain is one-half the amplification factor of the device. This amplifier operates well with supply voltages of only 4 V, but the voltage gain varies approximately in proportion to the supply voltage. By replacing the lower FET with a high drain current device 4.7(b), the source feedback resistor may be increased to provide significant feedback. Using a 2N3687 FET in combination with a 2N3086 FET and a 12-kΩ source resistor, we can reduce the voltage gain to 25 from a no-feedback gain of 125. With a 15- to 30-V drain supply—well above the device pinchoff voltage—the gain changes are reduced significantly by the feedback.

4.4 FET-TRANSISTOR AMPLIFIERS

Two amplifiers that use direct-coupled FET-transistor pairs with overall feedback are illustrated in Fig. 4.8. The phase inverter in Fig. 4.8(a) has a voltage gain of 1 from the input gate to each side and a 3-dB high-frequency cutoff at 120 kHz. The amplifier in Fig. 4.8(b) is obtained by reducing the feedback resistor and adjusting it to make the dc emitter voltage one-half the supply voltage. The gain is 50, and the 3-dB high-frequency cutoff falls inside the audio band unless, as shown, the source impedance is reduced to 240 kΩ. The phase inverter illustrates the need for a positive gate-to-ground bias voltage when an FET stage has a relatively large dc feedback voltage. With a low-impedance source the cutoff frequencies are 200 kHz for both amplifiers.

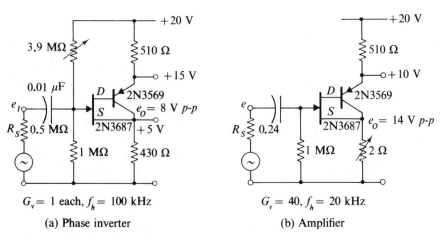

Figure 4.8. FET-transistor pair amplifiers

4.5 OSCILLOSCOPE PREAMPLIFIER

The input capacitance of an oscilloscope and the capacitance of an interconnecting lead produce a high-frequency cutoff that may limit the use of an oscilloscope in high impedance measurements. A 30 cm length of shielded cable may have a capacitance of 100 pF, which has a reactance of 1 MΩ at 1600 Hz. Even with an input impedance of only 10 kΩ the lead capacitance reduces the measured signal 3 dB at 160 kHz.

The oscilloscope preamplifier shown in Fig. 4.9(a) may be used to increase the upper cutoff frequency by a factor of 10, or the amplifier may be used, as in Fig. 4.9(b), with either a 10 times higher gain or a 10 times higher

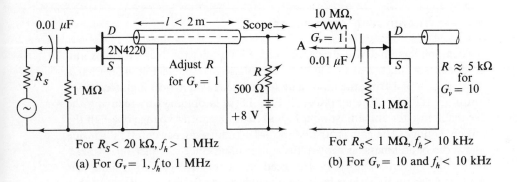

Figure 4.9. Oscilloscope remote preamplifier

input impedance. As shown in the figures, the JFET is placed at the input end of the cable, and both the dc power and the output signal are carried over a single-wire, shielded cable. The 2N4220 JFET has a high enough transconductance to provide a gain of 10 with R_L less than 10 kΩ, or a gain of 1 with R_L less than 1 kΩ. With zero bias the drain current is less than 2 mA, which may be supplied by a 9-V transistor battery with a battery life of approximately 150 hours, 300 mAh.

4.6 HIGH-IMPEDANCE REMOTE PICKUP

High-impedance pickups or transducers are sometimes located 30 m, more or less, from an amplifier. For low noise and a wide-band frequency response the first stage of the amplifier may be placed near the pickup with the input-stage power supplied over a 2-conductor line that is used also for the return signal. The circuit illustrated in Fig. 4.10 is useful for remote pickups from the low-audio frequencies up to 100 kHz, depending on the length of line between stages. The system is a form of FET-transistor compound, and, with the emitter bypassed, the second-stage input impedance is approximately 1 kΩ. The line used to interconnect the stages may be a twin-lead pair for high-level signals and low frequencies, but for low-level, high-frequency signals the line is preferably a coaxial cable.

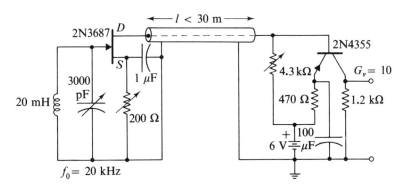

Figure 4.10. High-impedance remote preamplifier

The amplifier may be operated on a 6-V battery supply with a current drain of only 2 mA, and the gate-to-collector voltage gain is 10, 20 dB.

With a shielded 2-conductor line the collector load resistor may be returned to the FET source to provide feedback. However, for adequate feedback the amplifier should have a higher voltage gain and will require a 20- to 30-V supply voltage.

4.7 LOW-IMPEDANCE REMOTE PICKUP

For low-impedance transducers or signal sources a remote stage may be supplied power through a coaxial cable, as shown in the circuit in Fig. 4.11.

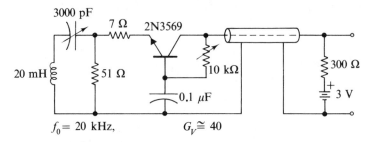

Figure 4.11. Low-impedance remote preamplifier

For illustration, a resonant circuit is shown as the pickup, and the transistor is used as a CB stage with an input impedance of approximately 8 Ω. The transistor power is supplied via the 300-Ω load resistor and the connecting line. The line used for the interconnection may be selected as described in the preceding section. The transistor base is shown grounded through a large capacitor, but a resistor with a shunt-compensating capacitor may be better for some applications. The overall emitter-to-collector gain is approximately 40, depending on the method used to bypass the base-to-ground. If the resonant circuit has a Q exceeding 5 with the emitter connected, the tuned circuit acts as an efficient transformer of energy received by the coil. The series-resonant system and amplifier have been used for remote reception and amplification of VLF radio signals at 15 kHz and above.

4.8 LOW-POWER, DC PREAMPLIFIER

The low-power preamplifier shown in Fig. 4.12 has a voltage gain of 100, 40 dB, and is useful where extreme simplicity and a low-power dc amplifier are required. The amplifier operates on 2 dry cells and requires only 100 μA, which D-size cells can supply for several years. The stage may be turned off by connecting the input gate to the negative battery. The gain may be made exactly 100, and the collector dc voltage may be reduced to zero for zero offset by adjusting the bias resistor and the collector load resistor. For the components shown in Fig. 4.12 the FET should have V_P less than 1 V, and the transconductance should be above 100 μmho with 0.1 mA zero-bias drain current.

The amplifier has the advantage of low noise, which with a 1-MΩ source impedance and a 25-kHz bandwidth is substantially thermal noise, 20 μV

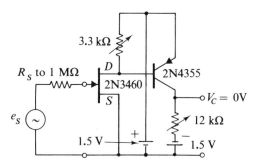

Figure 4.12. Low-power dc preamplifier, $G_V = 100$ to 25 kHz

rms. The bandwidth of the amplifier is determined by the Miller-effect feedback in the second stage, and for source impedances up to 1 MΩ the upper cutoff frequency is 25 kHz.

The first stage of the amplifier may be remotely located with the power supplied over a coaxial cable, as shown in Figs. 4.9 and 4.10. The preamplifier has the advantage that it may be used as a dc amplifier with zero dc offset from input to output, and the frequency response covers the audio range. However, because the amplifier uses only one FET, there is a small Q-point drift with temperature changes that may present drift problems when the preamplifier is followed by a high-gain dc amplifier.

4.9 TWO- AND THREE-STAGE FET-TRANSISTOR AMPLIFIERS

Multistage amplifiers often use FETs and transistors in alternate stages or direct-coupled as FET-transistor compounds. The FETs make it possible to use high interstage impedance levels—i.e., smaller coupling and filter capacitors—and allow the use of high-impedance potentiometers as controls. Transistors supply high transconductances at low dc collector currents. Thus, amplifiers combining FETs with transistors can be expected to exhibit the advantages of both devices. One of the most useful combinations uses a FET direct-coupled to a CE transistor amplifier.

The 2-stage amplifier in Fig. 4.13(a) illustrates the advantages of an FET-transistor pair amplifier. The FET input stage provides a high input impedance, and the transistor stage with a 5-kΩ load resistor provides a voltage gain of 80 with feedback. The FET provides an input impedance that is high enough for almost any purpose, but the signal source impedance should be 100 kΩ or less, for a flat frequency response. The relatively low gain in the first stage minimizes the Miller effect, and with a 100-kΩ source the 3-dB high-frequency cutoff is at 200 kHz. The amplifier has significant dc feedback,

70 Field-Effect Transistor Amplifiers   Ch. 4

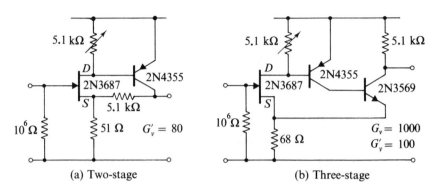

Figure 4.13. Direct-coupled FET-transistor amplifiers

12 dB, that keeps the gain from increasing more than 2 dB as the supply voltage is changed from 6 to 40 V. With the feedback bypassed with a 250 μF capacitor the gain is 50 dB up to 100 kHz when a 20-V collector supply is used.

The 3-stage FET transistor amplifier shown in Fig. 4.13(b) is similar in most respects to the 2-stage amplifier except for the advantages of additional dc feedback. With feedback the voltage gain is 100, 40 dB. With a 250-kΩ source the frequency response is constant up to a 70-kHz, 3-dB high-frequency cutoff. With a 100-kΩ source the response is constant up to 200 kHz. Thus, the amplifier with feedback has an equivalent 20-MHz gain-bandwidth product with an equivalent input capacitance of only 10 pF. The feedback may be partially bypassed for an additional 10-dB gain at small signal levels. With a fixed bias resistor the amplifier may be operated on a 12-to-30-V collector supply.

4.10 FET-TRANSISTOR PAIR AMPLIFIER FOR LOW-IMPEDANCE LOADS

The direct-coupled 3-stage amplifier in Fig. 4.14 offers a high input impedance, a voltage gain of 100, and a low output impedance. With a small adjustment of the bias the amplifier may be changed to drive a 200-Ω, capacitor-coupled load with an 8-V, peak-to-peak output. With a 500-kΩ source the amplifier delivers full output power up to 100 kHz. A 5-MΩ source may be used with full gain and output up to 20 kHz, provided a small capacitor of approximately 5000 pF is connected across the bias resistor to prevent instability or signal peaking. These direct-coupled stages are easy to construct if the FET has a zero-bias drain current less than 0.3 mA and a high figure of merit. FETs that require a high drain current must be loaded by a low-value drain resistor, and the overall voltage gain is reduced.

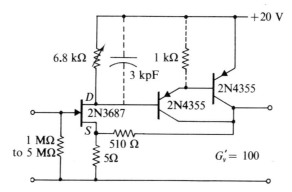

Figure 4.14. High-impedance dc to 100 kHz amplifier

4.11 TEMPERATURE-COMPENSATED, FET-TRANSISTOR AMPLIFIER

The direct-coupled amplifier in Fig. 4.14 is satisfactory for room temperature applications, but the high dc gain makes it unsatisfactory for service over a wide temperature range, as in industrial service. The amplifier in Fig. 4.15 is designed to be used in ambient temperatures from about $-25°C$ to $+60°C$. The wide temperature range is achieved by eliminating the direct-coupling between stages and by introducing temperature compensation in the transistor amplifier to control the Q-point drift. The diodes in series with the transistor base resistor reduce the base bias voltage as the temperature increases and in this way control most of the Q-point drift.

The temperature drift can be further reduced by using a positive temperature coefficient emitter feedback resistor and by increasing the feedback to reduce the voltage gain. Silicon resistors are available that have a 0.7 per cent/°C positive temperature coefficient. When the voltage across a silicon

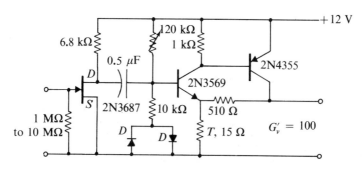

Figure 4.15. Temperature compensated FET-transistor amplifier

emitter resistor is 0.15 V, the net voltage change is 1.0 mV per °C. This voltage-temperature change can be adjusted easily by constructing the emitter resistor as a series or parallel combination of a carbon and a silicon resistor.

4.12 BROADBAND, DIRECT-COUPLED AMPLIFIER

The direct-coupled amplifier shown in Fig. 4.16 has a constant 30-dB voltage gain from dc to above 1 MHz. The amplifier is designed to operate on a 12-V battery supply in an instrument application that requires a high input

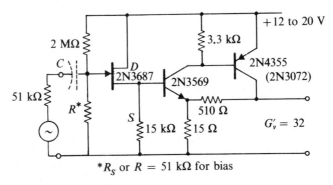

Figure 4.16. Broadband direct-coupled amplifier

impedance and a stable, wide-band frequency response. The gain is stabilized by feedback and increases only 1 dB as the supply voltage changes from 10 to 25 V. The amplifier delivers 4 V rms to a high-impedance load and 2 V rms to a capacitor-coupled 500-Ω load. With a 20-V supply the output voltage is 4 V rms with a capacitor-coupled load. Bias for the input stage must be supplied either through a resistor in series with the signal source, or else the source must be capacitor-coupled with a 51-kΩ bias resistor connected from the gate to ground. The amplifier may be made to cover the video-frequency band, 4 MHz, by reducing all resistors by a factor of 3 and by changing the output transistor to a metal-cased TO-5 device that dissipates 0.6 W. A 2N3072 transistor with a clip-on heat sink is recommended for the output stage of a video amplifier.

4.13 FET WITH BOOTSTRAPPED INPUT

The input impedance of a JFET is at least 100-MΩ at low frequencies, and in room-temperature applications the gate-to-ground resistor may be as high as 50 MΩ. The gate-to-source leakage current in a JFET is typically between 0.1 and 5 nA at room temperatures, but above 65° the current in the gate resistor may be high enough to forward-bias a JFET amplifier into cutoff. The effect of increasing gate current with temperature may be offset by reducing the gate resistor or by selecting a JFET with a low gate current. The JFETs with 2N numbers above 3000 usually have lower gate currents than the devices manufactured and supplied with a lower type number. Low-priced FETs are often inferior devices or manufacturer's rejects.

When the gate-to-ground resistor produces a lower-than-desired input resistance, the apparent ac input impedance may be increased by positive feedback, using a bootstrapped input. An amplifier with a bootstrapped input is illustrated by the circuit in Fig. 4.17. The amplifier has a direct-coupled FET-transistor pair with negative feedback to the source for stabilizing the collector Q-point. The amplifier input impedance is increased by connecting the ground end of the gate resistor to a point in the circuit that has an ac voltage approximately equal to the gate voltage. Thus, there is only a small ac current in the gate resistor, and the equivalent input impedance is higher than without feedback.

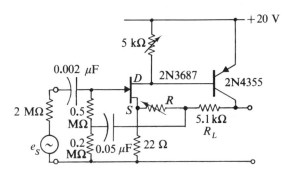

Figure 4.17. FET with bootstrapped input

In the bootstrapped input circuit shown in Fig. 4.17 the amount of positive feedback may be adjusted by the variable resistor R. The resistor R is between one and two times the source feedback resistor, depending on the amount of feedback required. High values of R produce a high value of the input impedance but narrow the amplifier bandwidth and may cause a 1- or 2-dB boost of the frequency response at 200 Hz. When $R = 11 \, \Omega$, the input impedance is greater than 8 MΩ, and the frequency response between 3-dB cutoffs extends from 40 Hz to 10 kHz. Without bootstrapping $C = 0$, the input impedance is 0.7 MΩ, and the frequency response extends from 30 Hz to 30 kHz. With a low-impedance signal source $R_S = 0$, the high frequency cutoff is at 100 kHz with the input bootstrapped, and there is a 5-dB peak in the response at 45 Hz. With higher values of the feedback resistor R the frequency response has the characteristics of a tuned circuit with a 500-Hz resonant frequency. The amplifier voltage gain is 40 dB.

Bootstrap circuits usually do not have the adjustable feedback resistor R and need considerable negative feedback to make the voltage at the feedback point follow the input signal. With additional stages and more feedback the input impedance may be increased and the high-frequency cutoff may be moved a decade higher or more. With $R = 0$ the feedback may be increased by making the collector resistor 2.4 kΩ. The bootstrapped input impedance is 2 MΩ, and the frequency response between 3-dB frequencies is from 30 Hz to 30 kHz. The voltage gain is 30 dB with $R_L = 2.4$ kΩ. Higher input impedances may be obtained by increasing the gate resistor. The upper cutoff frequency may be increased to above 20 MHz by using a high-frequency transistor and a low source impedance, and by careful layout and shielding of the circuit elements. The input impedance at high frequencies is determined by the equivalent input capacitance that is generally between 1 and 10 pF.

4.14 FET-TRANSISTOR, DC AMPLIFIER

The amplifier shown in Fig. 4.18 is designed for applications requiring a low-gain dc amplifier with high performance or selected FETs in the input stage. With a matched pair of FETs in a common enclosure the amplifier offers a low Q-point drift, a high input impedance, and an open-loop gain exceeding 1000. Where a minimum drift is not required, the upper FET may be replaced by an adjustable bias resistor, as indicated in the circuit.

The FET amplifier has an equivalent input capacitance of 10 pF, and, with source impedances less than 1 MΩ, the 3-dB high-frequency cutoff is above 10 kHz. Because the high-frequency cutoff is caused by the Miller effect in the

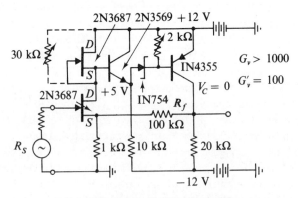

Figure 4.18. FET-transistor dc amplifier

input stage, the high-frequency cutoff is moved to higher frequencies by reducing the source impedance. With source impedances less than 100 kΩ the high-frequency cutoff is above 100 kHz, which is equivalent to at least a 10-MHz gain-bandwidth product.

4.15 FET MILLIVOLTMETER

A high-impedance voltmeter with a 20-mV full-scale sensitivity is easily constructed with an FET amplifier and a 10-μA meter. The amplifier illustrated in Fig. 4.19 has a pair of direct-coupled FETs that may be mounted on the back of a meter. The amplifier requires 0.6 mA at 3 V, and with two D-size cells the battery life is one year in continuous service.

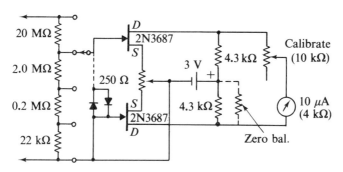

Figure 4.19. FET millivoltmeter—20 mV full scale

The sensitivity of the amplifier is high enough to show thermal noise in a 20-MΩ resistor as a just-observable movement of the meter pointer when the meter has 10-μA full-scale sensitivity. Because the sensitivity is noise-limited, there is no advantage in increasing the amplifier gain. The FETs should be mounted in thermal contact to minimize their temperature difference, and they may need to be selected for low noise in the $1/f$ region.

The meter may be zeroed with the potentiometer, and an initial balance is made at assembly by shunting a drain resistor to make the meter balance with the potentiometer at the center position. If necessary, the FETs should be interchanged to permit shunting the load on the inactive FET to keep the sensitivity of the active FET as high as possible. Once the amplifier is balanced, only an occasional adjustment of the potentiometer is required.

Except for an FET breakdown caused by a high input voltage the maximum meter current is 0.5 mA, which is low enough to protect a 50-μA meter. When a 10-μA meter is used, a push-to-read switch may be desired. For applications requiring less sensitivity the amplifier is capable of driving a

100-μA meter to full scale without linearity errors. If used with a 500-μA meter, the drain battery should be increased to 12 V.

Because 10 μA is produced with 20-mV input, or 1 nA through 20 MΩ, the meter may be used as a 20-mV full-scale voltmeter or as a 1-nA current meter. The simplicity of the circuit and the stability of the balance make the millivoltmeter useful in many applications. The meter may be used to balance a high-resistance bridge or to compare very low light levels with a photo-tube bridge. Important advantages are that the meter may be used off-ground, with source impedances up to 100 MΩ, and without a battery switch.

5 MOS FET Amplifiers

The insulated gate FETs have important advantages that are just now being utilized. The FETs are available as enhancement or depletion devices and as intermediate types that may be used either way. The insulated gate devices make possible some very simple direct-coupled circuits. The high gate resistance makes them useful in electrometer applications. The dual-gate types have exceptionally low input capacitance that makes the FETs competitive with transistors in UHF circuits. MOS FETs may be obtained to operate with 10-μA to 5-A drain currents. The high-current devices may have a transconductance of 0.1 mho and dissipate nearly 10 W. The low-current types have a transconductance of 1 to 10 mmho and dissipate 0.1 to 1 W.

The depletion-type MOS FETs have current-voltage drain characteristics very similar to the characteristics of JFETs and may be operated as a linear amplifier with zero bias. The enhancement types are cut off with zero bias, and the drain current rises abruptly at a threshold gate voltage. This threshold characteristic makes the enhancement devices useful as voltage-sensitive switches and provides a valuable noise immunity in switching applications.

The high resistance and small capacitance of the MOS-FET gate make these devices easily damaged by an electrostatic charge. A MOS FET should be picked up by the case rather than the leads and the leads should be shorted by foil or a wire wrap when stored. A person handling a MOS FET should be grounded by touching a grounded surface; the soldering iron should be grounded also. Some MOS devices are protected internally by a Zener diode.

5.1 BIASING SINGLE-STAGE MOS AMPLIFIERS (REFS. 4, 6)

Three ways of biasing an enhancement-mode MOS amplifier are shown in Fig. 5.1, and the options available to the designer are illustrated by the gain

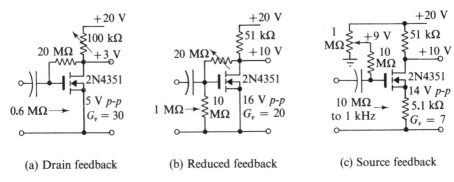

(a) Drain feedback (b) Reduced feedback (c) Source feedback

Figure 5.1. Single-stage enhancement MOS biasing

and impedance values given with each circuit. For a maximum voltage gain with resistance coupling an enhancement MOS FET should be biased approximately 1 V above the cutoff voltage, but, for a maximum peak output voltage, the drain should be approximately one-half the supply voltage. With drain feedback, as shown in Fig. 5.1(a), the gate and drain operate at the same voltage, so the load resistor required for either a maximum gain or a maximum voltage output depends on the drain-current characteristics of the MOS FET, and the resistor is best selected by trial.

With drain-to-gate feedback the input impedance of the single-stage amplifier is relatively low but may be increased or decreased by changing the feedback resistor. The gain should not change with the resistor value unless the source impedance is more than one-one hundredth the resistor value.

A collector Q-point voltage greater than the gate voltage may be obtained by connecting the gate to a voltage divider, as shown in Fig. 5.1(b). The voltage divider changes the dc Q-point, but the input impedance and the voltage gain are changed only because a lower drain resistor is used in the second example.

Figure 5.1(c) illustrates an example of fixed biasing with source feedback. By eliminating shunt feedback, the input impedance is increased to 10 MΩ up to the Miller cutoff frequency, 1 kHz. The gain without feedback is 20, and the stage has significant feedback that reduces the gain to 7.

The stages shown in Figs. 5.1(b) and (c) require a change of component values for a different supply voltage, but the stage in Fig. 5.1(a) is self-adjusting to give a maximum gain for a wide range of the supply voltage.

5.2 TWO-STAGE RESISTANCE-COUPLED MOS-FET AMPLIFIERS

The 2-stage MOS-FET amplifier shown in Fig. 5.2 requires relatively few components and is easily assembled. The amplifier gain may be varied over a wide range by using a variable feedback resistor, and a change of the feedback

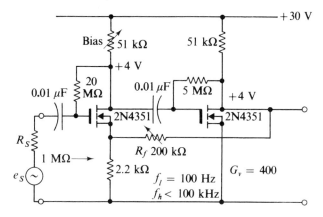

Figure 5.2. Two-stage MOS amplifier

does not adversely affect the Q-points. The low-frequency cutoff of the interstage coupling is placed at 100 Hz to reduce the low-frequency $1/f$ noise that may be objectionable in some applications. The high-frequency cutoff of the amplifier depends mainly on the first-stage Miller effect and the signal source impedance. The input impedance is approximately 0.6 MΩ in parallel with 35 pF, which with a high-impedance source makes the upper cutoff frequency only 10 kHz.

With closely matched MOS FETs the amplifier may be simplified by connecting the second-stage gate to the first-stage drain and eliminating the 5 MΩ bias resistor. Some adjustment of the operating Q-point may be obtained by changing the first-stage drain resistor, but the second stage lacks dc feedback, and satisfactory operation is achieved only by using well-matched semiconductors.

5.3 A WIDE-BAND MOS AMPLIFIER

The amplifier shown in Fig. 5.3 has a MOS FET input stage driving a transistor phase inverter. The circuit combines the high input impedance of the MOS with the low output impedance advantages and high power gain of a transistor. The MOS FET is operated with a fixed bias to avoid shunt feedback and maintain a high input impedance. The input capacitance that normally lowers the impedance at high frequencies is balanced out by positive feedback via the neutralizing capacitor C_N. The capacitance feedback is adjusted by using a potentiometer to vary the feedback voltage with the advantage that the capacitor is fixed and relatively large. With careful neutralization the amplifier input impedance exceeds 10 MΩ up to 30 kHz, or, with a 1-MΩ source, the upper cutoff frequency is as high as 100 kHz. The amplifier may be given various amounts of negative or positive feedback by

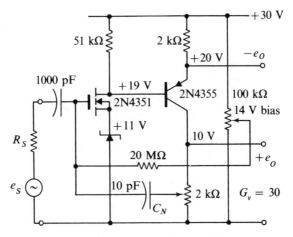

Figure 5.3. High-impedance wide-band MOS amplifier

connecting the bias control to either the inverted or non-inverted output. The circuit is simple and offers many interesting application possibilities.

5.4 HIGH-GAIN, 3-STAGE MOS-FET AMPLIFIERS

Enhancement mode MOS FETs may be operated with the gate and drain at the same potential, which makes possible direct-coupled amplifiers with few components. The amplifier in Fig. 5.4 has 3 identical enhancement-mode FETs with a feedback resistor connecting the output drain to the input gate. With like drain resistors the dc feedback holds the Q-points alike and independent of the supply voltage. The voltage gain of each stage is 20, 26 dB,

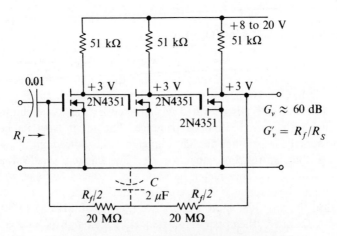

Figure 5.4. High-gain 3-stage MOS-FET amplifier

and the overall open-loop gain is 80 dB with a 10-V supply. The overall gain varies with the supply voltage and is 86 dB with a 20-V supply. The amplifier noise is approximately the resistance noise of the input loop, provided most of the transmitted frequency band is above the $1/f$ corner frequency that may be above 100 Hz.

The amplifier input impedance with feedback equals the feedback resistor R_f divided by the open-loop gain G_v. If the amplifier is driven by a source with an impedance less than the input impedance, the feedback is effectively bypassed, and the gain is 10,000, the open-loop gain. For source impedances exceeding approximately 2 kΩ, the voltage gain is the ratio R_f/R_S. For applications requiring a 20-MΩ input impedance the capacitor C may be used to eliminate the ac feedback, and the gain may be controlled by inserting a resistor in series with C.

The frequency response of the MOS-FET amplifier depends on the signal source impedance and on whether the capacitor C is used to bypass the ac feedback. With a low-impedance source the 3 dB f_h is 100 kHz, while with a 51-kΩ source the cutoff frequency is reduced to 20 kHz. With a high-impedance source and the capacitor C in place, f_h is determined by the Miller-effect capacitance of the first stage and by the 20-MΩ resistor $R_f/2$. Because the input capacitance is 30 pF, the high cutoff is at approximately 300 Hz.

The amplifier becomes a dc amplifier with a voltage gain of 100 if the input capacitor is replaced by a 40-kΩ resistor R_S. Dc operation requires that the input terminal be connected to a point 7-V positive with respect to ground, and requires a well-regulated dc voltage supply. As a dc amplifier the voltage gain may be adjusted by a resistor in series with the source. A gain exceeding 100 is generally impractical because of Q-point instability.

The simplicity and flexibility of the direct-coupled MOS-FET amplifier make it useful for a variety of ac applications. Because of the 80-dB dc feedback the amplifier has very stable Q-point voltages that are almost independent of the supply voltage. With lesser amounts of dc feedback the amplifier may require a well-regulated supply.

5.5 HIGH-GAIN, LOW-NOISE, LOW-DRIFT DC AMPLIFIERS

An amplifier that uses a depletion-mode MOS FET or a JFET in the input stage permits dc amplification from a grounded source, as shown in Fig. 5.5. Feedback is introduced by connecting the first and third FETs to a common source resistor, and the loop gain is maximized by an adjustment of the resistors R_L and R_A. The JFET should be a low-current type with a pinch-off voltage rating of 3 V or less. The drain resistor R_L is adjusted to make the JFET drain voltage as low as practicable for high gain and to permit a small value bias resistor R_A. The bias is adjusted to make the output stage drain voltage one-half the supply voltage.

82 MOS FET Amplifiers

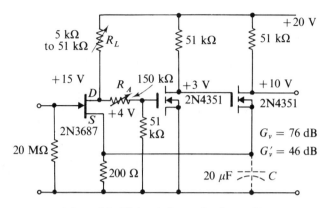

Figure 5.5. High-gain low-noise dc amplifier

The amplifier voltage gain with feedback is 46 dB, and the gain with the feedback bypassed by the capacitor C is 76 dB. Thus, the dc feedback is much lower than in the 3-stage MOS amplifier, and a well-regulated voltage supply is required.

With a low-noise JFET in the input stage the amplifier offers low noise for source impedances exceeding 100 kΩ, up to 20 MΩ. The bandwidth of the amplifier is determined by the source impedance and the Miller-effect input capacitance. With a 500-kΩ source the upper cutoff frequency is 20 kHz, and with a 5-MΩ source the cutoff frequency is reduced to 2 kHz, a factor of 10 lower, because the impedance is a factor of 10 higher. With feedback and a low-impedance source the upper cutoff frequency is above 100 kHz.

5.6 MOS-FET ELECTROMETERS OR PICOAMMETERS

A meter may be easily constructed to read currents of 1 pA or less, using the circuit shown in Fig. 5.6. The picoammeter uses a pair of FETs in a balanced bridge with a 1-mA or 100-μA meter, depending on the sensitivity desired. The resistor R is chosen to make the full-scale voltage drop across R approximately 1 V. Thus, for 1 pA full scale the resistor must have a resistance of 10^{12} Ω. The semiconductors should be an electrometer-type enhancement mode MOS FET with a low gate current and a high gate resistance. The RF types of MOS FETs are not usually satisfactory. The unit used on the input side may need to be selected by trial to obtain a sufficiently high gate resistance. To maintain a high input resistance the gate of TR-1 should not be connected to a terminal in the usual manner, and surfaces in contact with the gate should be treated with wax or a silicone grease to prevent surface conduction.

The 100-kΩ resistor and the neon lamp prevent breakdown of the gate if the input terminal is accidentally subjected to an over-voltage or an elec-

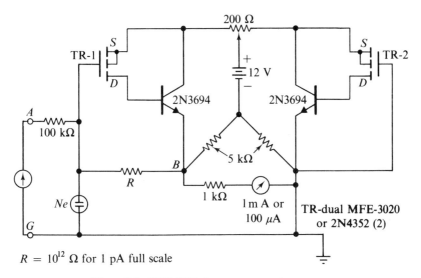

Figure 5.6. MOS FET electrometer or picoammeter

trostatic charge. The neon lamp should be a low-voltage type ne-86. The 1-kΩ resistor in series with the meter is to provide feedback and make the output current proportional to the input. A small adjustment of the series resistor may be used to calibrate the meter.

High-value resistors for R may be obtained from Victoreen, but a small capacitor may be used in place of R if the amplifier has adequate feedback. The amplifier has adequate feedback if the voltage BG is within 80 to 90 per cent of the input voltage AG when the input voltage is approximately 1 V. With adequate feedback and negligible gate leakage the input current is the product of the rate of change of V_{AG} multiplied by the capacitance C. Thus, if V increases 0.1 V/s and C is 10 pF, the input current is $0.1 \times 10^{-11} = 1$ pA. A measurement with R replaced by a capacitor assumes that the gate leakage is negligible, which is a safe assumption if the output voltage with the input open changes more slowly than when a current is measured.

5.7 MOS FET HIGH-FREQUENCY AMPLIFIERS

The exceptionally low gate-to-drain capacitance, 0.3 pF, of the single-gate MOS FETs offered for RF applications makes these devices useful in both tuned and resistance-coupled amplifiers for applications above the audio frequencies. Tuned and resistance-coupled amplifiers may be constructed easily, without neutralization and feedback problems, up to frequencies above the broadcast band.

A low-power resistance-coupled amplifier is shown in Fig. 5.7(a). The amplifier has the relatively low voltage gain of 10 because the supply voltage

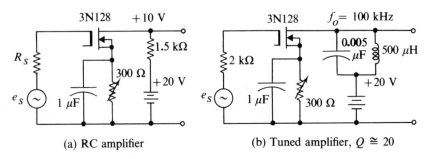

(a) RC amplifier (b) Tuned amplifier, $Q \cong 20$

Figure 5.7. MOS-FET high-frequency amplifiers

should not exceed the MOS FET breakdown voltage, 20 V, and the high drain current required by the FET, 15 mA, limits the drain resistor to 2 kΩ. Thus, the expected voltage gain is 10 to 20. With a voltage gain of 10 the input capacitance is 10 to 15 pF, depending on the capacitance between the external gate and drain circuits. An important advantage of the MOS FET is that the signal source impedance may be relatively high, up to a frequency where reactance of the input capacitance cannot be neglected. Thus, the resistance-coupled stage may be used with a 10-kΩ signal-source impedance at 1 MHz and 100 kΩ at 100 kHz. The maximum output voltage is 10 V $p\text{-}p$.

The tuned amplifier shown in Fig. 5.7(b) uses a readily available MOS FET and gives an iterated voltage gain of 30 up to frequencies as high as 10 MHz. With an electrostatic shield to minimize drain-to-gate feedback capacitance the tuned amplifier is stable and does not require neutralization. Because the internal resistance of the FET is 3 to 10 kΩ and g_m is between 5 and 10 mmho, the maximum gain of a stage is limited to $g_m R_o$, or approximately 20 to 50. The amplifier shown in the figure has a gain that exceeds 50; the maximum output voltage is 30 V $p\text{-}p$, and the circuit Q is approximately 20. High Qs are obtained only by making the reactances small compared with the 5-kΩ internal resistance of the MOS FET.

The voltage gain that may be attained in a practical MOS FET stage is usually less than 10, which is relatively low for low-frequency applications. However, the unique characteristics of MOS and CMOS (complementary MOS) devices makes them useful in RF and sometimes in low-frequency circuits. Both CMOS transistors and CMOS operational amplifiers are offered for low-frequency service. These devices offer high input impedance and over 50 dB voltage gain per stage, but have a high equivalent input noise.

5.8 MOS FETS IN RF AND UHF AMPLIFIERS
(REFS. 3, 4, 6)

Recent improvements of the insulated-gate, field-effect transistors make the MOS FETs suitable for many VHF and UHF applications. The MOS FETs offer low noise, low distortion, and a low feedback capacitance that permit their use as replacements for vacuum tubes in nearly any high-frequency, low-power amplifier.

The triode types of MOS FETs have drain-to-gate feedback capacities that are typically 0.1 to 0.7 pF with transconductance values of 7500 μmho. MOS FETs with dual gates have feedback capacities as low as 0.1 to 0.2 pF with the transconductance exceeding 10,000 μmho. These devices offer unneutralized power gains of 15 to 20 dB at 200 MHz and 10 to 13 dB at 400 MHz. The dual gate reduces the feedback capacitance, and one gate may be used for the input while the other is used for the local oscillator (LO) input of a mixer. The low-value leakage current of the insulated gate eliminates input circuit loading caused by high input signals. Thus, a MOS-FET amplifier tolerates large signals without causing either detuning or Q-reduction.

The circuit of a typical UHF dual-gate, MOS-FET amplifier is illustrated in Fig. 5.8. The input circuit is essentially the same as for a vacuum tube amplifier, while the output circuit has as high a load impedance as the stability permits. Bias is provided to the second gate by the voltage divider R_2 and R_3, or the gate may be returned to an AGC control voltage or a LO signal for mixing. Gate-1 is usually operated 0.5 V below the source voltage when gate-2 is 2 to 4 V above the voltage. Circuits with component values and electrode voltages required for a specific FET may be found in the manufacturer's data sheet.

The gate-to-drain feedback capacitance of JFETs is typically 1 to 3 pF, which generally makes these devices unsuitable for RF applications. JFETs may, however, be used to replace almost any low-power vacuum tube triode.

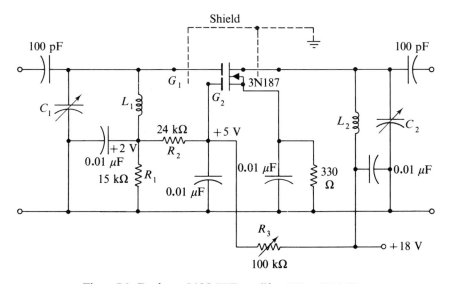

Figure 5.8. Dual-gate MOS-FET amplifier, RF to 200 MHz

Of course, such replacements require suitable changes of the bias and reduction of the *B*-voltage to a value suitable for the JFET collector. The replacement of vacuum tubes by MOS FETs requires similar changes, but the tuned circuits and their component values may be used without change. The dual-gate MOS FETs are equivalent to a cascode stage that may be used without neutralization and as a simple and inexpensive replacement for the active elements in many high-frequency circuits.

6 Power Supplies and Regulators

The dc power required for transistor equipment is usually obtained from a line-connected, ac power supply, or from a small battery. An ac power supply has a transformer for reducing the line voltage to a low-voltage ac that can be rectified and reduced by a smoothing filter to a steady dc. Transformers for reducing the line voltage are readily available, and the rectifiers may be almost any 50 to 100-V, 1-A, silicon diodes.

Regulators are used to improve the characteristics of a power supply. A regulator may reduce the ac hum output of a power supply, or reduce changes of load voltage caused by changes of the input voltage or by changes of the load demand. A regulator may be as simple as a resistor with a Zener diode or a transistor, or it may be purchased as a self-protected IC. Regulators are described near the end of this chapter.

6.1 AC POWER SUPPLIES (REFS. 3, 7)

Because of the simplicity of semiconductor rectifiers a full-wave or a bridge rectifier is used in preference to a half-wave rectifier unless the dc load current is relatively small. The full-wave rectifier produces a series of half sine waves at double the input frequency, so that the lowest frequency that must be removed by the hum filter has twice the supply frequency. The ac transformer has to supply an amount of ac power that is equal to the dc power delivered to the load plus a sizable component of reactive power that circulates in the filter capacitor and the transformer inductance. With a full-wave rectifier the volt-ampere (VA) input to the transformer can be conservatively estimated to be one-fourth to one-third larger than the dc load power. Half-wave rectifiers are seldom used because they require a transformer having about twice the VA rating needed for a full-wave rectifier. The full-wave rectified power, having double the ac input frequency, is more easily filtered than half-wave power at the input frequency. As a result, both the filter and the transformer for a full-wave rectifier are smaller, lighter, and less expensive than for an equivalent half-wave rectifier.

The VA rating of a transformer is usually specified by the manufacturer for either a 60 Hz or 400 Hz input. The ampere rating of a transformer is fixed by the wire size—i.e., the copper loss—and is therefore independent of frequency. The voltage rating is fixed by the reactance of the winding and is therefore approximately proportional to the operating frequency. In other words, a 400-Hz power transformer can be used at a lower input frequency if the input voltage is reduced in proportion to the frequency reduction. A 60-Hz transformer can be used at 50 Hz if the load current is reduced a little to offset the increased reactive input power. A low-frequency power transformer can be used at a higher frequency if the winding voltages are not increased quite in proportion to the frequency increase. Therefore, the power capability of a given transformer increases approximately as the square root of the operating frequency.

A rectifier is followed by a filter in order to reflect the ac components of the rectifier output back to the power line and transfer the dc component unimpeded to the load. The filter in a power supply is called a *smoothing filter*. The filter is often only a large electrolytic capacitor, in which case the dc voltage is about 1.3 times the rms ac input voltage.

The size of the capacitor and the dc load resistor determine the extent to which the rectifier ripple is reduced. The rms ripple voltage is measured as a percentage of the dc load voltage. For a 60-Hz full-wave rectifier the per cent ripple r is approximately:

$$r = \frac{200{,}000}{RC} \qquad (6.1)$$

where R is the equivalent dc load resistance in Ω, and C is the capacitance in μF (microfarads). (A half-wave rectifier has twice the per cent ripple given by Eq. 6.1.) If the load has 100 Ω and the capacitor is 1000 μF, the ripple is about 2 per cent. Because the large capacitor required for low ripple is expensive and causes high peak rectifier currents, we usually find that the input capacitor is chosen to allow about 5 per cent ripple. A further reduction of ripple is obtained more economically by adding a second-stage RC filter, a voltage regulator, or a capacitance multiplier.

A large capacitor across the load also presents the low ac output impedance needed to prevent undesirable crossover and feedback when high-gain amplifiers share a common power supply. The capacitor is continually discharged by the connected load but is recharged in brief current impulses whenever the peak rectifier output voltage exceeds the capacitor voltage. These current pulses can be 10 to 100 times the average load current and may easily exceed the peak current rating of the rectifiers. The current peaks can be limited by inserting a resistor anywhere in series with the transformer.

A resistance in series with the secondary winding that is equivalent to 10 per cent of the load resistance reduces the dc load voltage about 20 per cent, but reduces the peak rectifier currents by a much larger factor. Unfortunately, with a capacitor input the peak rectifier currents are not easily predicted, so it is difficult to generalize the prevention of peak current difficulties. Furthermore, the input capacitor itself should be conservatively rated when it is subjected to current peaks. If the capacitor has a low capacitance, the ac ripple voltage may be quite high. Hence, the capacitor should have a dc voltage rating at least one-third higher than the dc load voltage.

6.2 HALF-WAVE RECTIFIERS

Half-wave rectifiers are occasionally used in applications requiring very low dc power or for converting signal power to dc. For comparison 3 forms of half-wave rectifiers are shown in Fig. 6.1. In each example the dc voltage obtained with a given load resistor is shown, and the open-circuit or no-load voltage is shown in parentheses. The no-load voltage is approximately the voltage obtained with a transformer that is rated for 12-V ac at full load; it is not the theoretical value obtained with 12-V input.

The dc voltage obtained from a rectifier without a capacitor, Fig. 6.1(a), is approximately one-third that obtained with a coupling capacitor or a filter capacitor. A diode rectifier is sometimes used with a coupling capacitor, as in Fig. 6.1(b), and the diode must be connected across the load. While the dc voltage is approximately the peak signal voltage, the ac signal voltage appears across the load, and a separate filter may be needed. The filter capacitor cannot be connected across the rectifier and must be separated by a choke or a series resistor. The capacitor-input rectifier and filter shown in Fig. 6.1(c)

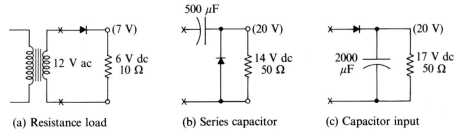

Figure 6.1. Half-wave rectifiers

may be used, provided the rectifier is driven with a transformer or a resistor at *x-x*. The rectifier cannot supply dc unless there is a dc connection from both sides of the rectifier to the load. The smoothing capacitor shown in Fig. 6.1(c) is sized to reduce the ripple to approximately 5 per cent of the dc voltage. The load resistors shown in Fig. 6.1 have the minimum values that can be used without overloading a transformer rated for 12 V and 1 A.

6.3 BRIDGE RECTIFIERS

The most commonly used forms of the full-wave rectifier are the full-bridge and the half-bridge illustrated in Fig. 6.2. The bridge rectifiers use a transformer rated for 12 V at 1 A and may be used with dc load currents up to 0.6 A. The rectifiers may be any 50-V diode rated for 1 A. With a 25-Ω load the 120-Hz ripple is approximately 5 per cent, less than 1 V. For higher load currents a 12-V, 2-A, transformer may be used with a 12-Ω equivalent load resistance, and the rectifiers should have a 3-A rating. Silicon power rectifiers are generally rated to withstand peak currents at least 10 times the average current rating. With small transformers the peak rectifier current is limited by the primary and secondary winding resistance, but a 1- or 2-Ω resistor in series with the secondary may improve the rectifier reliability. Transformers with ratings exceeding 25 volt-amperes are more efficient, and the peak rectifier currents may easily exceed 10 times the average current.

The smoothing capacitor for the filters shown in Fig. 6.2 should be rated for 25 V dc, regardless of the actual load voltage. Reducing the input capacitor lowers the load voltage and increases the per cent ripple.

The half-bridge rectifier requires a center tap on the transformer secondary and uses each half of the secondary on alternate half cycles. Thus, the transformer operates at a lower efficiency, and the dc load voltage is less than the full secondary ac voltage. Both the bridge and the half-bridge rectifiers are full-wave rectifiers, and for the experimenter the choice between them is mainly the opportunity to select the load voltage. The half-bridge with a 24-V, 0.5-A transformer secondary may be used for 15-V dc across a 30-Ω load.

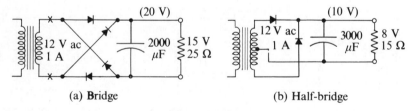

(a) Bridge (b) Half-bridge

Figure 6.2. Full-wave rectifiers

6.4 VOLTAGE DOUBLERS

A voltage doubler is a rectifier in which the transformer is used on both half cycles to double the dc voltage. The voltage doubler may be used to supply twice the dc voltage that can be obtained with a full-wave rectifier and a given transformer. The doubler has the disadvantages that the capacitors must be relatively large and the peak rectifier currents high, or else the load must be relatively small. Figure 6.3 shows a voltage doubler that may be used to obtain a dc supply of 50-to-70 V with load currents less than 40 mA. Figure 6.4 shows how the doubler may be extended to obtain three or more times the dc voltage obtained with a half-wave rectifier and the same transformer. The large total capacitance and the poor regulation of the voltage multiplier usually make the use of voltage multipliers less desirable than increasing the magnitude of the ac input voltage.

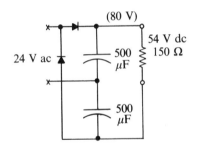

Figure 6.3. Voltage doubler

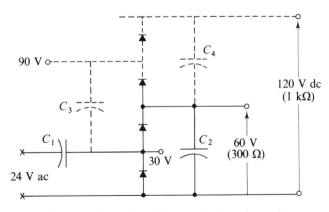

Figure 6.4. Voltage multiplier

6.5 REGULATED VOLTAGE DOUBLERS

The voltage doubler circuits in Figs. 6.5(a) and (b) show how a 6-V ac supply may be used to power low-current 8- and 12-V loads. In these circuits an 8- or 12-V Zener diode is used to reduce the ripple and improve the regulation, but the overall efficiency is relatively low. The internal output impedance of both circuits is approximately 6 Ω, and the ripple is approximately 50 mV for the 8-V and 200 mV for the 12-V circuit, independent of the load resistance. The rectifiers may be almost any low-current, low-voltage device, and the transformer may be a 6-V transformer rated for 0.5 A, or less.

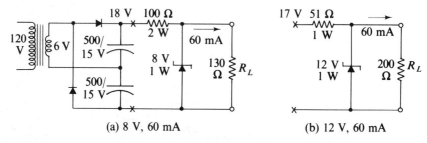

(a) 8 V, 60 mA   (b) 12 V, 60 mA

Figure 6.5. Voltage doubler with regulated load

6.6 REGULATED POWER SUPPLY FOR INTEGRATED CIRCUITS

A regulated plus-and-minus power supply may be constructed with a center-tap transformer and either a full-wave bridge or a half-wave, voltage-doubling rectifier. The bridge rectifier shown in Fig. 6.6(a) uses 2 series-connected capacitors with their common point connected to the transformer center tap.

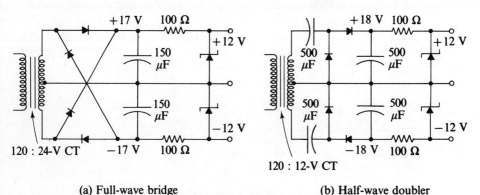

(a) Full-wave bridge   (b) Half-wave doubler

Figure 6.6. Regulated IC power supplies

The Zener regulators may be either the 8- or 12-V regulator that is shown in Fig. 6.5. The half-wave doublers shown in Fig. 6.6(b) may be used to obtain the same voltage and ripple output as with the full-wave circuit, Fig. 6.6(a). Comparison of the circuits (a) and (b) shows that the half-wave doubler may be used with a 12-V ac source, but for the same ripple the individual capacitors must be at least 3 times as large as for the full-wave bridge rectifier. When each of the power supplies shown in Figs. 6.5 and 6.6 is used with a 200-Ω resistance load, the ripple is 6 per cent of the dc voltage. The ripple frequency is 120 Hz for the full-wave bridge and 60 Hz for the voltage doublers.

6.7 TYPICAL LOW-POWER, 30-V DC POWER SUPPLY

A bridge-rectifier power supply typical of those used with a low-power, high-fidelity, class-B amplifier is shown in Fig. 6.7. This power supply is usually rated as supplying 34 V dc and may be used with peak loads as high as 1 A. The capacitor C may be between 1000 and 4000 μF, depending on the hum requirements and the need to sustain load peaks. The peak power capability of a class-B amplifier that is reproducing music is determined by the supply voltage existing with a low-power sustained load, and a well-regulated supply is not required.

For a class-B amplifier requiring ± 34 V a power supply may be constructed, as shown in Fig. 6.8, by using a transformer with a single 48-V, 1-A secondary, or two 24-V, 1-A transformers may be used in series. Two 24-V transformers may usually be obtained for the same or less cost than a 48-V transformer, and the former are more readily available. The dc-load curve shown in Fig. 6.7(b) may be used as the load curve for each side of the ± 34-V supply.

The capacitor nearest the rectifier should be rated for a dc voltage at least 30 per cent higher than the no-load operating voltage. A capacitor more

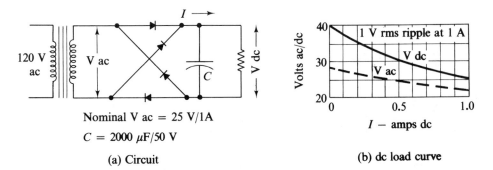

Figure 6.7. Dc power supply, typical 25 VA

94 Power Supplies and Regulators Ch. 6

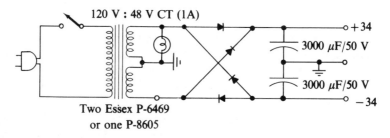

Figure 6.8. Typical power supply for 25-W high-fidelity PA

removed from the rectifier may be operated up to its rated voltage, but higher reliability is obtained by operating any capacitor at one-half the maximum rating. The simplest way to protect the transformer and diodes from a short-circuited capacitor is to connect a one-half or one-watt, 10- or 20-Ω resistor in series with the low-voltage winding of the transformer.

6.8 LINE-OPERATED POWER SUPPLY FOR A 2-W AMPLIFIER

The low-power audio amplifiers in recorders and radios are often supplied with dc power from a line-operated power supply similar to that shown in Fig. 6.9. The full line voltage is used without an isolating or step-down transformer, and for safety the line should be opened on both sides with a double-pole switch. The power resistor in series with the line is varied to adjust the dc voltage for the power stage, and the 15-to-30 V required by the preamplifier are adjusted by the resistor R between the high-voltage and low-voltage filter capacitors.

A line-operated dc supply for a 25-W amplifier, Fig. 6.10, may use a bridge rectifier and a relatively low resistance between the rectifier and the input filter. High-voltage transistors are available that may be operated up to 140 V, almost the peak ac line voltage. The voltage required for the preamplifier is reduced by a series resistor, as indicated in the figure.

6.9 ZENER-DIODE REGULATORS

A Zener diode is like an ordinary diode except that when it is reverse-biased there is a voltage at which the diode current increases abruptly. When the reverse current is limited by a series resistor, the voltage across the Zener diode is essentially constant for a wide range of currents. This is the control region where the diode acts as a constant voltage reference or as a control element useful for many purposes. Zener diodes are obtainable with voltage breakdown ratings upward from about 3 V. The diodes are available with

Sec. 6.9   Zener-Diode Regulators   95

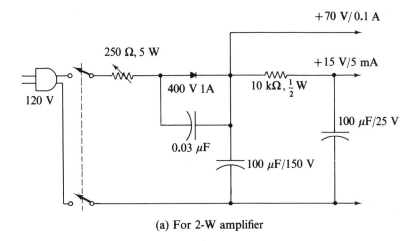

(a) For 2-W amplifier

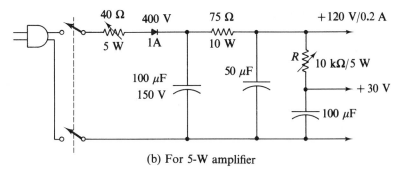

(b) For 5-W amplifier

Figure 6.9. Typical line-operated power supplies, 1 to 5-W amplifiers

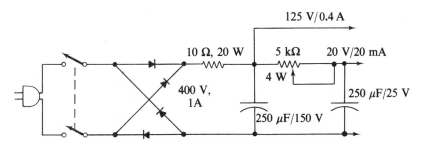

Figure 6.10. Typical line-operated power supply for 25-W amplifier

power ratings from 0.2 W to several hundred watts, but those with a 1-W dissipation rating are most readily available and are relatively inexpensive.

The volt-ampere characteristics for a typical 1-W Zener diode are represented in Fig. 6.11. The characteristics show the familiar forward conduction curves in the first quadrant. At the Zener knee in the third quadrant the reverse current abruptly increases when the current exceeds a minimum of about 1 mA. In the essentially constant voltage region the Zener voltage increases about 0.01 V per mA, thus indicating an equivalent internal resistance of 10 Ω. When the Zener voltage is 10 V and the current is 100 mA, the diode must dissipate 1 W—its maximum room temperature power rating. Low voltage Zener diodes have a more rounded knee than the Zener shown in Fig. 6.11, and high voltage Zeners of a given type are usually much like a 10-V diode.

A Zener diode operating in the breakdown region is like a battery being charged. For example, a 10-V, 1-W Zener diode with a positive current entering the + terminal, as shown in Fig. 6.12(a), is like the battery shown in Fig. 6.12(b). The diode has an equivalent 6-Ω series resistance that makes the Zener voltage V'_z increase 6 mV per milliampere of reverse current. The equivalence between the Zener diode and the battery exists only as long as the charging current is maintained. Because the diode tries to maintain a constant

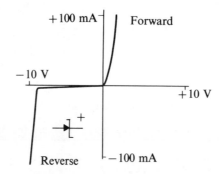

Figure 6.11. Zener diode current-voltage curves

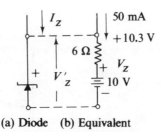

(a) Diode (b) Equivalent

Figure 6.12. A 10-V Zener diode and equivalent circuit

voltage, it must be protected from damage by currents exceeding the diode rating. Damage of a Zener diode is usually prevented by a series resistor R_S that should limit the product of the maximum reverse current and the diode breakdown voltage to no more than the diode power rating.

For reliability the power dissipated by a Zener diode should be less than one-half the maximum power rating. As our circuit examples show, the series resistance used to limit the diode current is several times larger than the internal resistance of the Zener diode. Hence, in selecting a series resistor we may represent the diode as a constant voltage (battery) and assume that the internal Zener resistance R_Z is negligible. The internal resistance of a Zener diode may be obtained from the diode data sheets and is used to calculate the change of the terminal voltage with a change of either the input or the load current.

The curves in Fig. 6.13 show how the terminal voltage of several 1-W Zener diodes varies with the current and indicates the corresponding internal resistance. The figure also shows the maximum and minimum currents that are used in design. The resistance of a half-watt diode is twice the resistance of a 1-W diode.

Two examples of Zener regulators are shown in Fig. 6.5, where a 20-V no-load voltage is reduced to a regulated 8 or 12 V. In Fig. 6.5(a) the dc input is 0.1 A at 18 V, and the load power is 0.06 A at 8 V. Thus, the load power is 27 per cent of the input power. For the 12-V regulator in Fig. 6.5(b) the load

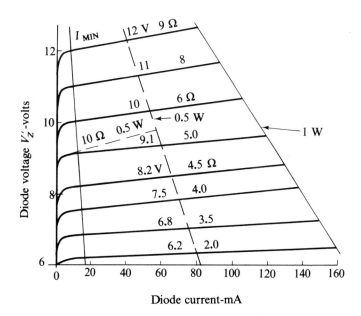

Figure 6.13. Zener diode voltage-current curves for 1-W diodes

power is 45 per cent of the input power. These examples illustrate the low efficiency that must be expected of a Zener regulator. Moreover, the regulated voltage cannot exceed 50 to 60 per cent of the input voltage, unless the input voltage is almost constant and the load current is less than 20 or 30 per cent of the Zener diode current.

6.10 REGULATOR FOR LOW-POWER, 9-V EQUIPMENT WITH A 12-V BATTERY

The regulator shown in Fig. 6.14 is designed for operating 9-V transistor equipment on an automobile 12-V storage battery. With a suitable regulator, transistor equipment may be plugged into the automobile cigarette lighter for a convenient power supply. The regulator supplies 9.1 to 9.7 V to a 50-mA load with the voltage depending on the battery condition and the Zener diode temperature. The circuit in Fig. 6.14(a) shows the Zener-diode voltage and current with the load removed and with a 16-V battery voltage. The diode is dissipating 3 W and should be rated for 5 W because the battery voltage can be higher than 16 V. The currents and voltages existing when the battery voltage is only 11.1 V are shown in Fig. 6.14(b). The maximum current available for the load is 50 mA because the minimum current in the diode should be 10 per cent of the maximum rated current.

A problem of some difficulty in the use of Zener diodes is the fact that the diode voltage varies with the load and temperature in addition to the 5 or 10 per cent tolerance that must be accepted when the diode is purchased. In the present example a 5 per cent tolerance may make the regulated voltage above 10 V or as low as 8.6 V. The author's experience indicates that the diode manufacturers withhold close-tolerance diodes and release high- and low-tolerance devices for the purchasers of small quantities. For this reason it may be necessary to select a diode that is 9.1 V minus 5 per cent if a regulated voltage near 9 V is desired. On the other hand, the difficulties in using a Zener diode with a directly-connected power load are an important reason for using either a transistor regulator or an IC regulator that may be adjusted to supply a particular and more stable output voltage.

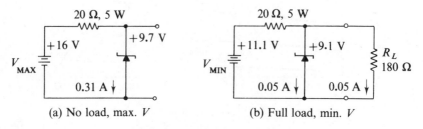

Figure 6.14. 9-V Zener regulator

6.11 TRANSISTORS AS ZENER DIODES

Many of the *pnp* planar transistors are useful as temperature-compensated Zener diodes. When the emitter is connected to the negative side of the supply, the emitter diode is reverse-biased and the breakdown voltage is relatively independent of the diode current for currents of 1 to 40 mA. The upper current limit is determined by the 300-mW rating of most transistors in the TO-105 plastic package. By connecting the diode collector to the positive side of the supply and leaving the base open, as shown in Fig. 6.15, the collector diode is forward-biased and provides temperature compensation.

A *pnp* transistor used as a reverse-biased emitter diode with the base open has a temperature-compensated Zener voltage that is between 6 and 8 V and an incremental resistance of only 15 to 30 Ω. Typically, a 1-W, 7-V Zener diode has an incremental resistance of 500 Ω at 1 mA and requires 10 to 30 mA to match the impedance of the transistor at 2 mA. There are many general-purpose *pnp* transistors that may be used as Zener diodes, and three examples of these are the 2N3638, 2N4355, and 2N4890. The emitter diode characteristics of *npn* transistors are not satisfactory as Zener regulators.

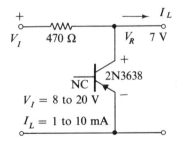

Figure 6.15. Silicon pnp transistor as a Zener regulator

6.12 LOW-IMPEDANCE REGULATOR

Zener diodes usually have an incremental impedance of several hundred ohms at low currents or must be operated near their maximum power rating to obtain impedances as low as 10 Ω. The circuit shown in Fig. 6.16 may be used as a low-impedance regulator with the operating voltage changed from 1 to 20 V by changing the resistor R_A. At operating currents of 1 to 5 mA the equivalent internal impedance at the output is less than 10 Ω, and at higher currents it may be 0 Ω or a negative resistance for small changes of the current. The transistors in Fig. 6.16 are the discrete equivalent of a breakdown diode.

The most useful characteristic of the regulator may be that the circuit operates as a Zener regulator with the voltage adjusted by changing a single resistor. The regulated voltage for the load is approximately 0.7 V more than the value of R_A in kilohms. Thus, with $R_A = 5.1$ kΩ the regulated voltage is approximately 6 V. For testing purposes the regulator should be supplied through a 470-Ω series resistor R_S, and the input voltage at which the voltage V_R becomes approximately constant with increasing input occurs when the input voltage is about 1 V more than V_R. The current at which the regulator is effective is thus 2 mA. The regulated current may be decreased to 1 mA by increasing all resistors in the circuit by a factor of 2, or to 0.5 mA by increasing them by a factor of 4. The value of R_A used to calculate the voltage V_R is decreased by the same factor, of course. When R_A is 100 Ω or less, the regulating voltage is about 0.7 V.

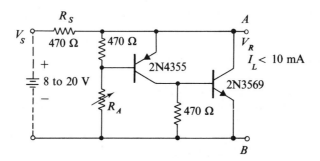

Figure 6.16. Low-impedance regulator

6.13 SHUNT TRANSISTOR REGULATORS

The power regulated by a Zener diode may be increased by a factor of 10 by adding a single power transistor, as shown in Fig. 6.17(a). We see that the transistor is made to conduct when the Zener diode conducts, and a silicon transistor tends to maintain the load voltage about 0.7 V above the Zener diode voltage. This regulator has a no-load to full-load regulation of about 1 per cent, or an equivalent internal output resistance of 0.2 Ω, and load changes

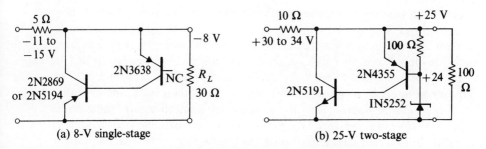

Figure 6.17. Transistor shunt regulators

are reduced to 7 per cent of the input voltage change. The disadvantage of the shunt regulator is that it continuously loads the power supply at full load and power is lost in both the series resistor and in the transistor. However, in some systems a shunt regulator may be useful in keeping a pulse-type load from producing noise or transient disturbances in an amplifier connected to the same power supply. Sometimes entirely separate instruments must operate on an ac supply that is connected through a long cable, and transistors in one instrument may load the line enough to produce noise in the other instrument, even when both have separate power transformers and rectifiers.

The circuit shown in Fig. 6.17(a) uses a selected *pnp* transistor as the Zener diode, and by careful selection the output voltage may be made 8 V for either a germanium 2N2869 or a silicon 2N5194 power transistor. The shunt regulator shown in Fig. 6.17(b) has a 2-stage transistor regulator and provides 1 per cent regulation for either input voltage or load changes. The regulator requires a low-impedance Zener diode, and for operation at higher output voltages the Zener diode should be a low-impedance type such as those in the 1-W, 1N5221 series. The power transistor must be selected to carry the no-load input current at the maximum input voltage without danger of second breakdown. The 2N2869 or the 2N5194 may be used safely with 1-A input currents and load voltages up to 30 V.

6.14 SERIES TRANSISTOR REGULATORS

A regulator that uses the power transistor as a variable series element is more efficient than a shunt regulator because the current into the regulator varies with the load current. For this reason an electronic regulator is usually a form of series regulator, as shown in Fig. 6.18. The Zener diode holds the base at a fixed voltage with reference to ground, and, in following the base, the emitter holds the load voltage 0.3 to 0.7 V below the Zener voltage, depending on the base-to-emitter voltage drop. The series resistor R_S must be adjusted to supply the base with about 1/20th of the maximum load current, when the input voltage is lowest, plus the current required by the Zener diode. The transistor must have at least 1 or 2 V from the collector-to-base at the minimum input. The maximum input is determined by the second breakdown characteristics of the transistor or by the amount of power that can be dissipated.

The operating characteristics of a single-transistor series regulator are better than those for a shunt regulator because the constant current characteristics of the collector help to keep the load current constant with a change of the input voltage. The single-stage series regulator reduces the load voltage changes to about 1 per cent of the input voltage change and has a full-load to no-load voltage change of approximately 1 per cent.

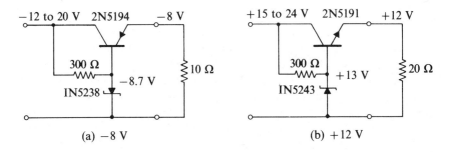

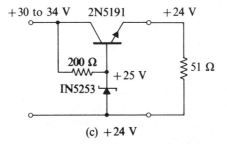

Figure 6.18. Transistor series regulators

In low-frequency, low-power amplifiers a series transistor regulator is more effective, more efficient, and requires less space than a resistor-capacitor filter. The simplicity of single-transistor regulators makes them less expensive than an RC filter and more widely useful than is generally realized. For these reasons several low-power regulators are described in the following section. For load currents of 1 A and over TO-3 packaged IC regulators may be purchased that are less expensive than a regulator made with discrete components.

6.15 LOW-POWER VOLTAGE REGULATOR (REF. 4)

A simple transistor regulator for low-power applications is illustrated by the circuit in Fig. 6.19. The regulator uses a low-power 2N1711 transistor, or an equivalent that is rated to dissipate up to 0.8 W in free air. The low-power regulator is recommended mainly for replacing the bulky and less efficient RC filter that is needed to separate a preamplifier or an equipment subsection from the high-power section of an electronic system. The transistor may be thought of as a way of increasing the power capability and efficiency of the reference diode V_Z. The regulator may be used with input voltages from 6 to 30 V, and for a collector power dissipation under 300 mW the load current limit is from 30 to 100 mA. The no-load to full-load regulation is about

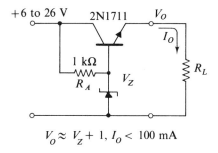

$V_O \approx V_Z + 1, I_O < 100$ mA

Figure 6.19. Low-power voltage regulator

1 per cent, and the output ripple is 10 mV peak-to-peak or less, with no more than 1 V peak-to-peak input ripple. The internal output resistance of the regulator is only 1 Ω with a 6-V Zener diode and increases to approximately 10 Ω with a 24-V diode. For use in high-frequency circuits the regulator may need a 0.1 μF capacitor across the output.

The voltage regulation is approximately 1 per cent of the load voltage change and 2 per cent of the input voltage change, and the output ripple decreases approximately in proportion to the reduction of the load current. The design of a transistor-regulator requires that the product of the collector-to-emitter voltage drop and the load current must not exceed the dissipation rating of the transistor. Preferably, the power dissipation should be a factor of 3-to-5 times lower if the voltage drift during the transistor warm-up is objectionable. The bias resistor R_A is shown as 1 kΩ, but lower values of R_A may be required if the collector-emitter voltage drop is low or the Zener diode requires a higher current for satisfactory regulation. In any event the voltage drop across the transistor should be at least 2 V, and higher values are acceptable as long as the transistor dissipation is not excessive. The regulator output voltage V_O is approximately 1 V above the Zener voltage V_Z.

Compared with an RC filter the transistor regulator is many times smaller and has the advantage of reducing low-frequency feedback or cross-coupling. Compared with a simple Zener diode regulator the transistor regulator may be operated with an input-to-output voltage drop of only 2 V where a Zener regulator requires a voltage loss that is 30 to 50 per cent of the input voltage. The transistor regulator is hardly more complicated than a simple Zener regulator and is usually better and more efficient.

6.16 ONE-AMPERE VOLTAGE REGULATORS

Regulators suitable for loads up to 1 A are illustrated in Fig. 6.20. A power transistor may be used to regulate high-current loads, provided the transistor has a current gain of 50 at the maximum load current and is mounted on a heat sink to dissipate the collector power $I_O(V_I - V_O)$. The bias resistor

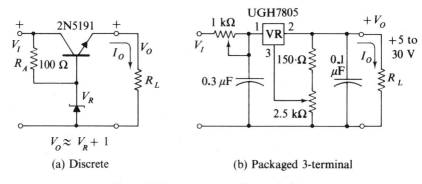

Figure 6.20. One-ampere voltage regulators

should be selected to make the current in R_A about 5 per cent of the load current, and a small capacitor may be needed across the load in high-frequency applications. For a negative voltage supply the Zener diode shown in Fig. 6.20(a) should be inverted and the transistor replaced by a *pnp* type 2N5194. Reasonably simple discrete component circuits that permit an adjustment of the output voltage usually require one or more additional transistors, and the performance characteristics are usually not much better than are obtained with a single power transistor.

For loads up to 3 A high-performance IC regulators are packaged as TO-3 or TO-220 devices and are available at low cost. On a suitable heat sink these devices dissipate up to 50 W and are capable of reducing a 30-V supply to 12 V regulated with a 1-A load. With some of the IC devices a variable voltage regulator may be constructed using the circuit shown in Fig. 6.20(b). A potentiometer is used to adjust the output voltage, or a 2-resistor voltage divider may be substituted when a fixed voltage is required. The input power may be obtained from a rectified supply as high as 35 V, provided the smoothing filter is capable of reducing the input ripple to 10 per cent or less at the maximum load current.

While the packaged regulators have output impedances of less than 1 Ω, a capacitor may be needed across the load, as shown in the figure. The type UGH regulator is self-protected from overload by limiting the maximum load current to 1 A. With the series resistor R the load current may be limited from a 30-mA maximum up to 1 A. The main limitation of the adjustable voltage regulator is that the output voltage cannot be reduced below 5 V. In other respects the regulators have excellent performance characteristics.

A positive and negative 1-A regulator may be constructed, as shown in Fig. 6.21, by using a transformer with 2 low-voltage windings and a pair of identical positive voltage regulators. This circuit illustrates also the simplicity with which the IC regulator may be used to obtain a well-regulated fixed voltage supply. The regulator may be purchased to supply a choice of output

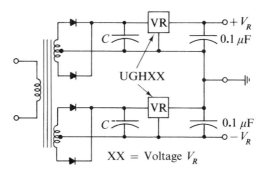

Fig. 6.21. Positive and negative 1A regulator

voltages, and the capacitor need only be selected to ensure that the ripple on the input side of the regulator limits the ripple voltage to about 10 per cent of the dc voltage. The value of the capacitor may be found by Eq. (6.1).

6.17 CURRENT LIMITERS

Transistor current limiters are easily constructed by using the 10-to-1 collector current change that is produced by a 60-mV change of the base-emitter voltage. A current limiter that may be used to limit load currents up to 20 mA is shown in Fig. 6.22(a). With a 25-Ω load-sensing resistor R_S the 2N3638 transistor turns ON when the voltage drop across R_S is approximately 0.5 V and cuts off the base current required to maintain current flow through the 2N4036 transistor. While the load current cutoff begins with a current of about 95 per cent of the maximum, the 2N4036 transistor may be required to dissipate a power that is $\frac{1}{2}V_{CC}I_L$. For the present low-power example the maximum dissipation is only 0.6 W with a 30-V supply, and the transistor does not require a heat sink. With higher load currents and supply voltages the designer may need to use a heat sink and to consider whether or not the $V_{CC}I_L$ product is within the transistor's second breakdown rating.

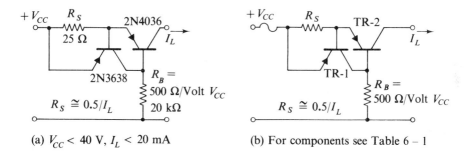

(a) $V_{CC} < 40$ V, $I_L < 20$ mA

(b) For components see Table 6 – 1

Figure 6.22. Current limiters, 20 mA to 1A

The circuit values shown in Fig. 6.22(a) are for use with a 30-V collector supply and a 20-mA maximum load current. For lower supply voltages the base resistor should be reduced in proportion to the V_{CC} reduction. To limit the load at currents below 20 mA the sensing resistor R_S is increased to make the voltage drop $I_L R_S$ equal to 0.5 V. With the components shown the limiting value of I_L may be reduced to 0.1 mA by making $R_S = 5$ kΩ.

Transistors and components recommended for higher load currents are shown in Table 6-1 for use in the current-limiting circuit, Fig. 6.22(b). With a

Table 6-1 Current Limiter Components for $V_{CC} = 30$ V—Fig. 6.22

| TR-1 | | TR-2 | | Case | R_B | | R_S (min.) Ω | I_L (max.) mA | Heat Sink °C/W |
pnp	npn	pnp	npn		kΩ	W			
2N4355	2N3569	2N4036	2N3108	TO-5	20	0.5	25	20	none
"	"	"	"	"	5	0.5	5.1	100	100
"	"	2N4235	2N4238	"	3	1	3.3	150	50
"	"	2N6125	2N6122	TO-220	1	2	0.5	1000	6
"	"	2N4398	none	TO-3	1	2	0.5	1000	3

positive supply the *pnp* transistors shown in Table 6-1 are recommended, and the *npn* transistors in Table 6-1 are complementary types for a negative supply. The table shows also the minimum recommended values for R_S and the thermal resistance of the heat sink needed with a 30-V supply and the maximum load current. For lower values of the input voltage and load current, the heat sink thermal resistance may be increased. For example, if the $V_{CC}I_L$ product is halved, the thermal resistance may be at least doubled, and with lower input powers a heat sink may not be required.

The need for a heat sink may be avoided by using in the regulator supply a fuse that is rated to open at one-half to three-fourths the cutoff current I_L. With a fast-blow fuse the transistor is protected from overheating, while the current limiter prevents the load currents from greatly exceeding the fuse rating in the time required to blow the fuse. Observe that replacing R_S by a fuse may result in a damaged TR-1. Circuits that protect a load from an abnormally high input voltage by shorting the load until a fuse blows are known as **crowbar circuits**.

6.18 FIFTEEN-AMPERE, SHORT-CIRCUIT-PROTECTED VOLTAGE REGULATOR

The circuit shown in Fig. 6.23 may be used as a voltage-regulated supply for output voltages up to 25 V with load currents up to 15 A. As with the low-current limiters the resistor R_S is selected to make the voltage drop $R_S I_L =$

Sec. 6.18   Fifteen-Ampere, Short-Circuit-Protected Voltage Regulator   107

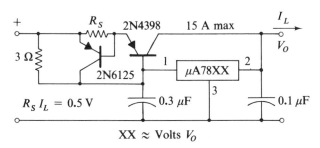

Figure 6.23. 15-A voltage regulator

0.5 V at the maximum load current. Because the power dissipated in R_S is 0.5 I_L watts, the power rating of R_S should be about $1 \times I_L$, or numerically equal to the magnitude of the maximum load current. Thus, if $I_L = 15$ A, $R_S = 1/30$ Ω, and the resistor should be rated for 15 W. A method of constructing such a low-value resistor is to connect fifteen 1-W one-half ohm resistors in parallel. For a 5-A load current only five 1-W, 0.5-Ω resistors are required, and five resistors may be paralleled without too much difficulty with space allowed for the dissipation of heat. A larger number of resistors for higher currents are best connected in parallel between a pair of 10- or 12-AWG wires formed as circles 2 to 5 cm in diameter. The resistors should be spaced to improve the power dissipation, and the external circuit may be connected to opposite sides of the circles to distribute the load current evenly to the resistors.

The resistors paralleled to make R_S may be rated for $\frac{1}{2}$ W if the maximum load current is expected to exist for relatively short times or only occasionally. Similarly, if 1-Ω resistors are used, twice as many resistors are needed, and, assuming equal currents, the maximum power is only 0.25 W in each resistor. Unfortunately, any resistance divider that eliminates the need for a low-resistance R_S also increases the voltage drop and the power loss in R_S.

The resistance values shown for R_S in Table 6-1 are minimum values that may be increased up to 100 times to reduce the maximum I_L. For example, with the components given in Table 6-1 for a 1-A, current-limited power supply, the current limit is adjusted by making R_S a 0.5-Ω, 1-W resistor in series with a 50-Ω, 1-W variable resistor. The current limit may be adjusted from a 1-A maximum down to a 10-mA maximum. A switch on the potentiometer may be connected to open the base connection of the current-sensing transistor TR-1 as a way to obtain full output from the power supply, provided the second breakdown ratings of the series transistor TR-2 are not exceeded. A separate switch that connects the load to the power supply is a better means of bypassing the limiting circuit. Unfortunately d.p.s.t. switches that may be attached to a potentiometer for the latter purpose are not often available.

6.19 CAPACITANCE MULTIPLIERS

A capacitance multiplier is a form of active filter that produces an equivalent capacitance that is β or S times larger than a reference capacitor. The circuits in Fig. 6.24 show two capacitance multiplying filters with the reference capacitor connected from the base to ground. In the capacitance multipliers the equivalent capacitance across the load R_L is approximately 20 times the base capacitor. Thus, the ripple across the load is lower than the input hum by a factor equal to the current gain of the transistor. The transistor is biased in the active region with a collector-emitter voltage of about 2 V. The transistor must be biased to ensure that the Q-point remains in the active region at any junction temperature, but the dc S-factor is low, and the filters are operable over a wide temperature range.

The capacitance multiplier in Fig. 6.24(a) offers an easy way of filtering a supply voltage or a low-frequency signal and may be used in place of a regulator where low hum is needed but a constant voltage is not required. The equivalent capacitance C, shown in Fig. 6.24(b), is 1000 μf in series with 2 Ω. The 2-Ω resistance component is the internal output resistance of the transistor. This resistance places a lower limit to the impedance than can be attained by increasing the base capacitor. The reactance of 1000 μF is less than 2 Ω at 120 Hz, so there is no advantage in increasing the base capacitor for a capacitor used at 120 Hz.

The capacitance multiplier shown in Fig. 6.24(c) is designed as a 120-Hz smoothing filter for a 15-V, 1-A load. The equivalent capacitor is over 100,000 μF, and the equivalent series resistor is 0.04 Ω. If either circuit is for use at a lower frequency, the base capacitor may be increased. For example, if the frequency is 12 Hz instead of 120 Hz, the base capacitor may be increased by a factor of 10 with a corresponding increase of the equivalent output capacitance.

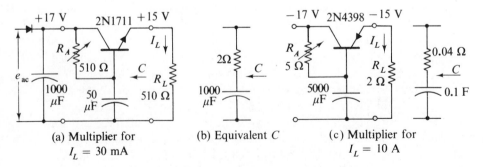

Figure 6.24. Capacitance multiplier filters

6.20 REPLACING FILTER CAPACITORS WITH ZENER DIODES

Zener diodes may replace power-supply filter capacitors or emitter capacitors in the low-power stages of a high-gain amplifier. In low-level stages where the low efficiency of a Zener regulator is unimportant, the diodes offer the advantages of high reliability and the ability to remove low-frequency noise and feedback. The circuit of a Zener-diode filter designed for a low-frequency preamplifier is illustrated in Fig. 6.25. The Zener diode is a low-current, 250 mW, 1N4112 that has 100-Ω ac resistance at currents as low as 1 mA. Moreover, the diode is a low-noise type in preference to a power type that may introduce unexpectedly high noise. If the amplifier has a wide band or is required to have a minimum noise output, Zener noise may be reduced by connecting a small capacitor across the diode.

When used as a smoothing filter, a Zener diode atttenuates the power-supply noise by a factor of 10-to-30, 20-to-30 dB, provided the resistor is sized to reduce the supply voltage to about two-thirds the input voltage. This reduction of the supply voltage is not often a disadvantage in a preamplifier. If the low-frequency response of the amplifier extends to 20 Hz, the diode in Fig. 6.25 is approximately equivalent to an 80-μF capacitor. For lower frequencies the diode is equivalent to proportionally larger capacitors and has the advantage of maintaining a constant voltage.

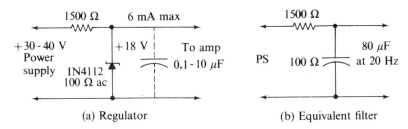

Figure 6.25. Zener diode smoothing filter

6.21 BATTERY POWER SUPPLIES

The most common reasons for operating transistor equipment with a battery power supply are the need for portability or the cost advantage obtained by eliminating an ac power supply. Other reasons for using batteries are the simplicity and convenience with which they can be used to eliminate most problems of power-supply filtering and stray ac ground currents. In some instruments an off-ground power supply is needed where the current drain is so low a battery without a cutoff switch may be operated for a year or even as long as the shelf life of the battery.

In estimating the life of a battery the most unpredictable factor is the amount of time the battery may be left ON unintentionally when the equipment is not in service. Unless the operating life in continuous service is several days or more, the problem of obtaining a replacement may make a battery impractical.

The voltage and milliampere-hour capacities of equivalent D-size cells of the most common battery types are shown in Table 6-2. However, when

Table 6-2 Approximate Characteristics of No. 2, D-Size Cells (3.2 cm. diam. × 6.1 cm. long)

Type	Initial Voltage	End Voltage	Milliampere Hours
Zinc-carbon	1.55	0.9	4000
Alkaline	1.55	0.9	7000
Mercury	1.35	1.25	12000
Nickel-cadmium	1.35	1.15	3000
Lithium	2.6	2.2	9000
Sodium-bromide	3.6		8000

different types of batteries are compared, the characteristics of a cell vary with the load and make difficult a prediction of the service life or the operating costs. For example, a NEDA 1604, 9-V transistor battery gives approximately 4 hours of service with a 20 mA load (80 mA-hr) or 150 hours with a 2 mA load (300 mA-hr). For loads less than 2 mA the battery supplies a constant 300 mA-hr with the voltage decreasing to 7.2 V at the end of life.

The cell types shown in Table 6-2 are listed in approximately the order of their availability and increasing cost. The end-point voltage shown for each cell is a value commonly accepted in designs that require a long service life. If equipment can be designed to operate down to a lower end-point voltage, the service life is increased by using a higher total energy from the cell. The main advantage of the alkaline cells, which cost up to twice as much as zinc-carbon cells, is realized only at high-current loads where the service life may be 2 to 10 times greater, although not substantially different under low-current standby service. The alkaline-D cell is more economical with loads exceeding 300 mA for an hour or more.

In applications where the cells are large enough to operate at relatively low currents the ordinary dry cell is the least expensive and most satisfactory. For motor loads or high load currents the alkaline cell may be more satisfactory but more expensive than using a larger zinc-carbon cell. The mercury cells offer a nearly constant voltage and are a reliable source of relatively high load currents. The mercury cells have approximately 3 times the ampere-hour capacity of zinc-carbon cells of the same volume, but they are expensive.

Lithium cells promise a 2-to-1 advantage over mercury cells, but most comparisons are not yet meaningful.

Storage batteries are generally the most satisfactory and least expensive power source for loads exceeding several amperes and requiring reliable operation for several hours a day. Storage batteries have a very low internal impedance, especially when being charged, and equipment that cannot share a common rectified dc supply usually can share a storage battery. Nicad storage batteries are readily available in a variety of shapes and sizes. A few sealed lead-acid storage batteries are manufactured, and one D-size cell has a 2500 mA-hr rating and 2.05 V.

For laboratory tests requiring a high-current power source an automobile battery may be used, sometimes with the battery left in the car. Even for frequent service a low-cost 12-V battery and an ac charger are more satisfactory and less expensive than a high-current, ac-operated, power supply.

The internal resistance of a storage battery is usually much less than the impedance of the leads, plugs, and switches of a connected load. Hence, instrument noise caused by an inverter or motor may be reduced a factor of 10 or more by connecting separate low-current and high-current circuits to the battery terminals. Additional isolation may be obtained with two parallel-connected storage batteries that may be recharged as a single unit.

The useful life of the dry cells used with transistor equipment is sometimes increased by connecting a large capacitor across the battery circuit. The need for the capacitor may be determined by observing how the equipment performs with a resistor connected in series with the battery. The resistor simulates the internal resistance of an old or depleted battery and should be a value that reduces the supply voltage by 10 to 20 per cent. A large capacitor may supply peak load currents, reduce noise, or prevent feedback. On the other hand, a battery connected across a filter capacitor or used to separate power supplies may resolve a difficult instrument problem or simplify amplifier tests.

7 Low-Noise Amplifiers and Preamplifiers

Amplifiers for some applications must be designed with the lowest possible amount of background noise because the smallest signal that can be used is fixed by the noise produced in the amplifier. Noise may include power-line hum and pickup of nearby signals, both of which can be eliminated by careful shielding or the relocating of circuits and components. However, when the hum and pickup are eliminated from a high-gain system, there always remains a random noise that cannot be reduced. This noise is the snow observed on a TV screen, the hiss heard in an audio amplifier, or the unsteady trace of an oscilloscope at maximum sensitivity.

A preamplifier is designed to shape and amplify a transducer signal and is generally used with a phonograph pickup or a magnetic tape head. Although a preamplifier should have low noise, the most important requirements of the amplifier are that it should be free of power-line hum and be capable of producing a distortion-free signal with even, relatively large input signals. A low-noise amplifier, on the other hand, is used to amplify very low-level signals and is designed to add the lowest possible amount of random noise. Because low noise is obtained by operating the transistors with a low-current, low-voltage Q-point, a low-noise amplifier is not usually capable of producing the high output signals needed in preamplifier applications.

The random noise observed in a high-gain amplifier may be produced in three or more different ways. The noise in a wide-band audio amplifier is produced by current flow in semiconductors and by the thermal agitation of electrons in conductors. The noise from these sources is reduced by reducing the current and the ohmic resistance in the input circuits. Random noise is proportional to the square root of the amplifier bandwidth, so the bandwidth should be no greater than is necessary to transmit the desired signals.

A narrow-band amplifier for the frequency spectrum below 100 Hz usually has a relatively high random noise produced by current flow in resistors and semiconductors because they have a granular structure rather than the uniformity of a metallic conductor. This excess, pink, or $1/f$ noise is particularly troublesome in poorly made resistors, MOS FETs, and high-frequency transistors. Low-frequency $1/f$ noise is reduced by reducing the current, by using low-noise resistors, and by limiting the low-frequency response of the amplifier in a following stage. For a low-frequency amplifier a transistor may be the best low-noise choice when the source impedance is less than 1 kΩ. A JFET may be better with source impedance of 10 kΩ or more, and a MOS FET may be best when the impedance exceeds 1 MΩ.

The noise characteristics of otherwise identical semiconductors vary considerably, and a worthwhile reduction of noise may be obtained by selecting the device used in the input stage. Low-noise semiconductors are produced by careful manufacture and testing, so there is often considerable value in comparing the products of several manufacturers. Transistors that have a high β at low collector currents generally have the lowest noise. The *pnp* transistors are better than *npn*s at low frequencies and with a low-impedance source. High-frequency transistors are always too noisy for low-frequency and wide-band audio-frequency applications. The noise in a UHF amplifier is determined by the high-frequency cutoff of the transistors and is reduced by using devices that have a high f_T value. The noise characteristics of monolithic ICs are definitely inferior to those of a discrete transistor or FET. Low-noise ICs are hybrid devices that have a monolithic IC preceded by a discrete semiconductor input stage.

7.1 NOISE IN CIRCUIT COMPONENTS (REF. 9)

The bias resistor and sometimes the emitter resistor are a source of $1/f$ and thermal noise that contributes to the amplifier input noise. The effect of thermally-generated noise is made negligible by making the bias resistors 5 to 10 times larger than the source resistance. The $1/f$ noise is negligible in wire-wound and in well-made composition resistors. Carbon film resistors and poorly made or defective composition resistors may be a source of noise in current-carrying circuits. Except where the ultimate in low noise must be attained, the better brands of composition resistors do not cause a noticeable increase of low-frequency noise when carrying relatively low currents.

Capacitors free of leakage are not a source of noise, although the input coupling capacitor used with a JFET may contribute a noise voltage when the reactance of the capacitor exceeds 1 MΩ. Batteries in normally good condi-

114 Low-Noise Amplifiers and Preamplifiers   Ch. 7

tion are not a source of noise. Because input stages may be operated with 100 μA or lower collector currents, a battery is often the simplest means of obtaining a noise-free collector supply. If the collector current is less than 10 mA, the collector battery may often be permanently connected in the circuit without a switch.

Because the signal source and the input stage both contribute noise, there is an optimum source impedance that should be used with any given input stage. The optimum source impedance is usually between 1 and 25 kΩ, but the noise increases only a little with impedances that are a factor of 10 smaller or larger than the optimum. Thus, in the construction of low-noise amplifiers the source impedance is of less importance than the selection of a good semiconductor.

7.2 LOW-NOISE AUDIO AMPLIFIERS (REF. 2)

The circuit of a wide-band audio amplifier, illustrated in Fig. 7.1, differs from that of an ordinary amplifier mainly in the choice of the transistor and the use of a low collector current. In the present example the collector current is made 100 μA by using a low collector-supply voltage and relatively large collector and emitter resistors. The bias resistor is selected to make the emitter and the collector-to-emitter voltages each approximately 4 V.

The values of the capacitors used in the circuit play an important part in reducing the amplifier noise. The emitter capacitor eliminates the thermal and $1/f$ noise produced by the emitter resistor. The input coupling capacitor must be large enough to ensure that the transistor sees the relatively low source impedance; and the reactance of the capacitor should be less than the source impedance if the frequency response extends into the $1/f$ noise region. The

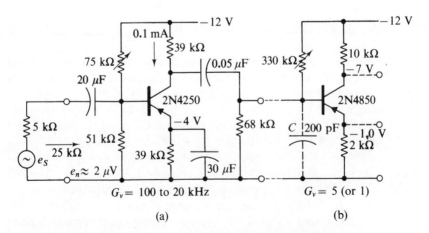

Figure 7.1. Low-noise wide-band audio amplifier

output capacitor is sized to eliminate low frequencies that are not required for a satisfactory response and thus helps eliminate $1/f$ noise.

The optimum source impedance for the wide-band amplifier is 5 kΩ, but, with a small increase in the noise level, the source impedance may be as low as 500 Ω and as high as 50 kΩ. For applications requiring a frequency response above the audio range the amplifier is useful with a 5-kΩ source up to approximately 100 kHz, provided the 68-kΩ load resistor is reduced to 10 kΩ. The bandwidth should be reduced as much as is practicable by reducing the interstage coupling capacitor when the low-frequency response is not required.

For a 10-kΩ or lower source impedance the voltage gain of the wide-band amplifier is approximately 100, 40 dB, and almost any audio stage with a 50-kΩ or higher input impedance may be used for the second stage. The stage shown in Fig. 7.1(b) may be used as a CE amplifier or as a CC output stage by removing the collector resistor and adjusting the bias resistor. The equivalent noise input voltage of the amplifier should be less than 3 μV rms. A higher noise may indicate the need to select a different transistor or to replace a noisy resistor.

The wide-band amplifier may be changed for narrow-band low-frequency applications by increasing the capacitors by a common factor and by adding a shunt capacitor to eliminate high frequencies and noise that are not needed in a narrow-band amplifier. For low-frequency applications the optimum source resistance is 5 kΩ and may be between 1 and 100 kΩ, with only a small increase of the noise. When the resistance component of the source impedance exceeds 100 kΩ, the input stage should be changed by increasing all resistors by a factor of 10. Increasing the resistors lowers the collector Q-point current and increases the optimum source impedance.

7.3 LOW-NOISE COLLECTOR FEEDBACK AMPLIFIER

The low-noise feedback amplifier shown in Fig. 7.2 has the advantages of considerable circuit simplicity and nearly 20-dB higher gain than the low-noise input stages usually recommended. Except for a slightly reduced bandwidth the collector-feedback amplifier has substantially identical input and noise characteristics as the wide-band audio amplifier of Fig. 7.1.

An FET is used in the second stage of the amplifier, and the high gain of the first stage makes the noise characteristics of the FET relatively unimportant. Because the bandwidth of the input stage is determined by the Miller effect, the upper cutoff frequency f_h is at approximately 10 kHz. However, f_h may be increased to 100 kHz by lowering the gate resistor to 10 kΩ, and the first stage gain is reduced to 30, 30 dB. With the increased bandwidth the equivalent input noise is increased by a factor of 3, unless the low-frequency

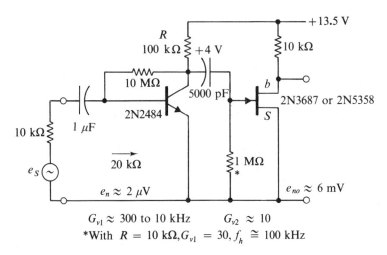

Figure 7.2. Low-noise collector feedback amplifier

cutoff is increased enough to maintain a fixed 10 kHz bandwidth. With the interstage resistor reduced just enough to increase f_h to 20 kHz, the amplifier can be used as a low-noise phonographic preamplifier. However, adequate dynamic range may not be available unless the FET supply voltage is increased to 30 V and the FET is selected to make the drain voltage approximately one-half the supply voltage. Ordinarily the record noise level is substantially higher than amplifier noise, so the input stage need not be designed for low noise. Hum induced in the pickup and inadequate power supply filtering are the most usual phonographic noise problems.

7.4 LOW-IMPEDANCE, LOW-NOISE, LOW NOISE AMPLIFIERS

Two amplifiers suitable for use with low-impedance transducers are shown in Fig. 7.3. Both input stages are essentially the same, and the different input noise levels are mainly a result of the difference in the overall bandwidth. The suggested impedances are for wide-band applications, and lower source impedances may be used for narrow-band applications with only a small increase of the equivalent noise voltages. The relatively large input capacitor, 100 μF, is needed as a means of reducing the $1/f$ noise by forcing the transistor to see a low impedance at frequencies a factor of 5 below the interstage cutoff frequency. A large capacitor is not needed when the interstage coupling is designed to eliminate frequencies below 1 kHz.

The single-stage may be coupled to almost any convenient audio amplifier, as shown in Fig. 7.3. The amplifier shown in Fig. 7.3(b) is designed for use with a 100-Ω transducer and may be used with transducer impedances from 0 to 1 kΩ with only a small change of the equivalent noise. The noise voltage

Sec. 7.5   JFET High-Impedance, Low-Noise Amplifiers   **117**

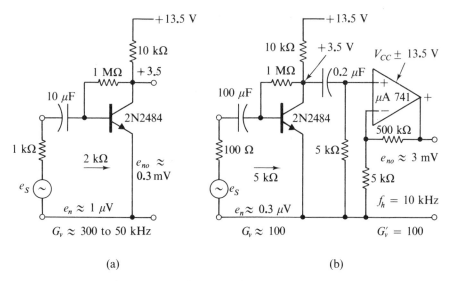

Figure 7.3. Low-impedance low-noise amplifiers

is 0.3 μV rms with the optimum source impedance 1 kΩ. Lower noise levels may be obtained with low-impedance sources by using an MPSA92, a *pnp* Fairchild device, in place of the 2N2484. With low impedances *pnp* transistors produce less noise than *npn*s.

The low-impedance input stage shown in Fig. 7.3(b) is coupled to a linear IC for additional gain. As a general rule, the equivalent input noise of an IC is at least 5 times the noise of a low-noise discrete transistor, and an IC amplifier should be preceded by a low-noise stage with a gain of at least 30. In the present example the input stage noise is amplified enough to make negligible the noise of a type 741 IC, and, by changing the IC feedback resistor, the bandwidth may be increased or decreased by a factor of 10 without reducing the accuracy of noise measurements. The simplicity of the amplifier and the ease with which the bandwidth may be changed make the amplifiers of Fig. 7.3 attractive for noise studies and low-impedance transducer applications.

7.5 JFET HIGH-IMPEDANCE, LOW-NOISE AMPLIFIERS

An amplifier used with a high source impedance usually has a relatively low upper cutoff frequency that is determined by the input capacitance and the Miller effect. The high-frequency cutoff to be expected with a 10-MΩ source is illustrated by the stage shown in Fig. 7.4. The FET selected for the amplifier has 75 pF input capacitance with a gain of 10, and the upper cutoff frequency

118 Low-Noise Amplifiers and Preamplifiers Ch. 7

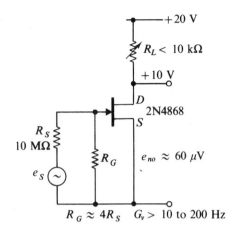

Figure 7.4. JFET high-impedance low-noise amplifiers

is at 200 Hz. With careful design and by using the FET as a source follower, the upper cutoff may be increased from 1 to 2 orders of magnitude.

The noise output of the high-impedance FET stage is almost entirely caused by thermal noise generated in the 10-MΩ source. With a lower impedance source the noise decreases as the square root of the resistance down to a source resistance of 100 kΩ. Because of the relatively high input noise the FET stage may be followed by any of the amplifiers shown in Figs. 7.1, 7.2, and 7.3. The input stage FET is operated without bias feedback, and the drain resistor R_L may require an adjustment because the drain current of the FETs tends to be quite variable. The correct adjustment of the drain resistor is with a value slightly less than the resistance that gives a maximum voltage gain. An alternate and essentially equivalent choice is to adjust the resistor to make the drain voltage 8 to 10 V. The signal source is direct-coupled, and the amplifier may be used to very low frequencies, depending on the frequency response of the succeeding stages. The 2N4868 FET is indicated because of its exceptionally low noise at 10 Hz.

The stage shown in Fig. 7.5 is intended for broadband applications that make the noise at 10 Hz relatively unimportant. The 2N3089 FET has an input capacitance of only 26 pF, which, with the lower source resistance, raises the upper cutoff frequency to above 20 kHz. The broadband noise output is again mainly thermal noise in the signal source, but the output noise is independent of the transducer resistance because the product $f_h R_S$ is essentially constant. Assuming that the $1/f$ low-frequency noise is removed in a succeeding amplifier, the output noise of the broadband stage is larger than that of the high-impedance stage only because the lower input capacitance has increased the $f_h R_S$ product.

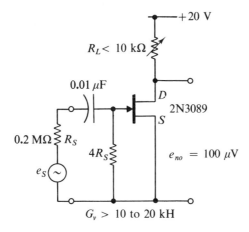

Figure 7.5. JFET low-noise audio amplifier

7.6 NONBLOCKING AMPLIFIERS

An amplifier connected to a capacitance transducer, or through a capacitor to a signal source, may be subjected to a large transient signal that is rectified by the input diode of a transistor or JFET. The rectified signal charges the capacitor and biases the amplifier beyond cutoff. Although the capacitor is charged rapidly through the forward-biased diode, the capacitor discharges slowly through the reverse-biased diode, and the amplifier is disabled, or blocked, for several seconds. While the input stage is recovering, the amplifier is unable to transmit small signals for a relatively long blocking time, and a disturbing thump is often produced when the amplifier recovers.

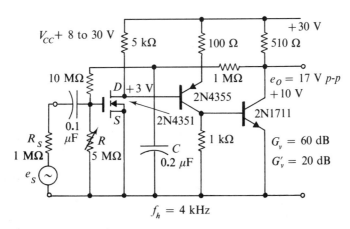

Figure 7.6. Nonblocking high-gain ac amplifier

The amplifier shown in Fig. 7.6 is nonblocking because the insulated gate is unable to rectify a large input signal. The amplifier has a voltage gain of 1000, 60 dB, without feedback; with the capacitor C removed, the feedback reduces the gain to 10, 20 dB. However, the gain with feedback depends on the source impedance and is 100, 40 dB, with a 100-kΩ source. The amount of feedback may be varied also by a resistor in series with the capacitor C. Several MOS FETs may have to be tried to find one that supplies the correct bias for the transistor amplifier, but there is 60-dB dc feedback and the amplifier requires only a minor bias adjustment for supply voltages from 8 to 20 V. The bias is adjusted by increasing or removing the resistor R.

The open-loop frequency response of the amplifier depends on the source impedance and on the care used to protect the input from overall capacitance feedback. With adequate shielding the upper cutoff frequency is at approximately 20 kHz, and the cutoff moves to higher frequencies when the gain is reduced by feedback. The output noise is 10 mV and is nearly independent of the amount of feedback, because the noise is predominately low-frequency $1/f$ noise. The noise may be reduced by a low-frequency cutoff in a later stage, but the amplifier cannot be recommended for low-noise, low-frequency applications.

The direct-coupled, insulated-gate preamplifier may prevent or reduce blocking in a succeeding amplifier because the peak output signal cannot exceed the drain supply voltage, which may be as low as 8 V. If a gain of 100 is sufficient, the preamplifier shown in Fig. 4.12 may be used to drive a power amplifier, and the output peaks cannot exceed 1.5 V. Blocking caused by energy stored in a reactive transducer is usually reduced by connecting back-to-back Schottky diodes across the amplifier input. When simpler methods fail FET switches or magnetic reed relays may be connected to discharge the coupling capacitors or short-circuit the transducer.

7.7 LOW-NOISE OP-AMP PREAMPLIFIERS

A low-noise, high-impedance OP-amp preamplifier is easily assembled and may be used as an inexpensive substitute for an expensive, though better, hybrid IC. With a pair of FETs and a μA741 amplifier the amplifier shown in Fig. 7.7 may be used to drive a pen recorder or as an oscilloscope dc amplifier. An advantage of the preamplifier is that it may be operated with a pair of 9-V transistor batteries with only a 3-mA current load. Thus, there is no problem with noise and pickup from a power supply, and the equivalent noise input is almost entirely thermal noise of the input circuit.

The input stage may be used as a single-sided amplifier or as a differential amplifier, and the source impedance may be from 0 to 20 MΩ. With a high source impedance the cutoff frequency is determined by the Miller effect at the input, but with source impedances less than 500 kΩ the bandwidth is 10 kHz

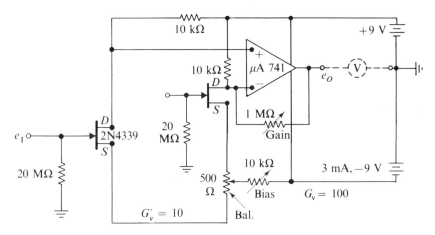

Figure 7.7. Low-noise OP amp preamplifier

and is determined by the OP amp. The voltage gain of the input stage is 10, which ensures that the noise in the input stage exceeds the noise of the OP amp as long as the input impedance exceeds 50 kΩ. The CMR of the input stage is 50 dB and is adequate for most applications.

The FETs should have a g_m of nearly 1000 μmho with a zero bias drain current that is no more than 1 mA. For ease of balancing the drain currents the FETs may need to be selected at assembly or purchased as a matched pair. With dual monolithic FETs for better temperature tracking and a 100-mV meter connected to the output, the gain may be adjusted to permit reading 100 μV full scale with a 10-MΩ impedance. Thus, the meter-amplifier has a sensitivity of 100 MΩ/V.

7.8 PHONOGRAPH PREAMPLIFIER (REFS. 2, 4, 7)

A low-noise preamplifier for phonographic uses may be built, using the circuit shown in Fig. 7.8. The amplifier has a low-noise JFET in the input stage and operates on a single + power supply. The amplifier has advantages when the input stage is to be placed very close to the pickup or is located at a distance from the main amplifier and power supply. A phono-preamplifier that uses an IC with a $\pm$ collector supply is described in the chapter describing IC applications.

The direct-coupled FET-transistor pair has a low-frequency gain of 100, 40 dB, and the capacitor C_1 provides the RIAA low-frequency equalization. Capacitors C_2 and C_3 are used to provide a high-frequency tone control with the RIAA high-frequency equalization obtained when the tone control is at the mid-position. If the tone control is placed in a later stage, the RIAA equalization is obtained by making $C_4 = 0.067$ μF and omitting C_2 and C_3.

122 Low-Noise Amplifiers and Preamplifiers　　　　　　　　　　　　　　　Ch. 7

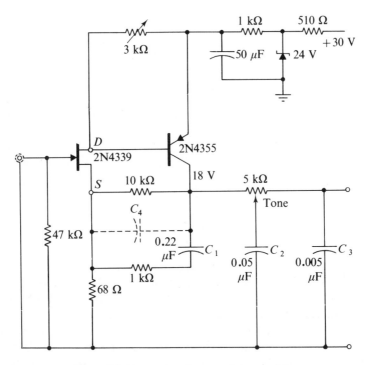

Figure 7.8. Low-noise phonograph preamplifier

For low distortion at large signal inputs the collector supply should be as high as practicable. With a 30-V supply that is shared with a power amplifier the preamplifier needs a well-filtered 22 V, easily obtained with a Zener diode, and a single RC filter to remove noise produced by the diode. The FET should be selected to operate with approximately 1-mA drain current when zero biased, and the bias resistor should be selected to make the collector voltage three-fourths the Zener voltage. The input FET may be remotely located near the pickup with a 3-wire interconnection. For a remote connection the drain, source, and the ground should be returned by a 3-wire cable, and the source resistor should be at the input end of the cable to prevent hum pickup in the input circuit. The resistor across the input is required only to terminate the pickup and should have the value recommended by the manufacturer.

7.9 TRANSISTOR PHONOGRAPH PREAMPLIFIER

A preamplifier that uses a single transistor is illustrated in Fig. 7.9. The amplifier has 36-dB gain at 30 Hz and is equalized to provide the RIAA frequency response needed with a magnetic pickup. The stage has collector

Sec. 7.9 Transistor Phonograph Preamplifier

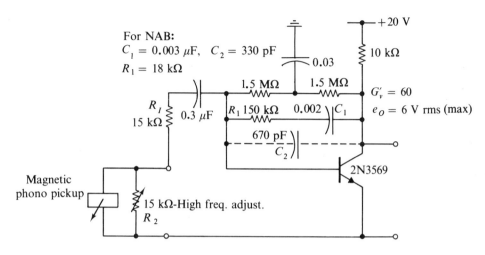

Figure 7.9. Transistor phono preamplifier (RIAA)

feedback bias that is bypassed at frequencies above 30 Hz. The maximum output voltage exceeds 6 V rms, and the performance characteristics are independent of temperature up to an ambient temperature of at least 60°C.

At 30 Hz the gain is determined by the transistor current gain, and the transistor should be selected to make the gain at least 60. At frequencies above 100 Hz the amplifier is like an OP amp with the frequency characteristic determined by the resistor R_I and the collector-to-base feedback network. However, above 2 kHz the impedance of a pickup increases and complicates the control of the high-frequency cutoff with feedback.

The RIAA high-frequency cutoff above 2 kHz is determined by the input resistance, R_I in parallel with R_2, and the feedback capacitor is not used unless the source impedance is less than 5 kΩ, as with an emitter follower. The resistance seen by the pickup should equal the reactance of the pickup at 2 kHz, which is 7500 Ω with a 600 mH pickup. Thus, the effect of the input resistance is to reduce the pickup voltage with increasing frequency, and the resistor R_2 offers a convenient means for adjusting the high-frequency response.

The transistor preamplifier has the advantages of a simple, uncomplicated circuit using small capacitors, having relatively little sensitivity to a change of the supply voltage and a bias circuit that does not require additional filtering. Because of the small amount of feedback at low frequencies, the equalization may not fit the RIAA curve exactly, but the deviation is small enough to be compensated by a minor adjustment of a tone control.

The preamplifier may be used for a tape recorder by changing the feedback network. The NAB frequency response used with a magnetic-tape pickup is obtained by making the feedback elements 18 kΩ in series with 0.003 μF, and the shunt capacitor C_2 may be omitted. The NAB frequency response is correct for a $7\frac{1}{2}$ inch per sec (ips) tape speed. With a 3 3/4 ips tape speed the series resistor R_1 may be 33 kΩ, or an adjustable 50-kΩ resistor. High-frequency tape noise may be reduced by using a shunt feedback capacitor C_2.

8 Audio Power Amplifiers

An amplifier is referred to as a *power amplifier* when the power output required by a load forces the designer to consider the voltage or current limits that can be withstood by the output transistors. Most audio transistors can safely handle the current and voltage required to deliver 50 or 100 mW to a load, so a medium power amplifier may be one that is required to deliver 0.25 to 1 W. Above 100 mW output transistors especially designed for power applications are required. An output transformer may be required when the current or the voltage demand of the load exceeds the capability of the output transistor, but transformers are almost never used in power applications for impedance matching.

Power amplifiers are designed to deliver a large amount of signal power with a low distortion and a high power efficiency. The last stage of a power amplifier should have as much power gain as practical in order to minimize the power required of the driver stage. Because the common-emitter amplifier has the highest power gain, from 30 to 40 dB, the output and driver stages of power amplifiers are almost always CE stages.

The power output of an amplifier is limited by the power capability of the transistors, and to develop maximum power a transistor must be loaded by a particular value of load impedance. Thus, a given power-amplifier design is not easily changed except by changing the load winding of the output transformer if the load is changed. Because power amplifiers are used in many different applications, a number of the circuits in this chapter are shown with output transformers that can have the output windings changed to meet load values different from those shown in particular circuits.

An amplifier is referred to as a class-A amplifier when the collector current flows continuously, both during the signal cycle and when there is no signal. Class-A amplifiers tend to dissipate considerable power and may need a large heat sink. If a single transistor is operated as a class-A power amplifier, there is usually an undesirably large amount of second and higher harmonic distortion, which may be eliminated by combining two single-sided amplifiers to form a push-pull amplifier.

A push-pull class-A amplifier offers about 3 times the power output of a single-sided amplifier with less distortion. It has the advantage that the dc collector currents flow in opposite directions relative to the transformer core, reducing the likelihood of saturating the iron core. The disadvantages are that the efficiency is low and the transistor dissipation is highest when there is no ac signal present. The push-pull class-A amplifiers are seldom used except to drive class-B amplifiers, because transistors operate almost ideally in class-B service.

A class-B amplifier is biased in such a way that the collector current is nearly zero when no signal is applied. With a signal applied, one transistor amplifies the positive side of the input signal and the other amplifies the negative side. These half signals are then combined by the transformer to restore the original waveform. The reader should observe that a class-B amplifier uses only one side at a time of both transformers, so the winding impedances must be calculated by using their value between one side and the center tap. The nominal full-winding impedance is 4 times the required side impedance.

Ideally, the transistor in class-B service should be biased to cut off the collector current under no-signal conditions. The transistor remains cut off at very low signal inputs because transistors have low current gain at cutoff and turn on abruptly with a larger signal. This nonlinear response gives the output signal a discontinuous waveform that is called **crossover distortion**. However, a small forward bias causes a small collector current to flow at low signal levels, suppressing most of the distortion. The residual distortion is further reduced by the negative feedback usually found in low distortion types of class-B amplifiers.

Class-B amplifiers are commonly used in portable radios and public address amplifiers because music and speech signals are highly intermittent, and small transistors will deliver the required power without needing heat sinks. The emitter resistors provide feedback and prevent thermal runaway at sustained high-signal levels. Although class-B operating efficiencies are somewhat higher than those of class A, the real advantage is that very little standby power is required under no-signal conditions.

Transistors are remarkably well adapted to class-B operation. When biased near cutoff, a transistor is linear and operating near its maximum gain, so that a class-B amplifier with feedback is remarkably linear at both large and small signal levels. Moreover, power transistors are such high-current devices that a voice coil may be driven without an output transformer. Thus, a high-fidelity audio amplifier is nearly always a direct-coupled amplifier that is also direct or capacitor coupled to the loudspeaker voice coil. A high-fidelity power amplifier usually has a complementary symmetry output stage operating class-B with 20 to 40 dB feedback.

8.1 LOW-POWER CLASS-A AMPLIFIERS (REF. 7)

A class-A power amplifier is usually coupled to a load by a transformer, which is the simplest way of connecting a load that must be completely isolated from the amplifier. A transistor may be easily coupled to a load with a transformer for loads of 0.1 W or less. For power outputs exceeding 1 W the transformer must be unreasonably large, and class-B amplifiers are more efficient and require relatively small output transformers.

For low-power applications the collector resistor of a resistance-coupled amplifier may be replaced by a transformer, provided the transistor CE breakdown voltage exceeds twice the supply voltage and the transformer is rated to carry the dc collector current. If the bias is not changed, the collector current does not change when the collector resistor is replaced by a transformer. With a low-impedance load a transformer provides higher gain than with capacitor coupling. Also, unless an efficient transfer of power is required, transformers are not used between stages, because an additional transistor gives greater gain at a lower cost, requires less space, and is much more reliable than a transformer.

The driver stages of class-B amplifiers in this chapter may be used as low-power class-A amplifiers by changing the output winding impedance to a value matching the load impedance. Although the amplifiers are generally described for low-power audio applications, they have many other industrial uses such as in servo systems, pen recorders, and controls.

The winding impedances of the input transformer for a class-A amplifier are usually selected to match the source impedance to the transistor input impedance. If the stage does not have significant feedback the input impedance in ohms is

$$R_I = \frac{26\beta}{I_E} \tag{8.1}$$

where I_E is the emitter current in mA and β is the minimum current gain of the transistor. If the stage has emitter feedback the input impedance is approximately the lower of the two base bias resistors.

The collector load impedance required for either a class-A or class-B

128 Audio Power Amplifiers　　Ch. 8

power amplifier is given approximately by

$$R_{cc} = \frac{V_{cc}^2}{P_o} \qquad (8.2)$$

where V_{cc} is the collector supply voltage and P_o is the desired output power. Equation (8.2) is a practical relation that takes into account the voltage and power losses that exist in low-power amplifiers and small transformers. In a practical design the collector load impedance may be as low as one-half the impedance given by Eq. (8.2). In a high-efficiency design the impedance may approach twice the value given by Eq. (8.2). Generally it is better to use a transformer that has an impedance too low than too high, and the maximum power output can be obtained if the bias is correctly adjusted.

8.2 POCKET-RADIO AUDIO AMPLIFIERS

The audio amplifier in a pocket radio is usually a transformer-coupled class-B output stage with a single CE stage for voltage gain. This amplifier generally uses low-power germanium transistors in a circuit similar to that shown in Fig. 8.1. The difference between a low-priced radio and a better radio is mainly in the use of larger and higher quality transformers with a more

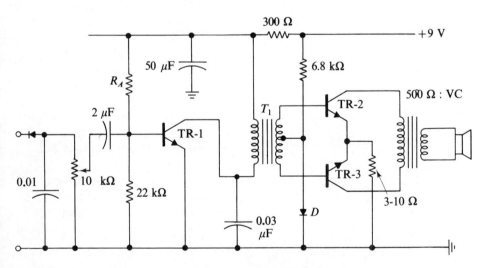

Note: TR-1: $\beta = 100$, set R_A for $I_C = 2$ mA
　　　TR-2, TR-3: $\beta > 40/100$ mA (200 mW)
　　　Diode D: $V_F = 1$ V/10-30 mA
　　　T_1: 2.5 kΩ : 2.5 kΩ CT (2 mA)

Figure 8.1. Pocket radio audio amplifier

efficient and larger loud speaker. The transformer manufacturers usually offer sets of input and output transformers with a suggested circuit for a given power output.

The diode D in the output stage serves as a low-voltage bias source that changes with temperature to keep transistors TR-2 and TR-3 slightly forward-biased, as required in a class-B stage. The diode is a low-current germanium or silicon device to match the transistor turn-ON voltage, and the transistors may share a common 3-Ω to 10-Ω emitter resistor to limit the current gain.

The driver stage may use a germanium transistor with an emitter resistor to limit the dc current gain and an emitter bypass capacitor to maximize the ac gain. If a silicon transistor is used, the emitter may be connected to ground with the transistor or the bias resistor selected during assembly.

8.3 LOW-POWER LINE-TO-LINE CLASS-A AMPLIFIERS (REF. 4)

The circuit of a low-power amplifier designed for use as an amplifier in a 500-Ω line is shown in Fig. 8.2. The amplifier uses a single CE stage operating on a 32-V power supply. The amplifier is intended for applications where 120-V ac is available, and the dc power may be obtained from a 24-V rectified ac supply. The amplifier may be used on a lower supply voltage by changing the bias resistor, but the gain and the maximum output signal are correspondingly reduced.

The base-to-collector voltage gain of the amplifier is 240, 48 dB, and with a pair of line-to-10-kΩ and 10-kΩ-to-line transformers the overall gain is at least 40 dB. The maximum output signal before overload is 3 V rms, and the

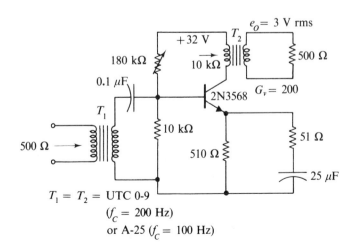

Figure 8.2. Line-to-line amplifier (46 dB gain, 1 W)

relatively high gain is obtained by operating the transistor with the partially bypassed emitter. The overall frequency response depends only on the characteristics of the transformers.

8.4 SINGLE-SIDED CLASS-A 2-W AMPLIFIER

The amplifier shown in Fig. 8.3 is designed for use in an auto radio operating on a 12-V to 14-V supply. The amplifier uses a direct-coupled transistor pair and gives a power gain of 36 to 40 dB with a 2- to 4-W output. The relatively small emitter capacitor provides high-frequency compensation, and the 3-dB cutoff is at 20 kHz. The low-frequency cutoff is determined by the inductance of the output transformer or choke, which must carry the dc collector current. If the load is connected across the choke or to a tap, the choke may be constructed by using a 2-cm square-tongue core and winding 22-AWG wire with a tap near the three-quarter point for a 4-Ω load. If the load is coupled using a 2-winding transformer, the core should be 1 or 2 sizes larger if low distortion and a good low-frequency response are desired. Unfortunately, the dc collector current tends to saturate the core, which should be assembled with a small air gap, and the amplifier should be biased to minimize the collector current whenever a lower power output is acceptable. The heat sink

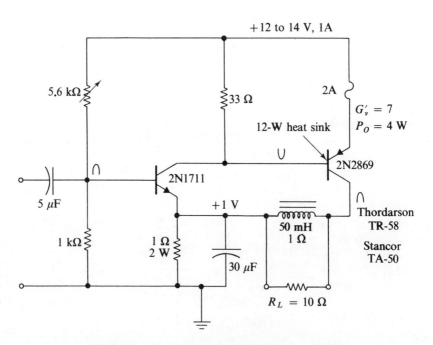

Figure 8.3. Class A audio power amplifier (12 V, 4 W)

should have a thermal resistance of 4°C/W or less; this requires a flat sheet with 300 cm² of exposed area.

The power amplifier shown in Fig. 8.3 may be used as a self-driven 4-W audio noise signal, for an intrusion alarm. Coupling the collector back to the input base provides positive feedback, and the signal frequency may be adjusted by a capacitor across the output inductor. A small reactor suitable for a single-frequency output above 400 Hz may be constructed by restacking the core of a 4-Ω output transformer or a 6-V filament transformer to provide a small air gap. The magnetic core with an air gap may be several times smaller than the core required for a satisfactory audio amplifier. However, the dc resistance of the winding should not exceed 1 Ω. Otherwise the dc current in the loudspeaker may be excessive.

8.5 LINE-OPERATED 1-W AUDIO AMPLIFIERS (REFS. 2, 5)

Some amplifiers such as those in small radios are operated on the 120-V ac line without an isolating power transformer, and for various reasons an audio-output transformer may be required. An example of a 1-W line-operated amplifier is illustrated by the direct-coupled 2-stage CE-CE amplifier shown in Fig. 8.4. In this amplifier an output transformer is required, not so much to isolate the load as to operate the transistor at a high voltage and low current, with the load voltage and current a factor of 10 lower and 10 higher, respectively. Amplifiers of this type are used in line-operated radios and TVs when 1-W output is acceptable without requiring the low-frequency response of a high-fidelity system.

The dc power for the amplifier is obtained from a half-wave rectifier and capacitor that are selected to make the collector supply voltage 70 V with an acceptable reduction of ac hum. With a capacitor input the dc input voltage is approximately 130 to 150 V, and a relatively large series resistor may be used in reducing the supply voltage to 70 V. A 200-Ω series resistor and a 50-µF capacitor provide adequate hum reduction for the output stage, and the additional 10-kΩ resistor and 10-µF capacitor comprise a low-current filter for the driver stage.

The power amplifier is a CE stage with the emitter resistor bypassed at audio frequencies. Because the second stage emitter supplies bias for the input stage, the driver operates very much like a combined collector and emitter-feedback amplifier. Full output is obtained with a 100-mV rms input signal from a 10-kΩ source. The overall power gain is about 64 dB, and the voltage gain is 40 from the input base to the load. The main limitation of the amplifier is that a relatively large output transformer is required, or a poor low-frequency response must be accepted at full output.

Output transformers for line-operated class-A amplifiers are not readily

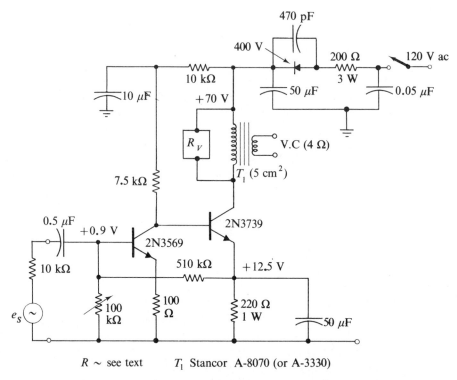

Figure 8.4. Line-operated 1-W amplifier

available. The primary winding should have an input impedance of 1 to 2 kΩ and be rated to carry an unbalanced 60 to 80 mA dc current. For voice-frequency applications a 12-V filament transformer may be used with an 8-Ω load, provided the core area is 5 cm² or larger.

Because the power transistor may be damaged by transient spikes, especially when the amplifier is operated without a resistor load, the transformer must be shunted by a Thyrite varistor, shown as R_V across the primary. Thyrite varistors are rated by a clipping voltage, at which the current increases as the fourth power of the voltage, and by the power that can be dissipated. The varistor for the 1-W amplifier should be rated for 60 to 100 V and 0.75 W. The amplifier may be safely tested without the varistor if the load is connected in a way that prevents accidental removal. The power transistor should be mounted on a chassis or a heat sink with an equivalent heat-radiating area of at least 50 cm².

Overall feedback may be applied when the transformer is correctly phased by connecting the load through a resistor to the input stage emitter. This feedback may be used to improve the frequency response at low signal levels, but, unless a larger output transformer is used, the 1-W output cannot be obtained below approximately 70 Hz. Unfortunately, the power loss in the output transformer cannot be removed by feedback, and the maximum obtainable power output is 1 W, reduced by the transformer loss for frequencies below the low-frequency cutoff.

8.6 TRANSFORMER-COUPLED FOUR-WATT AMPLIFIERS

A transformer-coupled, medium-power audio amplifier for operation on a 12-V to 14-V battery supply is shown in Fig. 8.5. The amplifier uses a germanium transistor, which is more efficient on low-voltage supplies, but a silicon transistor may be substituted by doubling both R_B and R_E and using a 20-Ω to 50-Ω secondary winding on the input transformer. The output transformer should present a 15-Ω to 20-Ω load impedance for the transistor, with the secondary winding impedance equal to the load impedance.

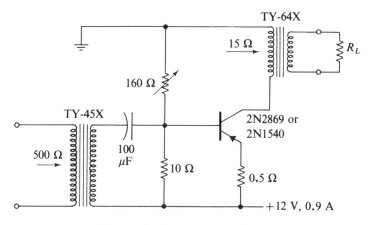

Figure 8.5. Class A 4-W amplifier

Because the collector voltage is fixed, the output power of the amplifier is determined by the bias and the collector current. For class-A operation the bias is adjusted to make the collector current 1 A for 1-W output. However, except for frequencies above 400 Hz, the output transformer requires a surprisingly large core area. For voice-frequency applications the collector current and the output power may be reduced to 0.25 A and 1 W, respectively. For higher frequencies the transformer requires fewer turns and a smaller core, so the amplifier may be entirely practical for a 4-W output. The distortion of a single-sided class-A amplifier may have 5 to 10 per cent even harmonics at full output. A push-pull class-B amplifier is usually much more satisfactory in audio applications.

8.7 POWER PAIRS

The amplifier in Fig. 8.6 is an interesting 5-W class-A power amplifier for audio and servo applications requiring an isolated load and a higher input impedance than the amplifier described in Sec. 8.6. The output stage uses a *pnp* germanium power transistor for a maximum power gain at low cost. The driver stage is a CE-CE pair direct-coupled to the output transistor. The input transistor is a high β silicon transistor used because it offers a very low I_{co} (leakage) current. The second stage is an *npn* 1-W silicon transistor that has a high current gain at 1-A emitter current. Each transistor offers the characteristics required for its place in the amplifier, and all are low-cost devices. The amplifier has 55 db power gain, unusually good waveform, and is thermally stable as long as the output transistor is mounted on a small heat sink. The amplifier operates equally well for low-power applications with TR-2 removed, provided the bias current is reduced and the output transformer is changed to present a higher impedance to the collector.

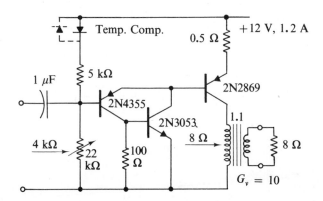

Figure 8.6. Class A 5-W power-pair amplifier

8.8 PUSH-PULL 2-W CLASS-A AMPLIFIERS

The push-pull amplifier shown in Fig. 8.7 is designed for low-power speech or music applications, using a 12-V automobile battery as the power supply. The input transformer is selected to present a 200-Ω source from base-to-base when the primary is driven from a 500-Ω generator. The output transformer provides a voltage step-down of about 3 to 1, and the maximum output is 12 V peak-to-peak across the 8-Ω load, or 2 W. The dc power input is 7 W, making the overall efficiency 28 per cent.

The output transformer is built on a small core, and the loss in the dc resistance of the windings and the emitter resistor is about 30 per cent of the

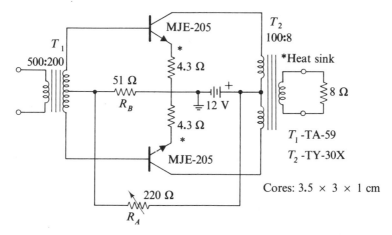

Figure 8.7. Push-pull 2-W class A amplifier

total output power. Thus, the collector circuit is developing about 3 W at over 40 per cent efficiency. The maximum theoretical efficiency is 50 per cent. With the transformers shown in Fig. 8.7, the amplifier has a power gain of 250 times, or 24 dB. Transformers built on a larger core would increase the power gain to about 30 dB. The collector winding of the output transformer should have a collector-to-collector impedance given by Eq. (8.2).

Equation (8.2) indicates correctly that high power output is obtained by reducing the load resistance. However, lowering the load resistance requires that the emitter resistance be lowered to keep the ac power loss in the resistor about 10 to 20 per cent of the total output power. When operated class A, the power transistors require a heat sink that will dissipate nearly 4 times the maximum output power.

8.9 CLASS-B 2-W AMPLIFIER AND DRIVER STAGE (REFS. 4, 7)

The class-B amplifier and driver stage shown in Fig. 8.8 is designed for use in an intercom or in similar voice-frequency applications requiring operation on a 12-V supply with a low standby current. The circuit and components used in the output stage are the same as those used in the class-A amplifier (Fig. 8.7), except that the transistors are a low-power TO-105 plastic type rated to dissipate 300 mW without a heat sink. The amplifier operates class B by increasing the bias resistor until the crossover distortion, Fig. 8.9, just appears and the standby current is approximately 65 mA. With the components shown in Fig. 8.8 the bias resistor is about 1 kΩ.

The maximum output power, 2 W, is obtained with a dc power input of

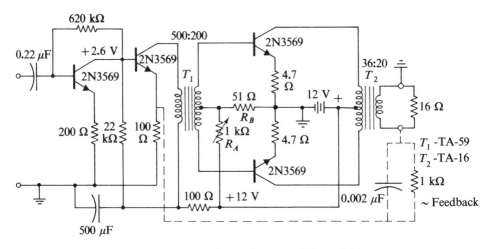

Figure 8.8. Push-pull 2-W class B amplifier and driver

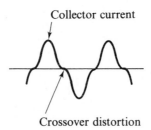

Figure 8.9. Crossover distortion

5 W, and the efficiency is relatively low because of power lost in the output transformer. The power gain of the push-pull stage is 100, 40 dB, and the voltage gain is 6 dB, with a frequency response from 100 Hz to 10 kHz. If the amplifier is to be used with a continuous output signal, the transistors should be changed to a medium power device that can dissipate 3 to 5 W with a small heat sink.

For applications requiring operation over a wide temperature range the resistor R_B is replaced by a low-power silicon diode that has a forward voltage of 1 V with 10 to 30 mA. With the diode the crossover distortion is removed by adjusting the bias resistor R_A and changing the emitter resistors if the adjustment cannot be obtained within the current rating of the diode. If the transistors are mounted on a heat sink, the diode compensates for temperature changes best when mounted in thermal contact with a transistor or a point on the heat sink near a transistor.

The driver stage shown in Fig. 8.8 has a voltage gain of 160, 44 dB, from the input to a 500-Ω load, and an input impedance greater than 15 kΩ. When used with a 10-kΩ source, the gain from the internal source voltage to the load is over 100, 40 dB, and the response between 3 dB points is from 80 Hz to 50 kHz. Thus, full output of the power stage may be obtained with an input signal of 30 mV rms or less. The driver stage develops 5 V rms across a 500-Ω load, or 50 mW, and with a suitable output transformer the driver may be used as a low-power output stage.

Feedback may be applied as indicated in Fig. 8.8, provided the transformers are correctly phased and the shunt capacitors are used in the output stage as indicated. The feedback resistor and the capacitor are adjusted by reducing the resistor until the amplifier just begins to oscillate and by adjusting the capacitor to stop the oscillation. The resistor left in the circuit should be twice the lowest value obtained with a stable amplifier, and the capacitor should be one-half the highest value obtained.

8.10 SINGLE-ENDED, CLASS-B, 10-WATT AMPLIFIER (REF. 4)

The single-ended amplifier shown in Fig. 8.10 has a pair of like power transistors that are driven class B by a transformer-coupled class-A input stage. The interstage transformer windings are polarized so that the series-connected

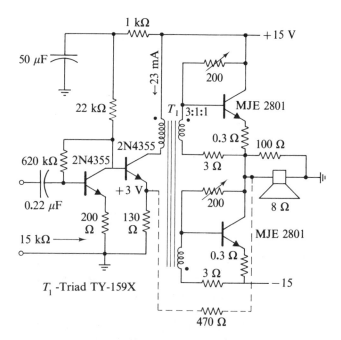

Figure 8.10. Single-ended 10-W class B amplifier

transistors conduct on alternate half cycles and cause the signal current to flow through the load in opposite directions on alternate half cycles. The load is driven as from a push-pull, class-B amplifier. The single-ended amplifier has the advantages that the circuit requires few components and is easily adjusted to operate over a wide temperature range. The transformer coupling gives a high power gain and high efficiency, but there is a limited choice of transformers that have three separate windings.

The power transistors are slightly forward-biased to minimize the crossover distortion, and for high-fidelity applications the residual distortion is eliminated by overall feedback. With the transformer correctly polarized, a feedback resistor may be connected between the load and the second-stage emitter.

Power transistors are sometimes damaged or destroyed by second breakdown caused by an accidental short-circuiting of the output load. Second breakdown is a sudden and destructive channeling of the collector current into a localized area of the transistor. This breakdown occurs when high currents and high voltages exist simultaneously in the collector circuit. The speed and destructiveness of the phenomenon complicate its control. Second breakdown is usually prevented by limiting the collector current-voltage product. In the single-sided amplifier short-circuit damage is prevented by limiting the peak output signal of the driver stage so that the power transistors are operated within the safe operating area shown in the data sheets.

8.11 COMPLEMENTARY-SYMMETRY AMPLIFIERS (REFS. 2, 4)

A complementary-symmetry amplifier uses complementary *pnp* and *npn* transistors that are connected in parallel and made to conduct on alternate half cycles, as in a class-B amplifier. These amplifiers use little standby power, and with speech and music the transistors need only a relatively small heat sink. Power outputs of 10 to 50 W are easily obtained, with harmonic and intermodulation distortion below 0.5 per cent. A quasi-complementary-symmetry amplifier uses series connected like power transistors driven by a complementary-symmetry phase inverter.

A low-distortion (high-fidelity) amplifier is usually some form of a direct-coupled complementary-symmetry power amplifier. The high-fidelity response is usually obtained by the combined use of direct coupling and considerable feedback. An advantage of the complementary-symmetry circuits is that coupling transformers are not required, and with direct-coupling it is relatively easy to use 20 to 40 dB of overall feedback.

A low-gain 3-W complementary-symmetry class-B audio amplifier is illustrated in Fig. 8.11. The power transistors are the series-connected complementary (*pnp* and *npn*) 2N5194 and 2N5191 silicon transistors. Both the

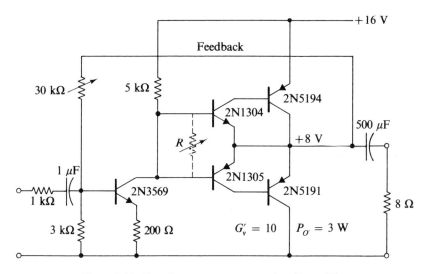

Figure 8.11. Complementary-symmetry class B amplifier

output stage collectors are connected to the load through a large capacitor. The power stage is driven by the parallel-connected 2N1304 and 2N1305 germanium transistors. Because the emitters of the driver stage are coupled to the load, the two-stage output section has a voltage gain of 1 and the load is driven as if by a low-impedance emitter follower.

The 2N3569 CE input stage provides the amplifier voltage gain, which is used partly for overall feedback. The bias resistor couples the output back to the input base and is adjusted to make the dc voltage at the output just half the supply voltage. Because the bias resistor supplies both dc and ac feedback, the Q-point is under feedback control, and additional ac feedback must be obtained by a separate capacitor-coupled resistor.

The amplifier shown in Fig. 8.11 has a power gain of 40 dB, and the input impedance is 1 kΩ. Since the current required of the power supply varies with the signal, the supply should be regulated or have a large filter capacitor.

The germanium transistors in the driver stage have a low turn-ON voltage, and with emitter feedback there is negligible cross-over distortion. If the germanium transistors are replaced by silicon devices, a forward-biasing diode or resistor must be connected between the two bases, and a satisfactory reduction of the cross-over distortion is not usually obtainable.

A quasi-complementary-symmetry amplifier uses alike power transistors (both *npn* or *pnp*). With this change one side of the driver and amplifier connects to the load as a Darlington pair and anti-cross-over bias is provided by a series of two or three diodes connected between the driver stage bases. Because complementary power transistors have become readily available the quasi-complementary-symmetry circuits are rarely used unless the output stage has *pnp* germanium power transistors.

8.12 COMPLEMENTARY SYMMETRY OP-AMP POWER AMPLIFIERS (REF. 5)

Many applications of linear ICs and OP amps require a power stage for low-impedance loads that cannot be directly driven by a low-power IC. For a number of reasons the amplifier shown in Fig. 8.12 offers a simple, trouble-free circuit for driving meters, pen recorders, or relays with impedances of 50 Ω or higher that cannot be driven directly by a low-power amplifier. The complementary-symmetry amplifier has these advantages: it may be used with supply voltages from 8 V to 30 V; there is a negligible voltage difference between the input and output; the input may be direct-coupled to an OP amp; and the amplifier has a high input impedance and a wide-band frequency response.

The 2-stage complementary-symmetry amplifier is a form of Darlington amplifier with a voltage gain of 1 and an input impedance of 100 kΩ up to 100 kHz. When this amplifier is driven by a 10-kΩ source, the frequency response is down 3 dB at 1.5 MHz. Thus, the amplifier may be brought inside the feedback loop of most low-frequency OP amp circuits.

The series-connected diodes in the output stage provide temperature

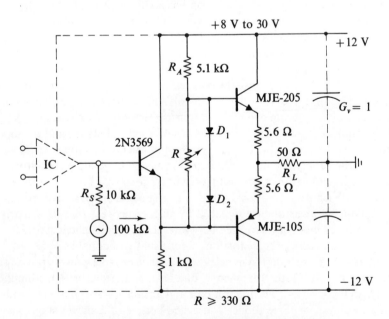

Figure 8.12. Half-watt power amplifier for linear ICs

compensation and make the bias adjustment independent of the supply voltage. The 5-kΩ bias resistor R_A may be used to adjust the offset voltage, as for centering a meter or recording pen, and the value of all resistors in the circuit need not be changed for supply voltages from ± 8 V to ± 30 V. With a ± 12-V supply the amplifier delivers 5 V rms to a 50-Ω load, which is 0.5 W, and with lower supply voltages the power is lower. With the emitter resistors lowered a factor of 5 in both stages and using a ± 30-V supply, the amplifier output is 6 V rms to a 10-Ω load, or nearly 4 W. However, lowering the emitter resistors reduces the local feedback, and a small crossover distortion appears that may not be acceptable unless the output stage is inside the IC feedback loop.

In Fig. 8.12, the resistor R across the diodes is entirely optional when used with a 50-Ω or higher load impedance. When R is 300 Ω, the amplifier operates as a class-B stage with negligible crossover distortion. The collector current is 10 mA with no signal and is 50 mA with 0.5 W output. If the resistor is omitted, the no-signal collector current is 25 mA, and it increases to 50 mA at full output. Thus, the amplifier operates class A at low signals, and the emitter feedback ensures a high degree of linearity up to full output. The fact that the output Q-point is approximately independent of the supply voltage makes the output ripple about 30 dB below the supply ripple, and the amplifier may be used on an unregulated supply.

8.13 LOW-POWER COMPLEMENTARY-SYMMETRY AMPLIFIERS (REFS. 4, 7)

The complementary-symmetry amplifier described in Sec. 8.12 may be simplified when the supply voltage is between 4 V and 12 V provided a reduced output power is adequate for the intended application. For low-power, low-voltage use the output transistors may be the 1-W TO-5 transistors shown in Fig. 8.13, and the driver transistor may be any conveniently available npn transistor. With a 10-Ω load and an 8-V supply, the output voltage is 1 V rms and the output power is 100 mW. With a 12-V supply the output voltage and power are 3 dB higher.

The output transistors do not require heat sinks for speech and music applications as long as the diodes are in thermal contact with the transistor case. With a sustained signal at full output the transistors should be equipped with a clip-on heat sink or be in good thermal contact with a metal chassis. The output stage operates class B, and on an 8-V supply the current increases from 10 mA with no signal to 40 mA at the maximum sine wave signal. On a 12-V supply the dc currents are 20 mA and 70 mA, respectively. If the load can be connected to a center-tap on the battery or to the mid-point of two capacitors across the supply, the input may be direct-coupled to an IC, as shown in Fig. 8.12.

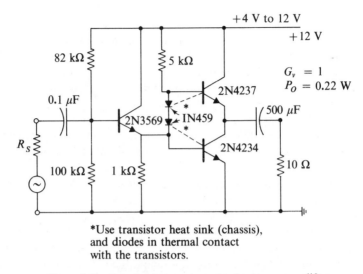

*Use transistor heat sink (chassis), and diodes in thermal contact with the transistors.

Figure 8.13. Complementary-symmetry low-power amplifier

The stability, simplicity, and low-voltage requirements of this circuit make the amplifier attractive for instrument applications where an ac or dc meter indication is required and the input signal is derived from a high-impedance source.

8.14 A 20-W QUASI-COMPLEMENTARY-SYMMETRY AMPLIFIER (REFS. 4, 7)

The quasi-complementary amplifier circuit is generally the designer's choice when germanium *pnp* power transistors are used because complementary *npn* devices are not available. The output transistors are series-connected, as shown in Fig. 8.14, with one driving the load as a CE stage and the other as a CC stage. The output transistors are driven by a complementary-symmetry CE pair, *npn* and *pnp*, and the overall feedback makes unimportant the fact that one side of the amplifier is a CE-CE pair while the other side is a CE-CC pair.

Bias for the output stage is obtained partly via the transistor TR-3 operating as a constant current source. This transistor supplies the current that drives the output transistors to the positive peaks without acting as a low-impedance load on the negative peaks. The constant-current bias has the advantage over the usual boot-strap bias circuit that one capacitor and its frequency effects are eliminated.

Bias to eliminate crossover distortion is determined by the voltage drop across two germanium diodes and a 7.5-Ω resistor connected between the bases of TR-4 and TR-5. The diodes should be mounted in close thermal

Sec. 8.14 A 20-W Quasi-Complementary-Symmetry Amplifier

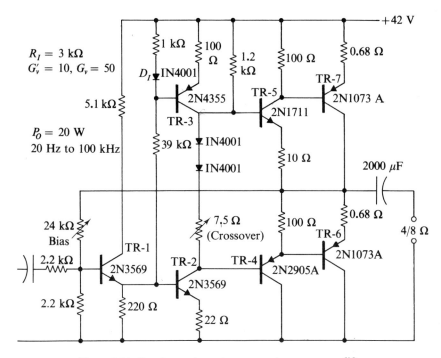

Figure 8.14. Quasi-complementary-symmetry power amplifier

contact with the power transistors to make the diode voltage drops follow the power-transistor voltage change with temperature. The series resistor is used to adjust the quiescent current and crossover distortion. The resistor between the output and the input base provides bias and feedback for the input and is used to adjust the output stage Q-point to one-half the supply voltage.

The power output of the amplifier depends on the supply voltage and the load impedance. With a 4-Ω load the rms power output is 10 W with a 30-V supply and is 20 W with a 42-V supply. With an 8-Ω load 20 W may be obtained with a 60-V supply. For speech and music and 20-W output the transistors should be mounted on a flat heat sink at least 10 cm × 10 cm. For 10-W output the heat sink may be as small as 7 cm × 7 cm.

The power that must be dissipated by the transistor heat sink is determined by the way in which the amplifier is used. The standby power in a class-B amplifier is usually 10 per cent of the rated power output. In speech and music applications the average power dissipated is hardly more than the standby power. Because an amplifier may operate accidentally with a sustained high-frequency signal, better reliability is ensured by providing a heat sink capable of dissipating 50 per cent of the maximum power. Most amplifiers are not damaged by thermal runaway and will operate normally after the transistors cool.

9 Linear Integrated Circuits

Integrated circuits simplify the design and construction of high-gain, low-power circuits and instruments. The linear ICs offer the advantage of high gain and high reliability in a small package, and they are generally designed with frequency characteristics and compensation circuits that simplify the use of feedback. Because the input stages of high-gain ICs are usually differential amplifiers with both positive and negative power supplies, the amplifier has zero dc input and output voltages, so the output is easily direct-coupled to the input or to other circuits. Except for special applications the ICs should be used in circuits that take advantage of the high available gain or input impedance, and discrete devices should be used where they offer lower noise, simpler circuitry, higher power, and greater reliability.

The manufacturers offer a number of different linear ICs with a confusing array of type numbers that may designate ICs that are identical for all practical purposes, except possibly in cost. The ICs used in the following sections are usually shown as the type 741 that is available from nearly all manufacturers. Unlike many of the earlier types the 741 IC has the advantages of high gain, low power requirements, a relative immunity to power supply voltage and ripple, and internal compensation that facilitates use of feedback. A remarkable advantage of the 741, as for many recently designed ICs, is that the amplifier is not usually damaged by improper hookups, short circuits, or by connecting the supply voltage to the wrong terminal. Because of the latter advantages the experimenter may be encouraged to use the 741 in place of earlier types like the 709.

9.1 BASIC IC CIRCUITS (REF. 2)

An IC is intended to be operated with overall feedback and may be operated only rarely as a high-gain amplifier without feedback. In order to make an IC useful with a variety of feedback circuits an operational amplifier usually has an internal frequency-compensating network that makes the frequency response decrease 6 dB per octave increase of frequency, beginning at a frequency between 1 and 10 Hz. Thus, an operational amplifier has a constant frequency response only with a resistance feedback network, and the closed loop bandwidth is the unity gain frequency divided by 1 plus the closed loop gain. The unity gain bandwidth of the 741 IC is 1 MHz, and with a gain of 1000 the frequency response is constant only from 0 to 1 kHz. A broadband IC that is not internally compensated requires an external compensating circuit, and with a gain of 100 it may have a constant frequency response up to 100 kHz or 1 MHz.

A linear IC intended for use as a high-gain amplifier generally presents 2 inputs, designated by $+$ and $-$, as shown in Fig. 9.1. Within the useful frequency range the output is 180° out-of-phase with the $-$ input and is in phase with the $+$ input. Thus, the output may be thought of as $+$, positive. When an IC is used for voltage gain, 3 external resistors are connected, as shown in Fig. 9.1. The input signal may be applied as e_1 on the inverting side or as e_2 on the noninverting side. The resistor R_f provides negative feedback; the signal source e_1 sees the input impedance R_1 as if the $-$ input is grounded; and the voltage gain is determined by the resistors and is R_f/R_1. For a signal applied to the noninverting side the gain is 1 more than R_f/R_1, and the input impedance is high, probably more than 1 MΩ.

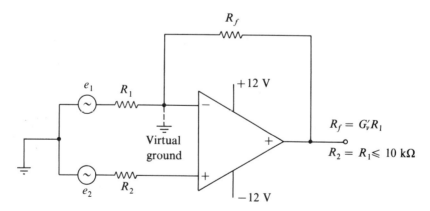

Figure 9.1. IC amplifier

The bias currents of the input transistors flow through R_1 and R_2, and the difference between these input voltages is amplified and offsets the output Q-point. The disturbing effects of the input bias currents are minimized by making the resistors R_1 and R_2 neither too different nor too large. The resistors R_1 and R_2 usually must be less than 10 kΩ for an IC with a transistor input stage, and the dc resistance of the signal source must be included as a part of R_1 or R_2. With high closed-loop gains the dc offset signal caused by bias currents is minimized by making R_2 equal to the parallel equivalent of R_1 and R_f, or the resistor R_2 may be changed to correct for a difference between the input and the offset voltages. Some ICs, including the 741, have terminals for connecting a potentiometer that may be used to compensate for offset errors.

For a high input impedance and a wideband frequency response an IC may be used with the source connected to the $+$ input, as illustrated in Fig. 9.2. The inverting input is connected to the output or through a resistor R_1 equal to the source impedance on the noninverting side to minimize the bias offset. An IC may require a particular compensating circuit for unity gain operation, and the voltage capability may be limited. However, if an IC feedback amplifier tends to be unstable and a wideband frequency response is not required, the gain at high frequencies may be reduced by connecting a small capacitor, 10 pF, across the feedback resistor.

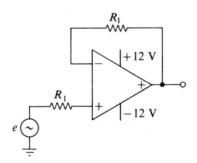

Figure 9.2. Unity-gain wide-band amplifier

Although the input requirements of a linear IC may be a limitation for many applications, the differential input offers important advantages in others. The circuit shown in Fig. 9.3 illustrates the use of an IC for the amplification of a balanced differential signal, $e_1 = -e_2$, while eliminating an even stronger common-mode pick-up signal e_{CM}. The resistor $R_2 = R_f$ makes the voltage gain the same R_f/R_1 for both input signals. The potentiometer R_b may be required to provide a return path for the bias currents or to balance the line when the signal source is grounded at the midpoint. The gain of this amplifier may be adjusted by changing the upper resistor R_f, provided the common-mode signal is not too large when R_f has the maximum value. However, the

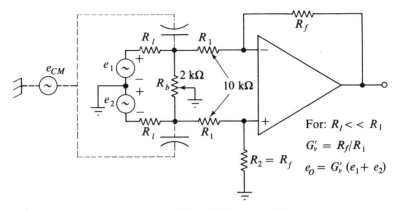

Figure 9.3. Differential-input amplifier

main limitation of the amplifier may be the low input impedance and the pickup signal caused by magnetic induction into the line as a single-turn loop.

9.2 SIGNAL ADDITION AND SUBTRACTION

When signals from two or more sources are to be combined by simple addition, there is often a need to prevent the signals from interacting by transmission from one channel into another. The preferred circuit for addition is usually the form that uses a differential input amplifier, as shown in Fig. 9.4(a). With a high-gain amplifier and with all resistors of equal magnitude, signal sources connected to terminals A, B, and C are combined and appear at the output in the form $-(A + B + C)$, provided the signal sources have low internal impedances compared with the series resistors. Each signal is amplified by the gain that is the ratio of R_4 to the equivalent source resistance R_1, R_2, or R_3 for that channel. When the inputs D, E, and F are not used, offset is minimized by making R_5 equal to the parallel value of all the other resistors, R_1 through R_4. With a high-gain amplifier the signal at the summing point S is 60 dB or more smaller than the input signals, so there is very little feedback from one channel to another. In the event a channel has an internal impedance that is not small compared with the 10-kΩ series resistor, the series resistor may be reduced.

Summing amplifiers with the feedback resistor equal to the input resistors are used in analog computers where there is no need for additional gain. However, when signals are mixed, as in a stereo system, the feedback resistor R_f may be increased to provide 20-to-30-dB gain. If high-impedance sources are to be combined, the series resistors may all be increased, but it may be necessary to keep the 10-kΩ feedback-resistor to prevent the bias currents from producing too large an output offset voltage.

Signals may be subtracted by connecting low-impedance sources to the

148 Linear Integrated Circuits   Ch. 9

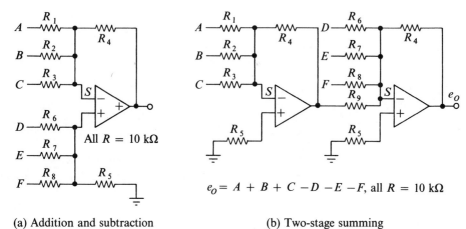

(a) Addition and subtraction   (b) Two-stage summing

Figure 9.4. Summing amplifiers

terminals D, E, and F on the $+$ input side of the differential amplifier. The output of the amplifier is $D + E + F - (A + B + C)$ as long as the number of resistors on the $-$ input equals the number on the $+$ input and $R_5 = R_4$. If there are fewer inputs on one side than on the other, the unused input must be grounded. Because the $+$ input has a high impedance, the signals connected to terminals D, E, and F may feed from one channel to another and cause objectionable crossover between channels. In such an event the plus signals may be added and inverted in the first stage shown in Fig. 9.4(b), and the inverted sum is added with the minus signals in a second stage. The inverted output is the sum $(A + B + C) - (C + D + E)$, and unused inputs are grounded only to reduce offset. The 2-stage summing circuit is usually preferred in analog computers.

9.3 INSTRUMENT AMPLIFIER

An instrument amplifier has a high input impedance and gain for differential input signals, while providing a high rejection of common-mode inputs. The instrument amplifier shown in Fig. 9.5 uses 2 noninverting buffer amplifiers for gain in the input stage, followed by a single IC amplifier that converts the differential input signal to a single-sided output. The gain of the input stage is 1 more than $2R_f/R_1$, except that for a common-mode signal the stage gain is 1, because the midpoint of R_1 is not grounded. Thus, the common-mode rejection of the output stage is improved by the signal gain advantage in the input stage. The output amplifier is usually operated with unit gain, and the feedback resistors are trimmed to ensure common-mode rejection.

Because the gain varies inversely with R_1, a linear potentiometer gives a very nonlinear gain control, but a potentiometer made for a logarithmic (dB) control may be used, except that the gain increases in dB steps when the control is rotated opposite to the direction ordinarily used. An instrument amplifier should have 10-to-20-dB gain in the input stage, and more gain, if needed, may be supplied by an amplifier following the CMR stage. An instrument amplifier with R_1 set for a fixed gain may be given a band-pass frequency response by inserting the capacitors C_1 and C_2. The low-frequency 3-dB cutoff is determined by $R_1 C_1$, and the high-frequency cutoff by $R_3 C_2$.

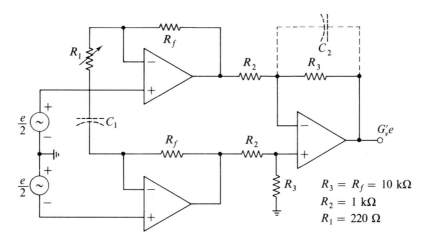

Figure 9.5. IC instrument amplifier

Instrument amplifiers are furnished as a single IC package with closely matched current and voltage offsets that track with temperature changes. The effects of bias currents are reduced by using IC amplifiers with FET or Darlington input stages. Also, since the feedback resistors in the input amplifiers do not affect the input impedance, these resistors may have relatively low values to minimize the effect of bias currents.

9.4 IC INTEGRATORS

An IC integrator is essentially a low-pass filter with a frequency response decreasing 20 dB per decade from the open-loop gain up to a frequency at which the input-to-output loss is at least 20 dB. The low-frequency cutoff of an integrator may begin at a frequency as low as 0.1 Hz or less. Integrators are usually made to perform a mathematical operation, which is the summing of a changing input signal over a specified time interval. An integrator may be used also to change signal waveforms, as in converting a square wave to a triangular wave or in converting a constant input voltage to a linearly increasing ramp.

The circuit of a typical analog integrator is shown in a simplified form in Fig. 9.6(a). The resistor R_o is usually between 1 kΩ and 10 kΩ, and the capacitor C_o is usually 0.1 μF or larger and should have a leakage current less than the amplifier bias current. For long-time integrating accuracy the capacitors are usually polystyrene or teflon insulated, while Mylar and mica capacitors may be used in high speed or less demanding applications. Similarly, chopper-stabilized, FET, or bipolar amplifiers are required respectively for long-time, medium-time, or short-time integration. The bipolar transistor types are usually satisfactory for signal generation and wave-form changing. The input impedance that loads the signal source is R_o.

Switches are required in an analog integrator to control the integrating time and to reset the integrator after use, as shown in Fig. 9.6(b). The output of the amplifier may be connected to a recording device, to other integrators, summing amplifiers, or to a voltmeter. The switch S_1 is first closed to discharge the capacitor when $V_C = 0$, or to give C an initial charge that is minus the voltage V_C. The capacitor charges until the output voltage equals minus the input voltage because $R_1 = R_f$. With S_1 opened, S_2 is closed and, after the desired integration time, is opened.

The output voltage existing with both switches open is minus the integral of the input voltage, multiplied by the scale factor $1/RC$. Thus, with $R = 10$

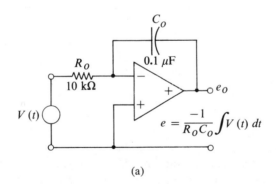

(a)

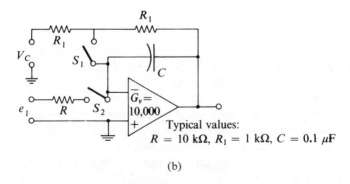

(b)

Figure 9.6. Analog integrator with switches

kΩ and $C = 0.1$ μF, the scale factor is $(1/10,000)(10^{-7})$, or 1000 reciprocal seconds. With 10 mV applied for 0.1 s, the integral is $10(0.1) = 1$ mVs, and the output voltage should rise to 1 mVs$(1/RC) = 0.001(1000) = 1$ V, but with 100 mV input a voltage rise to 10 V is too close to the amplifier overload voltage to give an accurate integration. Also, for accurate integration the integration time should be a factor of 10 smaller than the time constant of the integrator, which is $G_v RC$. For the integrator in Fig. 9.6(b), $G_v RC/10 = 10,000 (0.001)/10 = 1$ s. An equivalent statement is that the output amplitude should be less than $G_v E/10$, when E is a constant input voltage. For our example, $G_v(E/10) = 10^4(0.01/10) = 10$ V.

For experimental studies we may construct an integrator using a 10-kΩ input resistor with a 1.0 μF feedback capacitor. Thus, the integration scale factor is $1/0.01 = 100$, and the integration time constant is $G_v RC = 10,000 (0.01) = 100$ s. Hence, for a 1-V meter reading we need 1/100 volt-seconds input, which means that an input pulse that is on for 10 ms may have a 1-V peak without exceeding the 1-V output. The main error may be drift caused by the input offset that is minimized by an offset balance adjustment or by selecting a low-offset IC. We find that we may obtain accurate integration as long as the integrated input offset is small compared with the integrated input pulse. The meter may be used to study the closure time of manually- or pulse-operated switches and relays. Note that the amplifier gain determines a maximum useful integration time only, and a gain of 1000 gives an integration time that is adequate for most purposes.

9.5 ANALOG COMPUTERS

An analog computer uses operational amplifiers to represent the terms of a differential equation. The solution of the equation is obtained by observing the response of the analog circuit when the computer is driven by a signal that represents the independent variable. To avoid using differentiating amplifiers, the equation under study is successively integrated until the computer can be constructed as a series of integrating amplifiers. An analog computer uses summing amplifiers, inverters, integrators, and amplifiers or loss networks.

The analog computer shown in Fig. 9.7 may be used to study the steady-state and transient characteristics of a series-resonant circuit. The differential equation for the current i in a series resonant L, R, and C circuit is

$$L\frac{di}{dt} + Ri + \frac{1}{C}\int i\, dt = V'(t) \tag{9.1}$$

Integrating both sides of the equation and solving for the current i we obtain

$$i = -\int \left[\frac{R}{L}i + \frac{1}{LC}\int i\, dt - \frac{V'(t)}{L}\right] dt \tag{9.2}$$

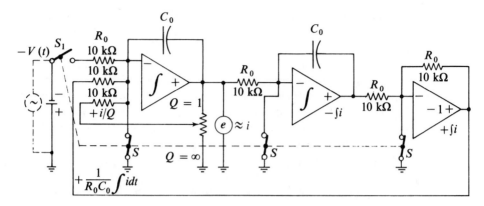

Figure 9.7. Analog computer for a series R, L, and C

Equation (9.2) shows that the current i may be obtained by integrating the sum of 3 terms inside the brackets on the right side of the equation.

In the computer circuit, Fig. 9.7, the integrator at the left has 3 inputs representing the terms in the brackets of Eq. (9.2). The output voltage e of the left integrator is i, the solution of Eq. (9.2). The middle integrating amplifier produces minus the integral of i, and the inverting amplifier at the right gives plus the integral of i. A voltage proportional to i is returned to the input as the first term in the brackets; and a voltage that is the integral of i is returned as the second term. The independent input voltage $V(t)$ is the third term in the brackets of Eq. (9.2). Because an integrator, as shown in Fig. 9.6(a), gives the time integral of the input voltage multiplied by $-1/R_0C_0$, the computer represents the equation

$$i = \frac{-1}{R_0C_0} \int \left[\frac{1}{Q}i + \frac{1}{R_0C_0} \int i\, dt - V(t) \right] dt \tag{9.3}$$

Equating coefficients in Eq. (9.2) with the equivalent coefficients in Eq. (9.3) we have

$$R_0C_0Q = \frac{L}{R}, \quad LC = (R_0C_0)^2 \quad \text{and} \quad V = R_0C_0\frac{V'}{L} \tag{9.4}$$

Substituting $\omega_0 = 1/R_0C_0$, we find that

$$Q = \frac{\omega_0 L}{R}, \quad \text{and} \quad \omega_0^2 = \frac{1}{LC} \tag{9.5}$$

Thus, the computer responds as a series R, L, C circuit resonant at the frequency ω_0. As a practical example, if $C_0 = 100\ \mu\text{F}$, $R_0C_0 = 1$, and $f_0 = 0.16$ Hz, the computer represents any series-tuned circuit resonant at

0.16 Hz and Q is adjustable from $Q = 1$ to ∞. If $C_o = 0.1 \ \mu F$, and $f_o = 160$ Hz, we may use for the series circuit $L = 1$ h, $C = 1 \ \mu F$, and $R = 1,000 \ \Omega$ for $Q = 1$.

The response of the computer is the amplifier output voltage e representing i that may be observed by connecting an oscilloscope or a pen recorder to record e. The switches S are used to ensure that the integrating capacitors are discharged when the voltage V is first connected. The current i may be calibrated by disconnecting the second and third amplifiers so that the computer represents the rise of current in a series R and L. When a step voltage $V' = 1$ V is applied to the series circuit, the current i rises to a maximum $i_{MAX} = V'/R$. In a practical computer setup the Q is more easily adjusted by removing the Q-potentiometer and changing the summing resistor.

9.6 IC AMPLIFIER WITH A SINGLE POWER SUPPLY

An IC amplifier may be operated on a single dc power supply using the circuit shown in Fig. 9.8. The amplifier is made operable with one side of the power supply grounded by returning the $+$ and $-$ inputs to the mid-taps of voltage dividers connected across the power supply. As shown in Fig. 9.8, a low-impedance signal source may be connected in series with one side of a voltage divider, or the source may be capacitor coupled to the center tap of either voltage divider. The resistors in both voltage dividers should be approximately equal with one of them adjusted to minimize the input voltage-offset. The fact that the voltage dividers are connected to the power supply may introduce input noise that may be eliminated by adding a Zener-diode noise filter or an RC filter between the voltage dividers and the power supply.

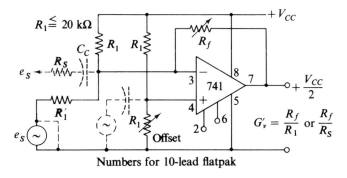

Figure 9.8. IC amplifier with single power supply

9.7 IC AMPLIFIER FOR LOW-IMPEDANCE LOADS

Low-power IC amplifiers are usually rated to operate with a minimum load impedance between 2 and 10 kΩ. The 741 amplifier may be used with a 2-kΩ

load and ±15-V supplies. With a complementary-pair emitter follower a low-power IC may drive load impedances that are a factor of 10 lower or more, depending on the power dissipation rating of the transistors. The circuit in Fig. 9.9 shows how load impedances as low as 240 Ω may be coupled to a 741 IC with the output stage inside the feedback loop. The series-connected diodes are needed to reduce the crossover distortion, and the distortion with feedback is negligible. The amplifier is shown with an input capacitor for ac applications, but it may be used as a dc amplifier by removing the capacitor.

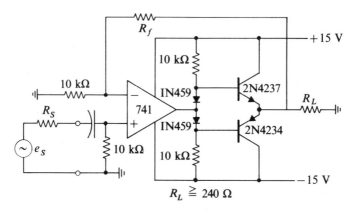

Figure 9.9. IC amplifier for low-impedance loads

9.8 R-C COUPLED IC AMPLIFIER

An IC amplifier may be resistance-capacitor coupled for use in low-frequency ac applications. The amplifier in Fig. 9.10 shows an IC OP amp with the input signal coupled to the noninverting input through an RC filter that removes low frequencies. The feedback capacitor C_f is used to remove high frequencies. As in R-C coupled amplifiers the bandwidth cannot be less than twice the center frequency, so the maximum Q is 0.5 and the cutoff frequencies are at approximately 0.4 and 2.4 times the center frequency. The minimum band is obtained when the input RC product equals the feedback RC product, and the gain at the peak frequency is one-half the ratio R_f/R. A problem with the tuned amplifier is that the high-frequency cutoff cannot be higher than the amplifier gain-bandwidth product divided by the gain R_f/R. With a gain of 1000 the 741 OP amp has a high-frequency cutoff at 10 kHz, which means that the peak frequency of an RC amplifier cannot exceed 4 kHz.

The amplifier shown in Fig. 9.10 may be tuned by inserting a capacitor (0.5 μF, or larger) in series with the resistor R. This capacitor gives the amplifier an additional low-frequency cutoff, but there is the disadvantage that the peak frequency and the cutoff frequencies are not easily estimated. The

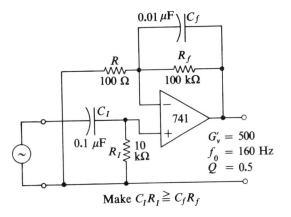

Figure 9.10. RC coupled IC amplifier

capacitor changes both the upper cutoff frequency and the gain at the center frequency and makes difficult the adjustments needed to produce a particular frequency response. As an example, 0.5 μF in series with R halves the gain, doubles the center frequency, and the Q remains 0.5.

The amplifiers shown in Fig. 9.11 are well suited for studies of the effect of changing the Q of a tuned circuit, as in an audio system. These amplifiers have a fixed peak frequency, and a potentiometer permits changing the Q by controlling positive feedback. The amplifier in Fig. 9.11(a) is a form of tuned active filter in which the Q may be varied from 1 to 10. The Q is difficult to adjust when Q exceeds 10, but the amplifier is stable with changes of the supply voltage, and the center frequency is independent of Q. The amplifier shown in Fig. 9.11(b) has similar characteristics, and the Q may be controlled up to

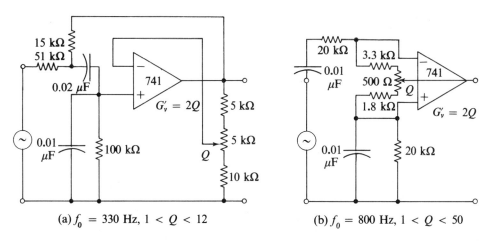

Figure 9.11. Variable Q, tuned IC amplifiers

$Q = 50$. Both amplifiers have a voltage gain that is approximately $2Q$, which makes the gain-bandwidth product constant, and gain measurements may be correlated with Q as an easy way of knowing the Q. The variable Q amplifiers have interesting uses in circuit studies, but high-Q tuned amplifiers without an adjustable Q are more satisfactory when built as narrow-band active filters.

9.9 BRIDGE AMPLIFIERS

An amplifier for detecting balance in a dc bridge may use a single low-impedance IC, as shown in Fig. 9.12. When the bridge is balanced, the voltage from B to D is zero and the amplifier output is zero. Because there is no feedback at balance, the bridge is loaded by R_{f_1} and R_{f_2}, which should be made equal or much larger than the resistors in the bridge. When the bridge is balanced, the voltage from B and D to ground is a common mode input, and the amplifier must have a high CMR rejection, or the balance is not correctly indicated. Because of the problems inherent in using a single amplifier, an instrument amplifier may be required for accurate bridge measurements. An IC amplifier is generally able to operate with a CMR voltage of one-half the supply voltage, so the resistor BC and CD should be smaller than AB and AD if the bridge voltage equals the amplifier supply voltage. Furthermore, series resistors may be needed between the bridge and the amplifier inputs to protect the amplifier when the bridge is off balance if the resistors AB and AD are too low to limit the input currents.

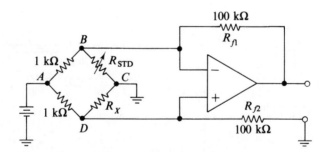

Figure 9.12. IC bridge amplifier

The half-bridge circuit of Fig. 9.13 is sometimes used for applications that require a voltage indication of small resistance changes from a reference value. Accurate bridge measurements require stable and equal magnitudes of the plus and minus voltages, and the noise and ripple of the voltage supply must be relatively low.

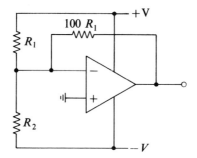

Figure 9.13. Half-bridge amplifier

9.10 HIGH-ACCURACY BRIDGE AMPLIFIER

The bridge amplifier of Fig. 9.14 requires a floating supply voltage for the bridge, but the amplifier indicates the bridge balance without loading any of the bridge arms. The amplifier gain is determined by the resistors R_1 and R_f, but the high input impedance of the inverting input makes the balance sensitivity independent of the bridge resistance, even for high values of the bridge resistors. However, the amplifier must have a low bias-current rating and an offsetting adjustment of the noninverting input voltage may be required. The circuit has the disadvantage of requiring careful shielding to prevent common-mode pickup.

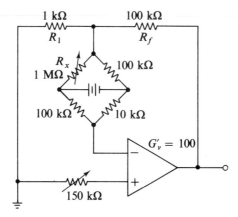

Figure 9.14. High-accuracy bridge amplifier

9.11 RESISTANCE-INDICATING AMPLIFIER

Ordinary bridge circuits produce an output voltage proportional to a resistance change only for small changes of the variable resistor. For measuring or

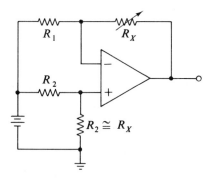

Figure 9.15. Resistance indicating bridge

matching resistors the amplifier circuit of Fig. 9.15 is recommended. With a fixed input voltage the output voltage varies linearly with the resistance of the feedback resistor, the unknown R_x, and the range of the resistors measured is changed by changing both of the input resistors R_1. This circuit is useful for indicating the resistance change of moderately high-resistance transducers, such as thermistors and high-resistance strain gages. The bridge is not satisfactory when the internal resistance of the voltage source is significant compared with R_1, nor is it satisfactory for values of $(R_1 + R_x)$ approaching the recommended minimum external load resistance for the IC.

9.12 IC FEEDBACK SWITCH

Positive feedback is useful in making sensitive switches that are either ON or OFF. The feedback switch shown in Fig. 9.16 uses an IC amplifier with positive feedback connected to the $+$ input. The IC inputs are connected to a balanced dc bridge, and the amplifier may be adjusted to turn full ON or OFF with only a 1 per cent change of a resistor. The output of the IC amplifier is

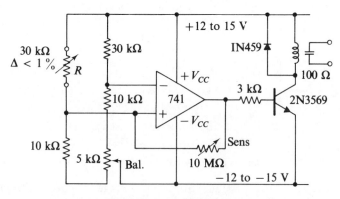

Figure 9.16. Feedback switch

connected to a switching amplifier that drives a 12-V, 100-Ω relay. An important advantage of the positive feedback is that the amplifier is unable to hold the transistor in the active region where the collector dissipation would be too high for a low-current plastic switching transistor.

The sensitivity of the feedback switch is adjusted by the feedback resistor, and the value of resistance at which the circuit switches is adjusted by the balance control. The variable resistor may be a thermistor or a photo resistor that need not have 30 kΩ at the switching point. A lower resistance device may be inserted in series with R, or a high-resistance device may be inserted across R. However, the sensitivity is greatest for a resistance value that can make up the entire resistance of one arm of the bridge. A second thermistor or photo resistor may be used in an adjacent arm of the bridge to offset changes of the ambient temperature or illumination. The diode across the relay is needed to protect the transistor from high-voltage inductive feedback from the relay.

10 Filters

Filters are used to separate desired signals from interference and noise. Filters often use only passive components, capacitors, resistors, and inductors, although inductors are rarely used in low-frequency circuits because iron-cored coils are heavy, expensive, nonlinear, and unreliable. However, the transistor amplifier has brought about improved filter circuits that use only resistors and capacitors—an especially valuable improvement at very low frequencies where inductors are useless. Active filters use relatively simple RC circuits, and nearly ideal filter characteristics are produced by amplifiers with positive and negative feedback.

The coupling circuits in FET and transistor amplifiers use resistors and capacitors to shape the frequency response where only a moderate selectivity is needed. Active filters are used in instruments and communication circuits where a precise frequency and phase characteristic or a high degree of selectivity is required. Active filters are widely advertised by the manufacturers of integrated circuits, but high-quality active filters can be constructed by using an emitter-follower in place of the integrated circuit.

This chapter begins with a description of high-pass and low-pass RC filters and shows how they may be combined to produce a band-pass or tuned characteristic. The design of RC-coupling circuits is described, and the use of capacitance feedback to control the high-frequency cutoff is discussed. Most of the chapter is concerned with active filters, and examples of high-Q tuned amplifiers and band-reject filters are included.

10.1 RESISTANCE-CAPACITANCE FILTERS

The most common way of removing unwanted signal frequencies is to use a simple resistance-capacitance filter. Low frequencies are removed by a series capacitor and a shunt resistor, as shown in Fig. 10.1. High frequencies are removed by interchanging the resistor and capacitor, as shown in Fig. 10.2.

Sec. 10.1 Resistance-Capacitance Filters

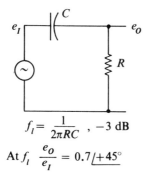
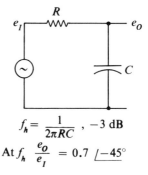

Figure 10.1. Low-cut RC filter

Figure 10.2. High-cut RC filter

The filter loss is 3 dB and the phase shift is 45° when the reactance of the capacitor equals the resistance R. This frequency may be found at the point on a reactance chart (see Appendix) at which the corresponding R and C lines intersect. On the other hand, the 3-dB frequency is doubled by using one-half the R or C values shown in either figure. Similarly, the 3-dB frequency is lowered by a factor of 2 by doubling either R or C in either figure.

In a single-section RC filter the loss is 20 dB and the phase shift is 84° at a frequency a factor of 10 removed from the 3-dB cutoff frequency. When the loss in the single-section filter is high, the slope of the frequency characteristic is 6 dB per octave, or 20 dB per decade of frequency change.

Simple RC filters are inadequate for many purposes because of their very gradual transition from the pass band to the cutoff region. A more rapid transition at cutoff is obtained by using a 2-section RC filter, but the improvement is relatively small except at frequencies a factor of 10 removed from the 3-dB cutoff. A 2-section filter may be constructed, as shown in Fig. 10.3, with the capacitors and resistors alike ($n = 1$) or with the resistance and reactance values increased by a factor of n in the second section. With both sections

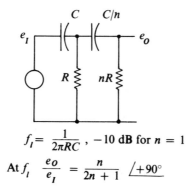

Figure 10.3. Two-section RC filter

alike ($n = 1$) the 3-dB loss falls at a frequency that is 3 times the low-frequency cutoff of a similar 1-section low-cut filter, and, as shown in Fig. 10.4, the cutoff rate does not increase significantly until the loss exceeds 20 dB. When more cutoff is required at relatively low frequencies, a 2-section filter with both capacitors or resistors larger by a factor of 3 places the 3-dB cutoff at the same frequency as for a single section. The phase shift is 90° at one-third the 3-dB cutoff frequency, and the loss is 24 dB at one-tenth the 3-dB cutoff frequency. These values show that a 2-section filter with $n = 1$ is only slightly better in the first frequency decade than a 1-section filter. Constructing the RC filter as two 1-section filters separated by an amplifier (or making $n > 4$) approximately doubles the loss and phase shift that are produced by a 1-section filter.

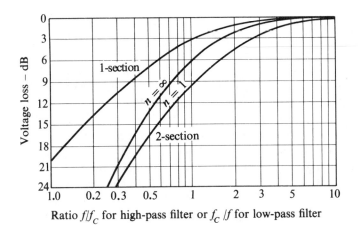

Figure 10.4. RC filter frequency response

The curves shown in Fig. 10.4 illustrate the fact that the 3-dB cutoff frequency increases when an amplifier is constructed as a series of stages with identical interstage coupling components. In order to maintain the same 3-dB cutoff frequency as calculated for 1 stage alone, a 2-stage amplifier must have the capacitors increased by 60 per cent, and a 3-stage amplifier must have the capacitors increased by a factor of 3. This change of the cutoff frequency is easily overlooked in the design of a multistage amplifier until the error is discovered at assembly.

10.2 AMPLIFIER INTERSTAGE COUPLING CAPACITORS

The coupling capacitors used between stages of a resistance-coupled amplifier act as a filter in limiting the frequency response. To obtain a particular frequency response, the interstage capacitors must be carefully selected. A

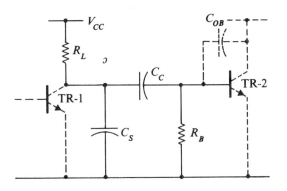

Figure 10.5. Amplifier interstage RC filter

For $f_l \ll f_h$ and neglecting transistor impedances,

$$f_l = \frac{1}{2\pi C_C (R_L + R_B)}$$

$$f_h = \frac{R_L + R_B}{2\pi C_S R_L R_B}$$

typical interstage RC coupling is represented in Fig. 10.5. The resistors and the coupling capacitor C_C determine the low-cutoff frequency f_l, while the resistors and the shunt capacitance C_S determine the high-frequency cutoff f_h. When, as is usually the case, the high-frequency cutoff is at least 10 times the low-frequency cutoff, the filters can be considered independently of each other.

If independent cutoff frequencies are assumed, the 3-dB low-frequency cutoff f_l of the interstage shown in Fig. 10.5 is the frequency at which the reactance of the capacitor C_C equals the series resistance $(R_L + R_B)$. Often the input resistance R_B is much larger than the collector resistance R_L, and the low-frequency cutoff may be obtained by finding the frequency at which the reactance of C_C equals the resistance R_B. In a similar manner the 3-dB high-frequency cutoff f_h occurs when the shunt capacitance C_S equals the resistance of R_L in parallel with R_B. When R_L is small relative to R_B, the high-frequency cutoff is approximately the frequency at which the reactance of C_S equals the resistance R_L. However, in most amplifiers the capacitance that determines the high-frequency cutoff is either the Miller input capacitance of transistor TR-2 when a shunt capacitor is not used, or else the input capacitance may be high enough to make the cutoff frequency much lower than that calculated by use of C_S. In general, the input resistance and capacitance of the second transistor cannot be neglected in calculating the cutoff frequencies, so a calculated cutoff frequency should always be checked by measurements.

When the collector-to-base capacitance of TR-2 is known and the wiring capacitance is included, the effective input capacitance is approximately the total collector-to-base capacitance multiplied by the base-to-collector voltage gain. For example, if the C_{OB} capacitance is 20 pF and the voltage gain of TR-2 is 100, the input capacitance is 2000 pF. With $R_L = 10 \text{ k}\Omega$ and $R_B = 30 \text{ k}\Omega$, the parallel interstage resistance is 7.5 kΩ, and the 3-dB high-frequency cutoff is at 100 kHz. On the other hand, if the interstage resistors are increased by a factor of 10 to 100 kΩ and 300 kΩ, the high-cutoff frequency is low enough to limit the response of a high-fidelity audio amplifier.

164 Filters   Ch. 10

The drain resistor used in a FET amplifier may be as high as 100 kΩ and the voltage gain of the following FET stage may be 50 but is not likely to be 100. Thus, a high-fidelity amplifier using FETs may be expected to have a limited frequency response unless the stage gains are less than 50 or the drain resistors are less than 50 kΩ.

10.3 ACTIVE FILTERS

An active filter uses amplifying elements to produce filter characteristics that are better and more easily built than with passive elements. In general, active filters eliminate the need for high-Q inductors that are bulky, expensive, and variable with temperature and signal. Active filters reduce the problems of insertion loss, critical tuning adjustments, and the effects produced by load changes. In the low-frequency range, 0.01 to 100 Hz, active filters are easily constructed with response characteristics that are practically unobtainable without gain elements. Between 100 Hz and 100 kHz active filters have accurate and stable frequency characteristics, provided there is adequate feedback to ensure that the response characteristics are controlled only by the resistors and capacitors of the feedback network. Tuned circuits and LC filters may excell above 20 kHz because active filters using ICs may not have adequate feedback to perform as well as the designer expects.

10.4 MILLER-FEEDBACK FILTERS

The simplest active filters use the Miller-effect feedback, as illustrated in the low-pass filter in Fig. 10.6(a). The signal source e_s is connected to the transistor base through a 10-kΩ resistor. A 160-pF capacitor is connected from the

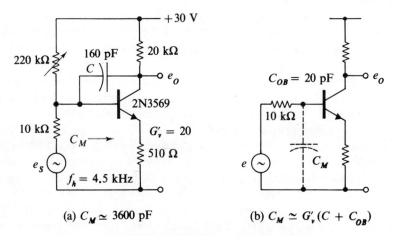

Figure 10.6. Miller-effect low-pass filter

collector to base, and the collector-emitter-resistor ratio indicates a base-to-collector voltage gain of 20. Feedback through the capacitor makes the circuit perform as a low-pass filter, as shown in Fig. 10.6(b). Because the ac collector voltage is 20 times the base voltage, the effective input capacitance C_M is approximately 20 (160 pF), or 3200 pF. The reactance of 3200 pF is 10 kΩ at 5 kHz, so the 3-dB cutoff frequency f_h is estimated to be 5 kHz. Measurements show that the cutoff frequency, 4.5 kHz, is 10 per cent lower because we neglected to include the transistor feedback capacitance C_{OB}, which is 20 pF.

This example of an active filter shows that with voltage gain a relatively small feedback capacitor or the transistor capacitance C_{OB} may produce a high-frequency cutoff at a surprisingly low frequency. As a practical filter this circuit has the disadvantage that f_h depends on the amplifier current and voltage gains, which change with the transistor selected, with temperature, and with the supply voltage. However, in amplifiers where the cutoff frequency is not of critical importance, instead of using the larger capacitor required from base to ground, a small collector-to-base feedback capacitor is often used to limit the upper-frequency response.

In applications using the Miller effect we must expect the cutoff frequency to be relatively variable, and it is always better to find the required feedback capacitor by observing the frequency response at assembly rather than by a calculation. The capacitor needed for a first trial may be calculated by the formula shown in Fig. 10.6(b). The present example emphasizes the fact that the collector capacitance C_{OB} of most audio-type transistors, which is 10 to 20 pF, is enough to limit the upper-frequency response. Stages with a 10-kΩ source or a 100-kΩ load impedance may have the frequency response limited to 50 kHz, or perhaps to a frequency within the audio band. If a stage must have a flat frequency response above the audio band, lower source and load impedances must be used, and the feedback capacitance must be reduced by shielding and using RF-type transistors.

10.5 PRACTICAL LOW-PASS FILTERS

The circuit (Fig. 10.6) illustrating a Miller effect, low-pass filter is not likely to be practically useful because the signal source must carry the transistor bias current. The stage illustrated in Fig. 10.6 is a form of the emitter-feedback stage described in Sec. 2.5, and the stage has adequate Q-point stability for practical applications. A coupling capacitor cannot be used in series with the signal source without eliminating the dc feedback and making the stage impractically variable with temperature and β.

However, a capacitor connected from the collector to the base in any CE stage illustrated in Chap. 2 may be used to produce a low-pass characteristic. As suggested above, the value of the capacitor required is best found by an experimental test. An important limitation, as with all active filters, is that the high-frequency loss cannot decrease indefinitely with frequency. When the reactance of the feedback capacitor is so low that the stage gain decreases with frequency, the amplifier gain falls off until the reactance of C_f approximately equals R_L/G_v. In the example we have used, Fig. 10.6, the high-frequency response decreases 6 dB per octave up to approximately 1.6 MHz, where the voltage gain becomes a constant -26 dB.

10.6 AN IC LOW-PASS FILTER

A simple low-pass filter that uses a high-gain IC amplifier and Miller-effect feedback is shown in Fig. 10.7. The input impedance seen by the source is R_1, which should be 10 kΩ or less, and the gain in the low-frequency pass band is R_2/R_1. The cutoff frequency f_h is the frequency at which the reactance of C_f equals R_2, and the high-frequency response conforms to that of a series resistor and shunt capacitor and falls 6 dB per octave at frequencies more than 10 times f_h.

Errors caused by the bias currents are minimized by making R_3 equal to the parallel value of R_1 and R_2, and the amplifier should be compensated to operate at unity gain. If the amplifier is an internally compensated operational amplifier (OP amp), the gain and the cutoff frequency should be chosen to make the response with external feedback at least 20 dB below the open-loop response of the OP amp. When the gain-frequency curve of the filter is too close to the open-loop gain characteristic, the filter may not have adequate feedback to ensure that the filter response is controlled by the external feedback elements. The presumed advantage that the response characteristic is determined precisely by the external feedback resistors and capacitors is defeated when the high-frequency gain of the IC cannot maintain adequate open-loop gain, high input impedance, and low output impedance. Most discussions of active filters tacitly assume that these conditions exist, whereas

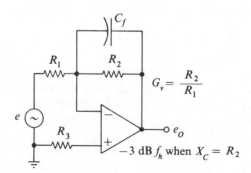

Figure 10.7. IC Miller-effect low-pass filter

an internally compensated OP amp may not have adequate gain above approximately 1 kHz.

The reader may have observed that the cutoff frequency of the IC low-pass filter was not found by using the amplifier gain, although the calculation may be stated in that form. If the feedback resistor R_f is removed, the filter has the form known as an *integrator*. The cutoff frequency is found by equating R_1 with the reactance of the capacitor $G_v C_f$, where G_v is the open-loop gain of the amplifier. Except for the effects of internal compensation, a simple IC low-pass filter has unity gain at the frequency above f_h when the reactance of C_f equals R_1. For applications requiring that the source be loaded by an impedance higher than R_1, the source may be capacitor-coupled to the $+$ side of the IC or may be connected in series with R_2, and the dc resistance, including the source, cannot greatly exceed the maximum value permitted in the IC bias circuit. If R_2 is too high, the dc bias current produces a voltage drop that is amplified and moves the output Q-point until collector saturation makes linear operation impossible.

10.7 ACTIVE FILTERS (HIGHER ORDER)

The active filters described in the preceding sections have used feedback networks with only one capacitor and a resistor. With a feedback network comprised of 4 or more RC components, low-pass and high-pass filters can be constructed that have different or better response characteristics than a simple RC filter. An important advantage of more complicated filters is that the active elements produce filter characteristics that would otherwise require inductances and negative resistances.

Active filters may be constructed that use a single high-gain operational amplifier with a complicated feedback network, or they may be constructed as a cascade of low-gain stages with 4 to 6 feedback elements. The cascaded stages with simple feedback circuits (2- or 3-pole) are stable, easily constructed, and the filter characteristics are independent of the collector supply voltage.

The manufacturers of integrated circuits generally promote active filters that use a cascade of ICs, although in many applications a single emitter-follower or CE stage is capable of providing essentially equivalent and equally useful characteristics. In the following sections we show what may be accomplished using emitter-followers or CE stages, followed by a few examples that require high-performance ICs. In any of the circuits each transistor may be replaced by an IC, provided suitable changes are made in the bias circuits. Similarly, the impedance level of any circuit may be changed, provided the emitter resistor and the bias resistors are scaled with the filter resistors. However, lowering the impedance level increases the power dissipated by the transistor, and increasing the impedance level tends to increase the effects of stray and feedback capacitance, which may limit the upper-frequency response.

10.8 FILTER TYPES AND NOMENCLATURE

The many types of active filters generally fall into one of the following classifications: low-pass, band-pass, high-pass, narrow-band, band-elimination, and phase-shift (all-pass) filters. Because the response characteristics of a low-pass filter determine its effectiveness in separating signals from interference and at the same time change the waveform of transmitted signals, the low-pass filters are further classified according to the way in which the filter may affect the transmitted signal. In a few cases the filter is designated by the name of the person credited with the first description of a particular circuit configuration. A Butterworth filter is designed to have a relatively flat response in the pass band, but the phase characteristics are unsatisfactory in pulse circuits, and the overshoot with a step input is large and increases with the number of filter stages. A Bessel filter has an approximately linear phase-frequency characteristic that is needed when pulses are transmitted and an output pulse is required to be of approximately the same shape as the input pulse. The Bessel filter has short rise and settling times that keep pulses from overlapping. Bessel filters are sometimes used to produce a time delay that depends on the slope of the phase characteristic. A Chebyshev (Tchebyshev) filter has a sharp cutoff characteristic that is used where a high degree of frequency discrimination is needed and a poor phase-frequency response is acceptable. The Chebyshev filter has a ripple-like frequency response in the passband that is denoted by the number of dB of ripple, and the step response approaches the final value erratically.

In each of the filters the frequency response, or the rolloff, at frequencies well above the cutoff f_c is $6n$ dB per octave for a filter with the equivalent of n sections of cutoff. A filter equivalent to n sections is referred to as an n-pole filter, and active filters are made usually with $n = 2$ for each amplifier, or stage of the filter. A 4-pole filter has 2 amplifiers, and to make the characteristics complementary each section is different from the others. As the number of sections is increased, the performance required of the amplifier is increasingly more severe, and a 6-pole filter may not perform as well as expected. Similarly, a 3-pole filter that uses a single amplifier makes a relatively severe demand of the amplifier, so it is usually easier and preferable to make odd-order filters by combining 2-pole stages with a 1-pole stage, especially in the high-frequency applications. Obviously, it is better to make each amplifier operate as a 2-pole stage than as a 1-pole stage, and active filters tend to be even order in preference to odd order.

Before beginning a description of Butterworth, Bessel, and Chebyshev filters, we shall compare the performance of several 2-pole filters to show how the characteristics of low-pass and high-pass sections are changed to shape the cutoff characteristics. As a matter of information we note that high-pass sections may be obtained by transforming a low-pass filter of any type to a high-pass filter. The resulting high-pass filter is given the name of the associated low-pass form. However, the transient response of a high-pass filter is not similar to nor has the desirable response characteristics of the low-pass filter from which the high-pass filter is derived.

The tables in the following descriptions of filters give the capacitor values required for each filter type when 1 through 5 poles are used. The capacitor values are for design cutoff frequencies of 1 kHz used with 10-kΩ resistors in a series of 1- and 2-pole filter stages. However, in some filters an emitter follower is inadequate, so the suggested circuit changes should be observed carefully. The Bessel filter is an exception in that the $f_0 = 1$ kHz used in the tables is a design constant and not a cutoff frequency, as is explained in the section discussing the Bessel circuit.

10.9 ACTIVE HIGH-PASS FILTERS

The active high-pass filter shown in Fig. 10.8 employs a 2-pole RC filter between the input and the base of the emitter follower. When the resistor R_1 of the input section is connected to the emitter, the feedback via R_1 augments the input signal and increases the output signal. As shown by curve A in Fig. 10.9, the filter and the emitter follower together have a voltage gain of 1.3, 2 dB, at approximately 120 Hz. Below 70 Hz the filter has a slope of 12 dB per octave, which shows this is a 2-pole active filter.

A study of the filter circuit shows that emitter feedback has the correct phase relations for an oscillator but lacks the required loop gain. In other words, the filter characteristics are obtained by a limited amount of positive feedback. Because the transistor voltage gain cannot exceed 1, these filters

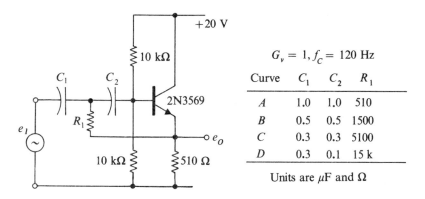

$G_v = 1, f_C = 120$ Hz

Curve	C_1	C_2	R_1
A	1.0	1.0	510
B	0.5	0.5	1500
C	0.3	0.3	5100
D	0.3	0.1	15 k

Units are μF and Ω

Figure 10.8. Active high-pass filter

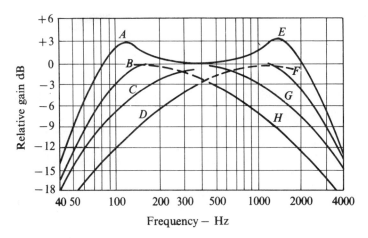

Figure 10.9. Active filter frequency response: high-pass, *A* through *D*; low-pass, *E* through *H*

are insensitive to the collector voltage, even though there is some positive feedback.

The filter represented by curve *B* has a monotonic frequency response with a relatively sharp break into the cutoff region. The curve-*B* circuit is approximately in the form of a 2-pole Chebyshev high-pass filter with an 0.5-dB ripple and a cutoff frequency at 100 Hz. Observe that with curve *A* moved to the left to place the peak response at 100 Hz the rise in curve *A* approximately offsets the loss in curve *B* at 100 Hz, but there is a 2-dB rise, or ripple, at a frequency slightly above 100 Hz. A filter made by connecting filter *B* to *A* has the circuit and response characteristic of a 4-pole Chebyshev 2-dB filter. The response below 100 Hz falls off at a rate of 24 dB per octave, and the transition from the passband to the cutoff is quite sharp with the -3-dB f_c at approximately 80 Hz.

A practical way of constructing the 4-pole Chebyshev filter with $f_c = 100$ is to construct both sections with the component values given in the table. The exact cutoff frequency and the ripple are easily changed by trial adjustments of the components while the effect on the overall frequency response is observed. This is the procedure usually recommended to obtain an optimum response characteristic. The cutoff frequency may be moved to a lower or higher frequency by increasing or decreasing all capacitors in the circuit by the ratio of the f_c in Fig. 10.9 to the cutoff frequency desired.

The curve *C* in Fig. 10.9 is obtained from a filter that has relatively little ac feedback and so is much like a 2-stage RC filter with $n = \infty$ and $f_c = 100$ Hz. The curve *D* is an example of the response characteristic obtained when the two cutoff frequencies are relatively far apart and the feedback is not effective.

10.10 ACTIVE LOW-PASS FILTERS

The low-pass filter shown in Fig. 10.10 has a 2-pole low-pass RC filter in the input, but is otherwise similar to the high-pass filters described in the previous section. The curves of both filters have similar characteristics, and an explanation of a low-pass filter is essentially the same as that of a high-pass filter. The curve F and the network components are approximately correct for an 0.5-dB Chebyshev filter, and, when combined with the network E, the 2-stage filter is a 4-pole Chebyshev filter with a 2-dB ripple and $f_C = 2$ kHz. The curve G in Fig. 10.9 is obtained with relatively little ac feedback and is a close approximation to a 2-stage RC filter with $n = \infty$ and $f_C = 1.6$ kHz.

An important difference between the low-pass and the high-pass filters is that the low-pass filters have a dc path between the input terminal and the transistor base. Because the low-pass filters require a dc bias voltage at the input terminal that is about one-half the collector supply voltage, they may be direct-coupled to the high-pass filters. The low-pass filter can be driven from a grounded source by changing the resistor R_1 to 20 kΩ and adding a 20-kΩ bias resistor, as indicated in Fig. 10.10. The suggested change increases the low-frequency loss from 2 dB to 8, but provides the bias and preserves the 10-kΩ Thevenin equivalent resistance of the input resistor R_1.

A cascaded connection of the filters B and F gives a band-pass response between -3-dB frequencies of 200 and 1800 Hz. A tuned-circuit characteristic may be obtained by adjusting filters A and E to peak at the same frequency, but similar characteristics are obtained more simply by a single-stage, active, tuned circuit. The tuned circuit has the advantage that the circuit requires about half as many components, and the Q of some active tuned circuits is adjustable.

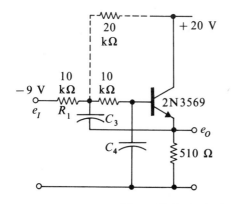

$G_v = 0.85$, $f_c = 1.5$ kHz

Curve	C_3	C_4
E	0.05	0.002
F	0.015	0.005
G	0.007	0.010
H	0.003	0.015

Units are μF and Ω

Figure 10.10. Active low-pass filter

10.11 CE STAGE FILTERS

The single-stage filters shown in Figs. 10.11, 10.12, and 10.13 use CE stages that have a voltage gain, without feedback, of about 20. Depending on the RC-feedback networks, the filters provide a high-pass, low-pass, or tuned-

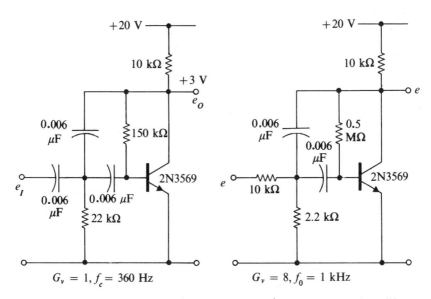

Figure 10.11. Active high-pass filter

Figure 10.12. Active tuned amplifier

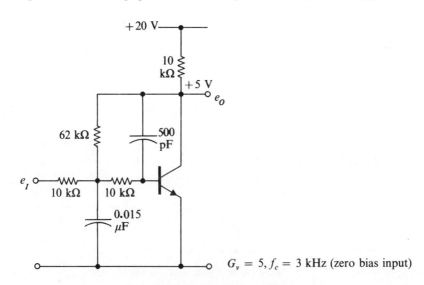

Figure 10.13. Active low-pass filter

amplifier gain characteristic. The curves in Fig. 10.14 show the normalized gain-frequency characteristics of each filter, with 0 dB representing the overall gain in the pass band, as given in the circuit diagrams. A band-pass characteristic is obtained by cascading the low-pass and the high-pass filters.

The cutoff frequency of all the active filters may be changed from 2 to 3 decades by changing the capacitors by a common factor. Changes of the cutoff frequency up to about a 3 : 1 change may be obtained without seriously altering the shape of the curves, by changing the resistors of the input network by a common factor.

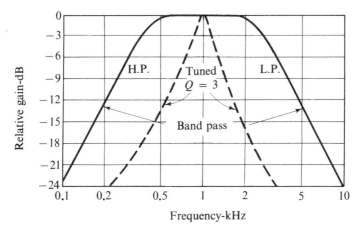

Figure 10.14. Active filter response curves

10.12 BUTTERWORTH FILTERS

The ideal low-pass filter, with unity transmission below and zero transmission above the cutoff frequency and with no phase shift in the pass band, is unattainable in practice. The choice of several approximations to the ideal is usually decided by examining the frequencies that must be separated and by selecting the filter that transmits the signal with an acceptable combination of amplitude-frequency change, waveform distortion, and time delay when required. A Butterworth filter is used when a flat frequency response is the most desired characteristic, although the phase characteristic makes the time delay vary with frequency, and the response to a square wave has an overshoot and some ringing.

The circuit of a 2-pole Butterworth filter is shown in Fig. 10.15. This filter uses an emitter follower, and the bias is obtained by dividing the 10-kΩ series input resistor into two 20-kΩ resistors. This means of supplying the bias introduces a 6-dB loss that is eliminated if the transistor is replaced by an IC. The frequency response is shown in Fig. 10.16(a), and the phase response is shown in Fig. 10.16(b). Observe that all the curves have a 3-dB loss at the

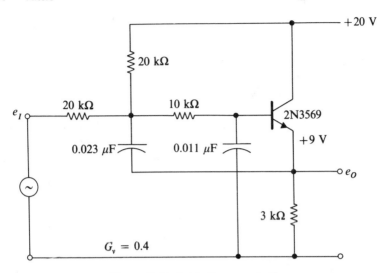

Figure 10.15. 2-pole Butterworth filter

cutoff frequency and are relatively straight lines in the entire cutoff region. By comparison, a series of RC filters with the same cutoff frequency has a loss at the cutoff frequency that is $3n$ dB, where n is the number of independent RC sections (with poles at the same f_C), and the curves above the cutoff frequency change slope relatively slowly in the first frequency decade. On the other hand, the phase curves of the Butterworth filter tend to be linear with frequency below cutoff, while the curves of the RC filters are more linear with frequency in the cutoff region.

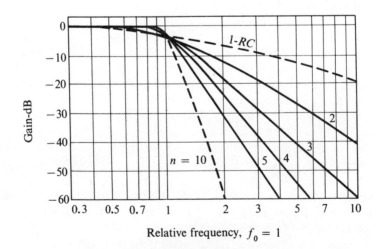

Figure 10.16a. Butterworth-gain vs. frequency

Sec. 10.12   Butterworth Filters   175

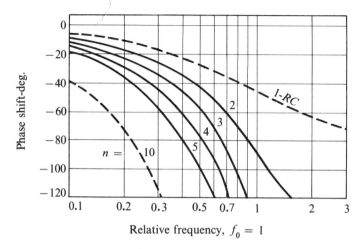

Figure 10.16b. Butterworth-phase vs. frequency

The circuit of a 3-pole Butterworth filter shown in Fig. 10.17 uses a single emitter follower in each stage, and the input stage is designed to use a low emitter current. A smoothing filter may be needed in the collector supply when more than 40 dB attenuation is required at high frequencies.

A circuit suitable for an n-pole low-pass filter is shown in Fig. 10.18. The circuit may be used for constructing filters with 2, 3, 4, or 5 poles by using only the first 2 stages for a 2- or 3-pole filter and using the entire filter with $C_0 = 0$ for a 4-pole filter. Table 10-1 gives the capacitor values required for a Butterworth filter with $f_c = 1$ kHz, and the cutoff frequency is changed by

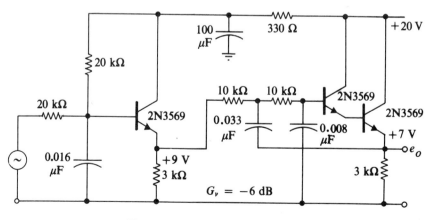

Figure 10.17. 3-pole Butterworth filter

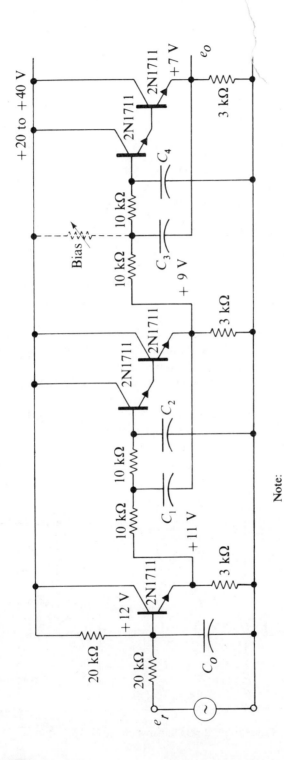

Note: **Butterworth filters use Darlington in last stage only. Bessel filters use emitter followers in all stages.**

Figure 10.18. Low-pass n-stage active filter

Sec. 10.13   Butterworth 3-Pole Filters with Equal Capacitors   177

Table 10-1   Capacitors for Butterworth 1-kHz Filters (μF)

n	C_0	C_1	C_2	C_3	C_4
2	0.000	0.023	0.011		
3	0.016	0.032	0.008		
4	0.000	0.017	0.015	0.041	0.0062
5	0.016	0.020	0.013	0.052	0.0049

dividing all the capacitors by a common factor that is proportional to the desired frequency change. Smoothing filters or Zener regulators that may be needed between filter stages are not shown, but a smoothing filter used with a 3-pole filter is illustrated in Fig. 10.17.

The reason the output stage of a 4- or 5-pole Butterworth filter needs a Darlington CC-CC amplifier is that the output section has considerable positive feedback to offset a loss at f_c that is produced in a preceding section. Positive feedback is used when the capacitor ratio exceeeds 2, and a Darlington CC-CC amplifier is needed whenever the capacitor ratio exceeds 4. A simple CC stage lacks the current gain needed to maintain both the input impedance and the low output impedance required by 2-pole sections with a high capacitor ratio. Insufficient current gain is indicated when a filter seems to have too much loss at the cutoff frequency and insufficient loss at 5 to 10 times the cutoff frequency. Darlington amplifiers are needed in all 2-pole sections of Chebyshev filters, in the output sections of Butterworth filters when n exceeds 3, and in Bessel filters when n exceeds 6.

10.13 BUTTERWORTH 3-POLE FILTERS WITH EQUAL CAPACITORS

The 3-pole Butterworth filter shown in Fig. 10.19 is interesting because the capacitors all have the same value and may be selected to provide a desired cutoff frequency by referring to a reactance chart. Both 3-pole filters (Figs. 10.17 and 10.19) have the same number of components when used with similar signal sources. To illustrate a simple means of providing the required input-stage bias, the filter with identical capacitors is shown connected to the collector of a preceding amplifier. Any of the active filters constructed entirely with CC stages may be direct-coupled throughout. If an emitter Q-point voltage is too low for the required signal amplitude the voltage may be increased by adding a bias resistor as illustrated in Fig. 10.10.

By removing the output capacitor and resistor, we may use the 2-pole section of the filter as a 2-stage low-pass filter with unit gain at 1 kHz and a 2-dB rise at 600 to 700 Hz. The 3-dB cutoff is at approximately 1.3 kHz, and the loss is 20 dB at 3 kHz and increases 40 dB/decade. With 0.02 μF capaci-

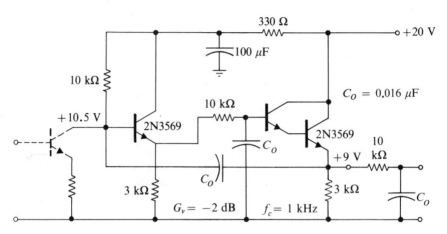

Figure 10.19. Butterworth 3-pole low-pass filter

tors the 3-dB cutoff is at 1 kHz. There is considerable advantage in constructing filters with identical capacitors, and in some cases the resistors may be changed to adjust the cutoff frequency when a particular capacitor value is not conveniently available.

10.14 CHEBYSHEV FILTERS

A Chebyshev filter offers a considerable sharper cutoff above the cutoff frequency than does a Butterworth filter. In exchange for the sharp cutoff the filter has an uneven ripple-like response in the pass band and a more non-linear and somewhat erratic phase response. The Chebyshev filter is useful for approximately sinusoidal signals, like speech signals, which tolerate the wave form distortion caused by a poor phase-frequency characteristic. Chebyshev filters are designed with a 3-dB, maximum-to-minimum amplitude change in the pass band, 2 dB, 1 dB, or less. With a low ripple the phase and amplitude characteristics are not very different from those of the Butterworth filter. In general, the Chebyshev filters have an amplitude response with a slope exceeding $6n$ dB per octave just above cutoff, where high selectivity is usually most desired. Greatest initial slope is obtained for a large allowed ripple and a large number of poles. In the filter circuit the sharp cutoff requires a large spread of capacitor values and imposes stringent amplifier characteristics. Hence, the Chebyshev 2-pole filters require a Darlington amplifier in each stage and a 3-pole Chebyshev filter using a single transistor amplifier is not recommended.

The amplitude and phase curves for a 3-dB Chebyshev filter are shown in Fig. 10.20. Capacitor values for 1-dB and 3-dB filters using the circuit shown in Fig. 10.18 are given in Tables 10-2 and 10-3. Values for filters with inter-

Sec. 10.14	Chebyshev Filters	179

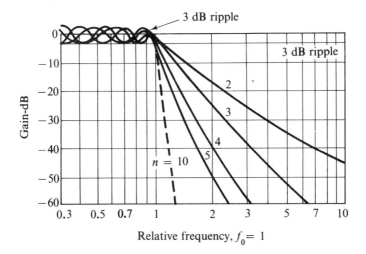

Figure 10.20a. Chebyshev-gain vs. frequency

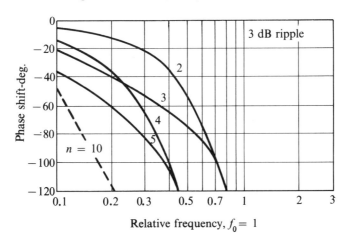

Figure 10.20b. Chebyshev-phase vs. frequency

mediate and less ripple may be obtained by interpolation, and Figs. 10.21 and 10.22 illustrate 2- and 3-pole, 1-dB and 3-dB ripple filters. The response of a filter at the cutoff frequency depends on the number of poles, and with a sharp cutoff the 3-dB cutoff frequency is not particularly meaningful. In a practical situation filter components with even a 5 per cent tolerance usually require adjustments at assembly if a precise cutoff frequency is desired.

Active filters that use the circuit shown in Fig. 10.18 produce a sharp cutoff by providing a peaked voltage rise in the output stage to offset the loss in an earlier stage. The expected voltage rise of a stage tested by

Table 10-2 Capacitors for Chebyshev 1 dB, 1 kHz Filters (μF)

n	C_0	C_1	C_2	C_3	C_4
2	0.000	0.029	0.0080		
3	0.032	0.064	0.0040		
4	0.000	0.047	0.0190	0.11	0.0023
5	0.055	0.068	0.0088	0.18	0.0014

Table 10-3 Capacitors for Chebyshev 3 dB, 1 kHz Filters (μF)

n	C_0	C_1	C_2	C_3	C_4
2	0.000	0.050	0.0072		
3	0.054	0.106	0.0028		
4	0.000	0.078	0.0168	0.19	0.0015
5	0.090	0.112	0.0061	0.29	0.0009

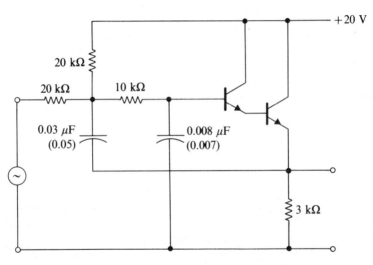

Figure 10.21. 2-pole Chebyshev filter, 1 dB (3 dB) ripple

itself is approximately one-half the square root of the capacitor ratio, C-odd/C-even. Thus, in the 5-pole, 1-dB Chebyshev filter the second stage capacitor ratio is 128, and the expected voltage rise is 5.7. To obtain the necessary voltage rise the 4- and 5-pole 1-dB and the 3- and 4-pole 3-dB Chebyshev filters must be operated with a 40-V supply. The 5-pole 3-dB filter requires a 10-to-1 voltage rise that is usually unattainable with a Darlington stage. The sharp cutoffs produced by a Chebyshev filter with more than four poles may be closely approximated by cascading 2 or more 3-pole filters that are more easily adjusted than is a single higher order filter. On the other hand, when

Sec. 10.15   Bessel Filters   **181**

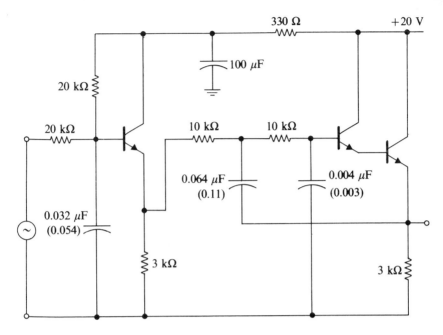

Figure 10.22. 3-pole Chebyshev filter, 1 dB (3 dB) ripple

the capacitor ratios are low, as in a 5-pole Bessel filter, simple emitter followers are adequate in all stages.

10.15 BESSEL FILTERS

A Bessel filter has a linear phase-frequency characteristic and is used as a low-pass time-delay network. The cutoff frequency f_0 is used only to design a network that gives a time delay t_0. The f_0 characterizing the design is given by $2\pi f_0 = 1/t_0$. Thus, when $f_0 = 1$ kHz, the time delay is 160 μs, and a time delay of 1.6 s requires a filter with $f_0 = 0.1$ Hz. The phase shift of a Bessel filter is approximately linear up to $n(45°)$, or $45n°$, where n is the number of poles in the filter, and the time delay is approximately constant for frequencies up to $\frac{1}{2}nf_0$. For example, with a 1.6-s time delay that uses a 4-pole Bessel network the phase shift is about 180° at a frequency of 0.2 Hz, or twice the design frequency f_0. With $f_0 = 1$ kHz the phase shift is linear with frequency and is 180° at about 2 kHz. If a square wave is to be delayed and the first 10 harmonic components of the filter are needed to preserve the waveform, the fundamental frequency of the square wave cannot exceed approximately $f_0/5$ when a 4-pole filter is used.

Because the frequency response of a Bessel filter is determined by the phase characteristic needed to produce a time delay without changing the signal waveform, a Bessel filter does not have a flat frequency response or a particular 3-dB cutoff frequency. However, it is useful to know that the response decreases with frequency less rapidly than with a Butterworth filter, and the 3-dB cutoff frequency is approximately at $1.2\sqrt{n}f_0$. At the frequency $10 f_0$ the signal is attenuated 30 to 40 dB for any value of n, and the slope is $6n$ dB per octave for frequencies above $10 f_0$.

Bessel gain and phase curves are shown in Fig. 10.23, and circuits for 2 and 3-pole filters are illustrated in Figs. 10.24 and 10.25. Because the capacitor ratios are relatively low and because a filter rarely needs more than 40-dB attenuation, Bessel filters may use emitter followers throughout. A Darlington stage is not required unless a Bessel filter has 6 or more poles. Table 10-4 gives

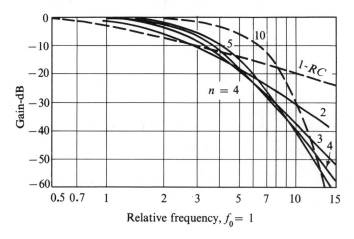

Figure 10.23a. Bessel-gain vs. frequency

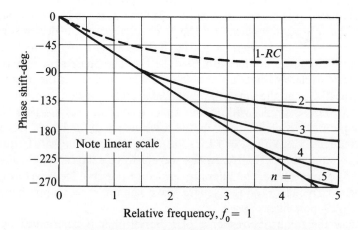

Figure 10.23b. Bessel-phase vs. linear frequency

Sec. 10.15 Bessel Filters 183

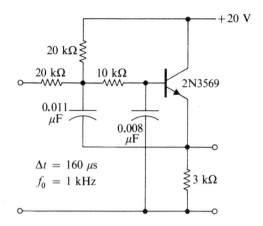

Figure 10.24. 2-pole Bessel filter

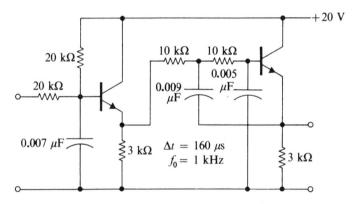

Figure 10.25. 3-pole Bessel filter

Table 10-4 Capacitors for Bessel 160 μs Filters (1 kHz), (μF)

n	C_0	C_1	C_2	C_3	C_4
2	0.000	0.0107	0.0080		
3	0.0069	0.0087	0.0045		
4	0.000	0.0055	0.0050	0.0076	0.0029
5	0.0044	0.0048	0.0037	0.0068	0.0020

the capacitors used in Bessel filters (Fig. 10.18) when the resistors are 10 kΩ and $f_0 = 1$ kHz.

The transient responses of 4-pole Butterworth and Bessel filters are shown for comparison in Fig. 10.26. The input to the filters is assumed to be a square wave with a frequency one-fifth the filter cutoff frequency. The transient response of a Chebyshev filter is not shown because a filter with more than

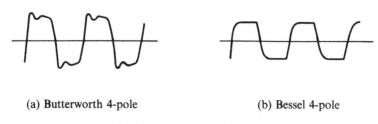

(a) Butterworth 4-pole   (b) Bessel 4-pole

Figure 10.26. Filter response with square wave at $\frac{1}{5} f_c$

1-dB ripple has a large overshoot, and a 4-pole filter with only 0.5-dB ripple has approximately twice the overshoot of a 4-pole Butterworth filter.

10.16 OP AMPS IN ACTIVE FILTERS

Operational amplifiers are often used in active filters because they offer a low output impedance with 60 to 80 dB voltage gain at low frequencies. With these advantages reasonable values of resistors and capacitors may be used in filters operating at frequencies as low as 10^{-3} Hz. With high gain in a feedback loop an amplifier has the low output impedance required to drive the relatively low-impedance elements of a filter and may still present a low internal output impedance that minimizes tuning interaction between stages. Narrowband and tuned circuits with a Q of 10 are generally practical in active filters, and higher Qs require a high stability of the circuit elements and of the amplifier. For high-Q applications an OP amp may have to be selected that has exceptional gain stability, and an uncompensated amplifier may be needed to obtain adequate gain for high-Q circuits at high frequencies.

Active filters are sometimes designed with an input-to-output voltage gain, but there are disadvantages in this practice, especially when the filter is required to have high-performance characteristics at 1 kHz and above. The difficulty is that an OP amp is stable with the feedback circuits used in active filters because the amplifier is compensated to reduce the gain linearly with increasing frequency above 10 Hz. At 1 kHz and above the OP amp may not have adequate high-frequency gain for a filter with overall gain without degrading the response characteristics. For the purposes of this book it seems better to exhibit active filters that use unit gain OP amps or an emitter follower and to assume a separate amplifier may be used for gain. On the other hand, when the OP amp has a plus input terminal, any filter designed for unit gain may be easily changed to have an overall gain by a circuit described in Sec. 10.17, provided there is more gain than is needed for the filter alone.

Active filters may be complicated by the bias network and bias currents needed by the amplifier, and an OP amp that requires low bias currents may be needed for an active filter with high resistances. However, as long as the amplifier bias requirements are satisfied, the input impedance of an OP amp and its effect on the filter need not be considered when the amplifier is used as a voltage follower.

10.17 ACTIVE FILTERS USING INTEGRATED CIRCUITS

Active filters may use a variety of feedback circuits. For example, a 3-pole Butterworth filter may be made with a single IC amplifier, as illustrated in Fig. 10.27. An IC has the high gain needed to drive feedback networks with a high capacitor ratio, and the filter in Fig. 10.27(a), using the IC in the positive feedback mode, has a voltage gain of 2 at low frequencies. The Rauch filter [Fig. 10.27(b)] uses an extra feedback component that lowers the capacitor ratios. In this example the feedback resistors show that the low-frequency response has a 6-dB loss, but with different component values the filter may be designed with voltage gain.

Voltage gain may be obtained easily in an active filter that uses an IC in the positive feedback mode. As illustrated for a 2-pole filter in Fig. 10.28(a), the voltage gain of the IC amplifier is adjusted by selecting the resistors R_f and R_A. For a voltage gain G_v we make $R_f/R_A = G_v - 1$, and 1000 Ω is a good choice for R_A. The feedback resistors R_3 and R_4 take the place of the filter resistor R_1 and make the feedback gain $= 1$. The ratio R_4/R_3 equals G_v, and the parallel value of R_4 and R_3 must equal the resistance R_1 of the filter designed for the unit-gain amplifier. The reader may easily confirm that the parallel value of R_3 and R_4 equals R_1 when we make

$$R_3 = \frac{G_v + 1}{G_v} R_1 \quad \text{and} \quad R_4 = (G_v + 1) R_1 \tag{10.1}$$

This method of providing voltage gain may be used when the amplifier has a low output impedance, and the feedback elements can be replaced by a

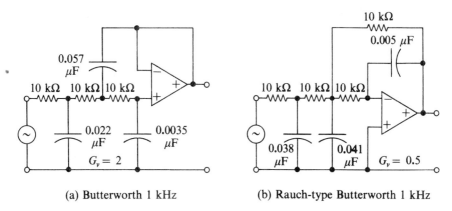

(a) Butterworth 1 kHz   (b) Rauch-type Butterworth 1 kHz

Figure 10.27. Butterworth 3-pole L.P. filters

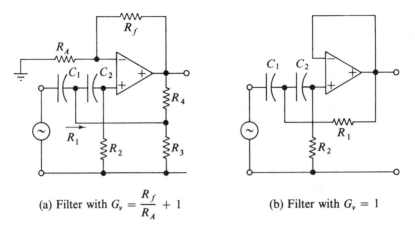

(a) Filter with $G_v = \dfrac{R_f}{R_A} + 1$ (b) Filter with $G_v = 1$

Figure 10.28. Active high-pass filter with gain

Thevenin equivalent to reduce the loop gain. If the feedback element is a capacitor, as shown in Fig. 10.17, the voltage divider uses the capacitors C_3 and C_4, as shown in Fig. 10.29. Because the reactance of a capacitor varies inversely with C, we make $C_3/C_4 = G_v$, and the parallel equivalent of C_1 is $C_3 + C_4$. From these relations we make

$$C_3 = \frac{C_1 G_v}{G_v + 1} \quad \text{and} \quad C_4 = \frac{C_1}{G_v + 1} \tag{10.2}$$

The addition of voltage gain by this method has the advantage of considerable simplicity, and a small adjustment of the feedback components permits an easy adjustment of the gain and the Q of the filter. However, at high frequencies the compensated gain of an IC may well be too low to provide both the overall gain and the feedback gain needed to produce the desired filter response. As a general rule it is probably better to use IC or discrete component amplifiers without gain in the filter and to place the gain where

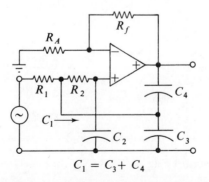

$C_1 = C_3 + C_4$

Figure 10.29. Active low-pass filter with gain

needed and in an amplifier designed for gain only. Separation of the filtering and gain sections gives greater flexibility in making changes and usually gives better performance. In a difficult filter application designers may find that one or another of the unit gain circuits has advantages. An experimental study of a particular objective is the best way to make a choice.

As a final reminder, the cutoff frequency of any of the filter circuits may be changed up or down by changing all the capacitors by the same factor, except that the high-frequency response of transistors and ICs generally limits the cutoff frequency of low-pass filters to below 20 kHz, while the Q and the practical size of capacitors generally limit the cutoff frequency to about 0.01 Hz on the low side. In high-pass filters the cutoff frequency limits are approximately 1 or 2 decades higher.

10.18 HIGH-PASS FILTERS

Any of the Butterworth filters may be converted to high-pass filters with the same cutoff frequency by a simple transformation. The high-pass circuit and the conversion relations are given in Fig. 10.30. Observe that the shunt resistor in the high-pass filter is the same as the series resistor of the low-pass filter. Both series capacitors in the high-pass filter are the same as the first capacitor of the low-pass filter. The only calculation required in the conversion is making the feedback resistor smaller than the shunt resistor by the capacitor ratio used in the low-pass filter. A 2-pole high-pass Butterworth filter with a 110-Hz cutoff frequency is shown in Fig. 10.30(b). The high-pass filter was obtained by converting the low-pass filter illustrated in Fig. 10.30(a). The phase and amplitude characteristics of either filter are images of the other reflected in the vertical line at the cutoff frequency, 110 Hz.

The cutoff frequency is the same in all stages of a Butterworth filter and may be obtained from

$$f_0^2 = \frac{25{,}300}{R_1 R_2 C_1 C_2} \qquad (10.3)$$

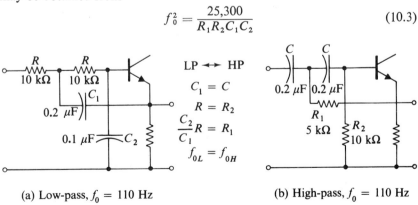

(a) Low-pass, $f_0 = 110$ Hz (b) High-pass, $f_0 = 110$ Hz

Figure 10.30. LP-to-HP filter change with constant f_0 (Butterworth)

where C_1 and C_2 are equal capacitors in the high-pass filters and R_1 and R_2 are equal resistors in the low-pass filters. The units used in Eq. (10.3) are μF, $k\Omega$, and Hz. For the high-pass filter shown in Fig. 10.30(b) we obtain by the conversion $C = 0.2\ \mu F$, $R_2 = 10\ k\Omega$, $R_1 = 0.1/0.2(10,000) = 5000\ \Omega$, and $f_0 = 110$ Hz.

The component values for a high-pass Chebyshev filter may be obtained by using the conversion relations given for a 2-pole Butterworth filter. However, the cutoff frequency of each 1- or 2-pole stage may differ from the cutoff frequency f_0. If the cutoff frequency of a low-pass filter is a factor of m below f_0, the cutoff frequency of the corresponding high-pass stage should be adjusted to a frequency that is f_0/m, above f_0. The adjustment is made easily by multiplying both capacitors in a stage by m^2 during the conversion. The squared frequency factors for 1-dB and 3-dB Chebyshev filters are given in Table 10-5 where N is the order of the filter and the subscript of m is the stage number.

Table 10-5 Frequency (Square) Factors for Chebyshev Filters

	1-dB LP Filter				3-dB LP Filter		
N	m_0^2	m_1^2	m_2^2	N	m_0^2	m_1^2	m_2^2
2		1.10		2		0.71	
3	0.24	0.99		3	0.089	0.84	
4		0.28	0.99	4		0.20	0.90
5	0.084	0.43	0.99	5	0.031	0.38	0.94

The 4-pole Chebyshev filter shown in Fig. 10.31 has a 1-kHz f_0, but the cutoff frequency for the first stage is 470 Hz and for the second stage is 970 Hz. Thus, $m_1^2 = (0.47)^2 = 0.22$ and $m_2^2 = (0.97)^2 = 0.94$, and, after multiplying C_1 and C_3 by m^2, the capacitors in the high-pass filter (Fig. 10.32) have the values 0.014 μF and 0.14 μF, as shown. It is interesting to see that the 2-dB Chebyshev filter may be closely approximated by using capacitor

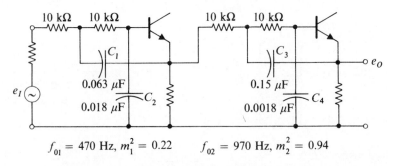

Figure 10.31. LP Chebyshev 2 dB, 1 kHz filter (to Fig. 10.32)

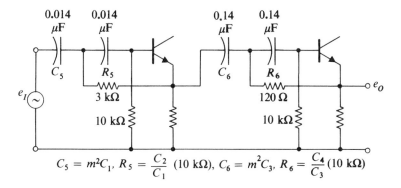

Figure 10.32. HP Chebyshev 2 dB, 1 kHz filter (from Fig. 10.31)

values obtained by averaging the capacitor values given in the 1-dB and 3-dB tables.

10.19 BAND-PASS FILTERS

A true band-pass characteristic has a flat response in the band and a sharp cutoff at both sides of the pass band. Unfortunately, a true band-pass characteristic cannot be obtained with a simple circuit. Band-pass circuits may be constructed that use relatively complicated feedback circuits described in the literature, or they may be purchased already adjusted for the most common applications, as for an IF amplifier. When a simple tuned circuit cannot be used in the low audio-frequency range, two or three tuned circuits may be constructed with the peak frequencies staggered to obtain a flat-band characteristic in the pass band. At frequencies above the audio band, tuned LC circuits are often over-coupled and staggered to produce a band-pass characteristic. The staggered amplifier network has the advantage of providing gain. If gain is not required and relatively high selectivity is desired with a flat response in the band, low-pass and high-pass filters may be series-connected to produce a band-pass characteristic, as illustrated in Fig. 10.14. As an example of a narrow-band band-pass amplifier the low-pass Chebyshev filter of Fig. 10.31 may be series-connected with the high-pass Chebyshev of Fig. 10.32, and the pass band may be adjusted by changing the resistors in either filter. A 10 per cent reduction of the resistors in the low-pass filter produces a nearly flat response with 100 Hz between the 3-dB cutoffs.

10.20 BAND-PASS FILTERS USING OP AMPS

A narrow-band filter may be constructed as a series of high-Q tuned circuits with the peak frequencies staggered to give a flat response near the center frequency. The OP amp circuit shown in Fig. 10.33 is easily adjusted to have a

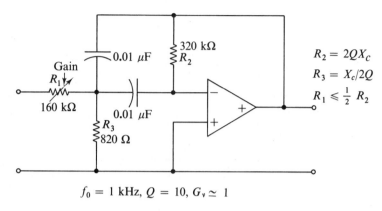

$f_0 = 1 \text{ kHz}, Q = 10, G_v \simeq 1$

Figure 10.33. Bandpass OP-amp filter ($BW = f_0/Q$)

desired peak frequency and bandwidth. The ratio of the peak frequency f_0 to the bandwidth, f_w, f_0/f_w, is the Q of the tuned circuit. Usually, the center frequency f_0 and the Q are given. The circuit shown in Fig. 10.33 has an overall peak-frequency gain of 1, 0 dB. The values shown in Fig. 10.33 are for a filter with $f_0 = 1$ kHz and $Q = 10$. The capacitors are always of equal value and may be changed to move the peak frequency without affecting the Q.

The filter is designed by selecting a convenient size for the capacitors, which should have a reactance of approximately 10 kΩ at the desired peak frequency. The 0.01-μF capacitors have a reactance of 16 kΩ at 1 kHz. The resistor R_1 is Q times the reactance, or 160 kΩ. The resistor R_2 is $2Q$ times the reactance, or 320 kΩ, and the resistor R_3 is the reactance divided by $2Q$, or $16{,}000/20 = 800$ Ω.

The tuned circuit filter has the remarkable characteristic that the voltage gain may be increased up to gains of 5 or 10 by decreasing R_1 without changing either the center frequency or the Q. The Q of the filter may be changed by changing the feedback resistor R_2 without affecting the center frequency or the gain. The center frequency may be adjusted by changing both capacitors, but, for small adjustments, either capacitor may be changed 5 or 10 per cent.

10.21 BAND-REJECT FILTERS

Band-reject filters almost always use a parallel-T network that is adjusted to reject a single frequency. Within a narrow frequency range the signal transmitted through one T network is cancelled by the signal through the other T network. For a high degree of rejection the Ts need to be adjusted carefully to produce exactly equal and out-of-phase output signals. The general form of the parallel-T is shown in Fig. 10.34(a) with the amplitude characteristic shown for a Q of 0.3 in Fig. 10.34(b). The parallel-T gives a deep null at the notch frequency f_0, regardless of the source and load impedances. However, for a narrow-band and equal low- and high-frequency responses well away from the null frequency the series resistor R should be 1.4 times the geometric mean value of the source and load impedances.

The parallel-T filter has the disadvantages that the Q is always less than 0.5, and, with equal source and load impedances, the loss a decade away from the null frequency is 14 dB. A load impedance 8 times the source impedance is about the minimum practical impedance ratio, and the loss is 9 dB. For a load 8 times the source impedance the shunt resistor is 2 times, and the series resistors are each 4 times, the source resistor. These impedance values are easily remembered, and the null frequency is always the frequency at which

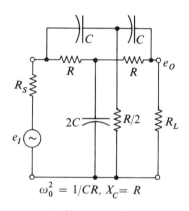

(a) Circuit relations

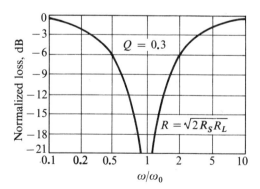

(b) Frequency response

Figure 10.34. Twin-T filter

the reactance of C equals R, regardless of the source and load impedances. Similarly, the null frequency occurs when the reactance of the shunt capacitor equals the shunt resistance.

Because of the relatively high pass-band loss, the twin-T filter is really only practical with a load impedance at least 100 times the source impedance. The filter shown in Fig. 10.35 is used to represent a practical configuration with a null at 100 Hz. When the load-to-source impedance is so high, a twin-T filter may be driven by an emitter follower and coupled into an FET amplifier. However, a more practical twin-T filter used as a component in an active filter is described in Sec. 10.22.

A little-used but very practical bridge-T rejection filter is shown in Fig. 10.36. The filter requires an inductor with a Q of at least 10 and is, therefore, practical for removing a single frequency at 400 Hz and above. A deep null is obtained by adjusting the resistor R, which is approximately 4 times the ac resistance of the inductor, and the null frequency is adjusted by changing one or both capacitors. The advantages of the bridge-T filter are that the pass-band insertion loss is relatively small and the filter Q is determined mainly by

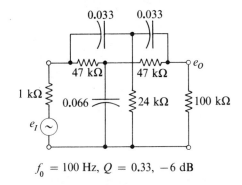

$f_0 = 100$ Hz, $Q = 0.33$, -6 dB

Figure 10.35. Practical low-loss twin-T filters

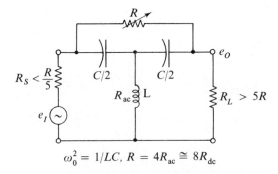

$\omega_0^2 = 1/LC$, $R = 4R_{ac} \cong 8R_{dc}$

Figure 10.36. Bridge-T rejection filter

the Q of the inductor. With a 10-to-1 load-to-source impedance, the circuit Q is reduced only to approximately one-half the Q of the inductor, and for higher impedance ratios the Q approaches the inductor Q. The filter is most easily adjusted by finding the null frequency with $R = 0$ and increasing R to obtain the desired null. Because the inductance and ac resistance of an iron-core inductor change with the signal level, the bridge-T circuit is more satisfactory at high frequencies, when an air-core inductor may be used.

A practical form of the bridge-T null filter is shown in Fig. 10.37. The 30 mH air-core inductance has a Q of 20, which implies a 100-Ω equivalent series resistance. A 40 to 60 dB null is obtained at 10 kHz when the resistor R is 250 Ω, and the pass band loss is only 2 dB at frequencies 10 per cent away from the null frequency. The filter has the advantages of simplicity and of not requiring an amplifier. Suitable air-core inductors may be hand wound or

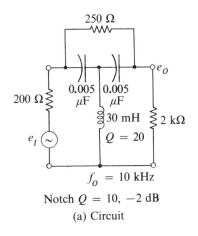

(a) Circuit

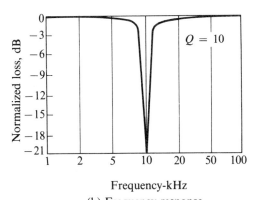

(b) Frequency response

Figure 10.37. Practical bridge-T filter

obtained by removing the core from an iron-cored choke. The filter is useful in distortion and noise measurements for removing a fundamental signal while transmitting noise and distortion with negligible loss.

10.22 ACTIVE BAND-REJECT FILTERS

The rejection characteristics of a twin-T filter may be improved greatly by combining the twin-T with an emitter follower, as shown in Fig. 10.38. With two variable resistors the null and the Q may be adjusted almost independently of each other; the amplifier produces a gain at the high and low frequencies; and the Q and gain are essentially independent of the collector supply voltage. If an emitter follower is used in place of the Darlington, as shown, the overall voltage gain is reduced to 3, and the Q is 5 with a 50-dB null.

When the source is capacitor-coupled, as indicated in Fig. 10.38, the output voltage is 7 V rms with $V_C = 17$ V. Small adjustments of the null frequency may be made by changing the series resistors, and large adjustments are made by changing all capacitors by a common factor. The emitter follower may be replaced by an IC, and voltage gain may be obtained if the point A is connected to the IC output with a voltage divider to offset the gain, as shown in Fig. 10.28 for a 2-pole filter.

If the signal source can carry the bias current, the input coupling capacitor may be removed, and the 20-kΩ resistor is not needed. For most of the active filter circuits the bias resistor is correct when the collector voltage is between 17 V and 20 V.

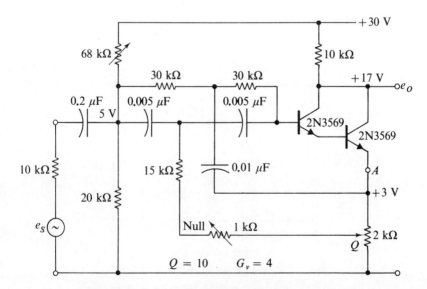

Figure 10.38. Twin-T rejection filter with gain and $Q = 10$

11 Tuned Amplifiers

Tuned amplifiers are used when the desired band of signal frequencies is a small percentage of the center frequency and when there are advantages in rejecting all frequencies outside the band. Tuned circuits are commonly used at radio frequencies to eliminate interfering signals or the harmonic distortion of oscillators and class-C amplifiers. At low frequencies tuned-circuit amplifiers are used to select a single frequency, as in impedance bridge measurements, or to select a particular telemetry signal. Examples of telemetry signals are the tones used in telephone switching and the supersonic signals used for the remote control of a TV receiver.

Tuned circuits offer important advantages in high-frequency circuits. The transistor and circuit capacities are made a part of the tuned circuit instead of becoming a low-impedance problem for the designer. Tuned circuits provide a simple means for coupling a high-impedance collector circuit to a low-impedance base without a power loss. The power gain obtained by using a tuned collector circuit as an interstage transformer substantially increases the gain of a multistage RF amplifier.

An important problem in the design and application of tuned amplifiers is caused by their tendency to oscillate when there is even a relatively small amount of positive feedback. Accordingly, there are frequent references in this chapter to the requirements that must be met to obtain a stable tuned amplifier. The chapter begins with an explanation of the Q factor, which is always used to describe the selectivity of a tuned circuit.

11.1 TUNED CIRCUITS AND THE Q FACTOR (REFS. 3, 9)

A discussion of the relative bandwidth of a tuned circuit is simplified by using the term Q to describe the sharpness of tuning. ***Q is defined as a measure of the***

196 Tuned Amplifiers Ch. 11

quality of an inductor or capacitor. However, the quality of a tuned circuit can be measured by comparing the peak response frequency to the bandwidth at points 3 db down from the peak response. As indicated in Fig. 11.1, the Q of a tuned circuit is given to a close approximation by the relation:

$$Q \cong \frac{f_0}{BW} \tag{11.1}$$

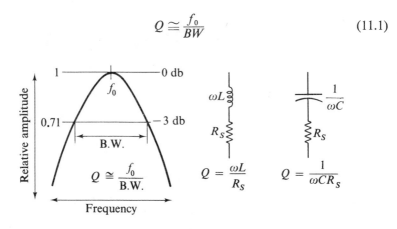

Figure 11.1. Q curve and formulas

where f_0 is the center frequency and BW is the bandwidth in Hz between the 3 dB points. If the center frequency is 1000 Hz and the bandwidth is 50 Hz, the tuned circuit Q is 20. The Q factors of tuned circuits are ordinarily between 2 and 100.

The Q of a tuned circuit is usually controlled by the energy loss in the inductance. The ratio of the inductive reactance ωL to the equivalent series resistance R_S of the inductance is known as the Q, or figure of merit, of the coil. Algebraically:

$$Q \equiv \frac{\omega L}{R_S} \tag{11.2}$$

The Q of an inductance varies with frequency but usually has a broad maximum, relatively independent of frequency, and is a useful design parameter. The Q of inductances varies from 1 to about 400. The Q of capacitors varies from about 100 to 10,000. The Q of an electromechanical resonator may be much higher. Because the rms voltage of electrical noise is usually proportional to the bandwidth of a signal channel, the noise is minimized by limiting the bandwidth as much as possible. On the other hand, a given signal cannot be transmitted in too narrow a band without destroying part of the signal. For these reasons the bandwidth of a tuned circuit is determined by the needs of the signal channel.

11.2 TUNED AUDIO-FREQUENCY AMPLIFIERS

A narrow-band tuned amplifier for applications in the audio-frequency range is illustrated in Fig. 11.2. The amplifier is suitable for detecting the balance signal of an impedance bridge, a metal detector, or telemetry signals. The amplifier operates on an 8-to-15 V supply and has a voltage gain of 56 dB with a 24-V p-p, no-load, output signal. With the additional FET-transistor feedback amplifier shown in Fig. 11.2 the overall gain is 85 dB and the output signal is 10 V p-p. With both amplifiers the collector supply uses only 3 mA at 12 V.

The Q of the tuned amplifier is approximately one-half the Q of the inductor, which indicates that the inductance must be less than 100 mH to take advantage of a coil with a Q exceeding 30. For frequency changes of 2-to-1 the capacitor may be changed, but both the capacitor and the inductor should be increased or decreased by the same factor for a greater change of the peak frequency. The amplifier may be used for narrow-band, audio-frequency applications by shunting the tuned circuit with a resistor to adjust the bandwidth. A Q of 2 or less is usually the optimum when high selectivity is needed in a voice-frequency amplifier,

With an effective Q of 16 at 1 kHz the output signal is down 20 dB at 0.8 and 1.3 kHz, 30 dB at 0.5 and 2 kHz, and 60 dB at 20 Hz and 50 kHz. The simplicity, selectivity, and high gain make the amplifier attractive for many single-frequency applications where the peak frequency must be independent of the supply voltage. The amplifier has an input impedance of approximately 5 kΩ and operates a little better when the emitter is used as the input terminal,

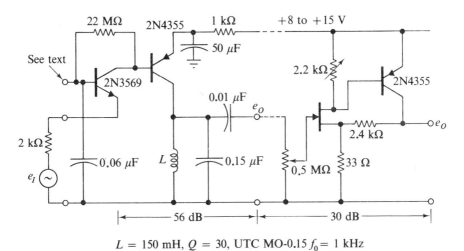

$L = 150$ mH, $Q = 30$, UTC MO-0.15 $f_0 = 1$ kHz

Figure 11.2. Tuned audio-frequency amplifier

as shown in Fig. 11.2. When a high input impedance is required, the input stage may be converted to a CE stage by using a 5-kΩ emitter resistor. When the signal is capacitor-coupled to the base, the input impedance exceeds 100 kΩ and the voltage gain is reduced approximately 10 dB. The tuned amplifier operates over a wide temperature range, and when a 9-V supply is used the emitter resistor may be omitted.

11.3 TWIN-*T* TUNED AMPLIFIERS

One of the most interesting developments brought about by the use of transistors is found in the construction of tuned amplifiers where the need for a high-Q inductor is eliminated by using an active feedback circuit. A considerable variety of circuits is offered for tuned circuit applications, and many of these prove disappointing. Some circuits are relatively complicated, others are difficult to adjust, and some do not have a stable Q or voltage gain. The problem with tuned active circuits is, of course, that positive feedback is required, and positive-feedback circuits tend to be unstable, especially with high Q. Many of the high-Q tuned circuits require a regulated or relatively fixed supply voltage.

The tuned amplifier represented by the circuit shown in Fig. 11.3 has a Q that is adjustable from 5 to 50 and is tunable over a small frequency range without producing a large change of the Q. The amplifier uses a form of parallel-T feedback, and the peak frequency f_0 and overall voltage gain are relatively independent of the supply voltage. Except that the Q varies with the supply voltage, a regulated supply is not usually needed.

The resistor and capacitor values shown in Fig. 11.3 are correct for a tuned frequency of 1 kHz. For a lower frequency all the resistors may be increased by a common factor, as much as 10 to 20 times, and the capacitors

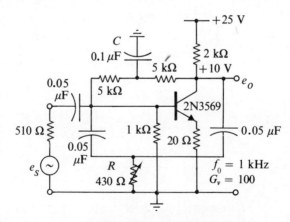

Figure 11.3. Twin-*T* tuned amplifier, $Q < 50$

may be increased by a factor of 100 or more. Frequencies through the upper audio range may be obtained by reducing the capacitors by a common factor. Above audio frequencies tuned LC circuits are simpler and more stable, and high-Q inductors may be obtained in a relatively small package.

The peak frequency of the amplifier may be changed over a ± 20 per cent frequency range by changing the resistor R from 200 Ω to 1 KΩ. The Q of the circuit is selected by the value used for the shunt capacitor C. Values of the peak frequency, Q, and the stage gain are shown in Table 11-1. The circuit is useful for Q values up to about 40, and the emitter resistor may be changed if lower or higher supply voltages are used. With a 10-V supply the emitter resistor may be omitted.

Table 11-1 Tuned-Filter Characteristics—Fig. 11.3

$C(\mu F)$	$R(\Omega)$	$f_0(Hz)$	Q	$G_v(dB)$
0.05	680	1000	5	27
0.05	330	1300	12	35
0.05	51	2000	6	33
0.1	680	900	20	42
0.1	430	1000	35	40
0.1	220	1200	15	40

11.4 NARROW-BAND HIGH-Q TUNED AMPLIFIERS

A remarkably stable high-Q tuned amplifier is shown in Fig. 11.4. The circuit differs from the one shown in Sec. 11.3 mainly by the use of a variable emitter resistor as the means of adjusting the Q, and also by the isolation provided to

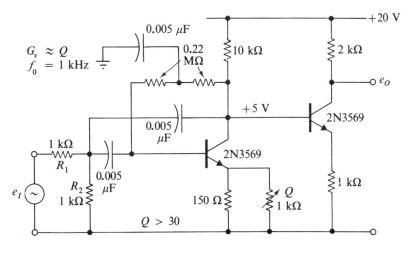

Figure 11.4. Narrow-band stable high-Q amplifier

eliminate output loading and changes of the source resistance. However, the performance is not independent of the values used for either R_1 or R_2. Small adjustments of f_0 may be obtained by shunting R_2 to increase the peak frequency, and the Q is adjusted by changing the emitter resistor. The frequency control and the Q are remarkably independent of each other, and the circuit may be operated with a Q approaching 1000. However, high Qs require a stable power-supply voltage, and the resistors and capacitors should be selected for a low temperature-coefficient over the temperature range in which the amplifiers may be used. One disadvantage of the circuit, perhaps, is that the gain varies with the Q and is approximately equal to Q. However, the circuit does have a desirably high loss at off-peak frequencies. For a Q of 50 the loss is 20 dB at 0.88 and $1.2 f_0$, 30 dB at 0.75 and $1.5 f_0$, and 40 dB at 0.45 and $4.5 f_0$. With higher Qs the skirt losses are correspondingly higher. Because the voltage gain is approximately equal to the Q, the Miller effect produces a significant input capacitance. Thus, the high-Q amplifier is mainly useful for frequencies below 1 kHz.

The Q-control may be adjusted to produce a stable oscillator with an excellent wave form. The frequency stability is approximately 0.5 per cent per volt change of the supply voltage, but is, unfortunately, sensitive to transistor temperature. Because of the temperature sensitivity, the amplifier and oscillator are useful mainly in room-temperature applications where the temperature effect can be offset by a frequency adjustment.

11.5 LOW-IMPEDANCE, TUNED, CB-CE AMPLIFIERS

A tuned amplifier with a series-resonant, impedance-transforming, input circuit is shown in Fig. 11.5. When the amplifier is driven by a signal source with a resistance less than 1 Ω, as with an electromagnetic coil pickup, the Q

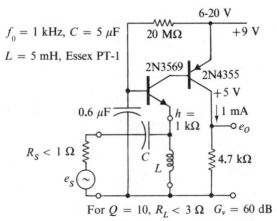

Figure 11.5. Low-impedance, tuned CB-CE amplifier

of the tuned circuit is 10, and the overall voltage gain is 1000, 60 dB. The components shown in the figure give the amplifier a 1-kHz peak frequency that may be moved a factor of 4 up or down by changing the capacitor C. Because of the low impedance of the tuned circuit the capacitor C should be a good quality mylar or paper capacitor. For a 10-kHz peak frequency the inductor L may be a hand-wound, air-core coil made with 20 meters of 26-gauge wire wound in an approximately square cross section on a 1-cm diameter wood dowel. A 0.25-μF tuning capacitor is used with a 1-mH inductance for 10 kHz.

The amplifier requires only 1 mA when operated on a 9-V supply, and the peak output signal is nearly 3 V. When additional gain and a very narrow frequency response are required, the amplifier shown in Fig. 11.2 may be used with the series-resonant, L-C-input circuit of Fig. 11.5 and a 1-Ω signal source. With the resonant input circuit and the tuned collector circuit the amplifier in Fig. 11.2 has adequate gain and selectivity for a metal detector that operates with a 1-kHz electromagnetic field generator.

For broad-band applications the amplifier in Fig. 11.5 may be used with source impedances up to 500 Ω connected in series with the emitter and with the tuned circuit removed. With source impedances up to 500 Ω the amplifier has a voltage gain of 100, 40 dB, between the 3-dB frequencies, 350 Hz and 25 kHz. The emitter current is low enough to permit connecting a dynamic microphone without an input transformer, or a 500 Ω-to-voice coil transformer may be used for additional gain and isolation. The low-frequency cutoff may be moved down to 35 Hz by making the base capacitor larger by a factor of 10.

11.6 TUNED-AMPLIFIER GAIN LIMITATIONS (REF. 2)

As in any high-gain system, an amplifier becomes an oscillator and is useless when the forward gain exceeds the loss from the output back to the input. At frequencies just below resonance, capacitance feedback returns a signal in phase with the input signal and effectively shunts the transistor input with a negative component of resistance. If the magnitude of the negative resistance is low enough to offset the positive resistance of the input circuit, the stage is forced to oscillate. As in any feedback system, oscillations begin when the loop phase shift is 180° if the base-to-collector voltage gain exceeds the feedback loss.

The effects produced by feedback through the collector-to-base capacitance and through internal resistance can be offset by providing additional feedback in the opposite sense. If an external feedback circuit cancels the effects of all resistance and capacitance feedback, the amplifier is said to be *unilateralized*. If the circuit cancels the effect of only the reactance feedback, the amplifier is said to be *neutralized*.

Generally amplifiers are neutralized by connecting a small capacitor (C_N in Fig. 11.6) from the transistor base to a winding on the output transformer, which provides a phase reversal. The neutralized condition is determined by adjusting the capacitor C_N until the signal at the base does not change when the collector circuit is tuned. With low available gain where neutralization is most needed, this adjustment is easy and effective. Residual effects that may be observed usually indicate additional stray coupling, and the stage gain should be reduced if the amplifier cannot be satisfactorily neutralized.

The unilateralization of tuned stages is generally considered impractical because the circuit requires a delicate balance of several feedback effects, which makes the adjustments difficult, and the amplifier is unstable for small changes away from the balance condition.

For similar reasons the neutralization of cascaded high-gain amplifiers is generally costly, time-consuming, and is optimized for only one frequency. Neutralization and tuning problems are particularly troublesome in cascaded stages, because the tuning of a collector circuit affects the tuning of the previous base circuit and may even interact from stage to stage. However, these difficulties may be removed by deliberately reducing the gain per stage enough to ensure a desired degree of stability and ease of tuning. The price paid for the improved performance is lower stage gain, but we obtain ease of

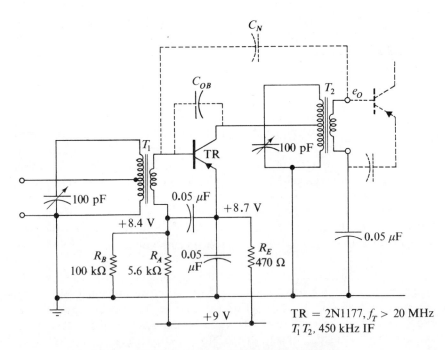

Figure 11.6. Typical tuned IF amplifier, −450 kHz

design, ease of alignment, and a simpler circuit. When the cost of an additional stage is important, the designer may use a combination of reduced gain and neutralization, a technique generally used in the IF amplifiers of transistor radios.

11.7 TUNED IF AMPLIFIERS

The problems encountered in working with multistage tuned amplifiers are essentially the same as those found in the IF stages of a broadcast-band radio. A typical tuned IF amplifier, as shown in Fig. 11.6, has tuned coupled circuits in both the input and the collector circuit. The resonant impedances are selected to provide suitable input and load impedances, and the inductances as so selected that the resulting Qs provide the desired selectivity. The capacitors may be found from the resonant frequency and the given inductance, except that the transistor input capacitance usually supplies an appreciable part of the capacitance.

Amplifiers at IF, RF, and UHF frequencies generally require tuned coupled circuits because adequate gain is obtained in the β-cutoff region only by making the transistor capacitance part of a tuned circuit. By this means the amplifier input impedance may be increased by almost the Q of the tuned circuit. If the transistor has more capacitance than can be permitted in the design, the base may be connected to a tap on the input circuit, and the tuned circuit will operate as a step-down transformer.

RF transformers are tuned either by capacitors or magnetic cores, and one winding usually has so many turns that the collector or the base connects to a tap on the winding. The excess winding is used either to accommodate the tuning slug better or to increase the impedance so that the capacitor is physically small. The collector and base connections are tapped down on the windings to limit the effect of the internal collector resistance on the Q or to reduce the stage gain and avoid the need for neutralization.

For small-signal applications a tuned stage is biased class A the same way a stage is biased in an audio-frequency amplifier. As shown in Fig. 11.6, a tuned stage usually has an emitter resistor and a base-circuit voltage divider so that the dc collector current is controlled by the bias resistor and the dc S-factor. In RF applications the emitter resistor is usually bypassed to make the ac current gain as high as possible.

11.8 TUNED AMPLIFIER EXAMPLE

The amplifier shown in Fig. 11.7 may be constructed for an experimental study of the problems encountered with tuned amplifiers. The CE stage has a tuned input and a tuned collector load. The amplifier operates at the 450-kHz IF frequency often used in broadcast-band radios, and is useful for demon-

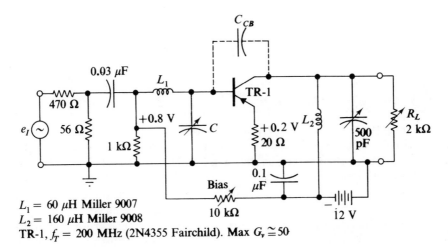

$L_1 = 60\ \mu H$ Miller 9007
$L_2 = 160\ \mu H$ Miller 9008
TR-1, $f_T = 200$ MHz (2N4355 Fairchild). Max $G_v \cong 50$

Figure 11.7. Experimental 450 kHz tuned CE amplifier

strating the stability problems met in an IF stage. The input circuit is a series-resonant circuit that can be connected to a signal source. Regardless of the source impedance, the Q of the tuned circuit at 450 kHz is approximately 3.4, so the equivalent parallel resistance is approximately $50(Q^2)\ \Omega$. When corrected for the transistor input resistance, the equivalent base impedance R_b is 450 Ω at resonance. The collector load impedance may be adjusted by the variable resistor R_L.

Feedback through the collector-to-base capacitance C_{OB} makes the stage oscillate with high values of the load resistor and gains exceeding 150. A study of the performance with lower gains shows that although the stage is stable for a voltage gain of 100 without neutralization, a higher stability, as that required for manufacturing, may be obtained by reducing the collector load. Moreover, unless the stage gain is reduced to about 20, which is a severe reduction, the stage shows interaction between the output and input that makes tuning difficult and unstable. The stage illustrates the fact that neutralization, which reduces the interaction, may be needed, even though not required for stability.

The stable gain that may be attained in a tuned amplifier varies inversely with the frequency as long as the feedback capacity remains fixed. Thus, when tuned at 50 kHz the experimental amplifier has a stable gain of 500 and oscillates when the gain exceeds 1500. A tuned stage operating at radio frequencies above 50 MHz may have a voltage gain of only 5 and would be inoperative without neutralization. However, with a voltage gain less than 10 neutralization is relatively easy and most effective.

Neutralization and tuning problems are particularly troublesome in multistage tuned amplifiers, because the adjustment of one stage may affect the tuning of the next or the second stage removed. For these and other reasons the popularity of the multistage tuned IF amplifier has declined in favor of a high-gain amplifier preceded or followed by a high-performance mechanical or crystal filter. Since the mid-twenties the superheterodyne receiver has been used to move the signal to lower frequencies where a high-gain tuned amplifier is constructed more easily. Thus, the multistage amplifier is a low-frequency IF amplifier unless the signal requires a broad-band amplifier as in video and radar systems.

In summary, amplifiers with tuned base and collector circuits are unstable with high values of the collector load resistor. At high frequencies where stage gains are less than 10 neutralization is easy and effective. At low frequencies where the critical gain is high neutralization is more difficult and a stable amplifier may be obtained by using a relatively low value of the load impedance. Sometimes a combination of loading and neutralization is most effective.

Because the instability of a tuned amplifier is caused by collector-to-base feedback the transistor should be selected for low capacitance C_{OB} and the base and collector circuits should be shielded carefully and separated to prevent inductive feedback. Tuned IF amplifier circuits may be found in the transistor handbooks, and a good working example may be obtained from a transistor radio for less than the cost of the components. An RF transistor with 2 pF feedback capacitance in a 455 MHz IF amplifier gives a power gain of approximately 30 dB/stage. Thus, the voltage gain measured from base-to-base is 30.

11.9 UHF TUNED AMPLIFIERS (REF. 3)

A UHF tuned amplifier is similar in many respects to an IF amplifier with tuned input and output, except that the circuit must be simplified with as many components grounded on one side as possible. In the UHF spectrum a circuit is usually placed inside a copper enclosure to provide short low-impedance grounds and for shielding.

A typical low-noise, broad-band, 200-MHz amplifier circuit is shown in Fig. 11.8. The amplifier has tuned input and output circuits with neutralization to increase the stability and ease of tuning.

The amplifier uses a CE stage that is biased as if it were a CB stage in order that the input inductor and one tuning capacitor may be grounded. In addition the bias circuit is bypassed by a single feedthrough capacitor so the emitter may have a very short ground connection, which is necessary to prevent regenerative feedback. Similarly, one collector tuning capacitor is grounded and the collector supply is brought in through a feedthrough capacitor. The amplifier has a 20-dB power gain between a 50-Ω source and a 50-Ω load, but a 300-Ω source and load (a 300-Ω line) may be used with similar performance characteristics.

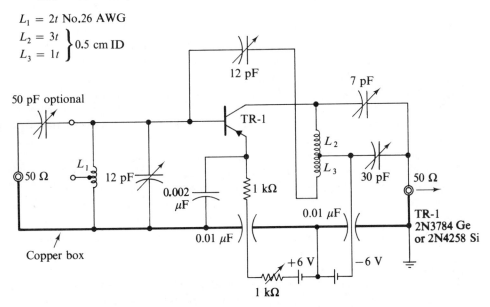

Figure 11.8. Wide-band UHF 200 MHz amplifier

Each of the tuned circuits has two tuning capacitors which permit an adjustment of both the resonant frequency and the impedance match. The input is connected in series with a tuning capacitor to provide an impedance step-up, while the load is coupled across a capacitance voltage divider, which has the advantage of bypassing the harmonic frequencies to ground. When the tuned circuits are designed with a Q of 10, the amplifier bandwidth is 10 per cent of the center frequency, and the 20 MHz frequency band makes this a broadband amplifier.

The transistor may be a high-frequency germanium device which has an f_T exceeding 1 GHz. A germanium transistor generally offers a high f_T or a lower cost than is available with a silicon transistor. Neutralization is required because of the relatively high power gain, but may be omitted by using a transistor with an $f_T = 200$ MHz and by accepting about 15-dB gain. The transistor is operated with a 3-mA emitter current that is adjusted by the variable resistor in the +6-V emitter supply. At 200 MHz the transistor current gain is about 5, so 20-dB power gain requires a collector load impedance that is at least 1000 Ω.

11.10 JFET TUNED RF AMPLIFIERS (REF. 3)

Junction field-effect transistors that are designed for high-frequency applications have a drain-to-gate feedback capacitance of 1 to 3 pF. With a feedback capacitance of this magnitude the gain of a tuned RF stage cannot exceed about 5 without requiring neutralization. A single-stage JFET amplifier may be used at frequencies as high as 150 MHz, but, where higher gains are needed, amplifiers generally use a MOS-FET stage or an FET-FET cascode. Above 30 MHz high-gain multistage amplifiers are avoided by frequency conversion for amplification in an IF amplifier.

Amplifiers that use an FET-FET cascode have the advantages of both the grounded-source and the grounded-gate FET amplifiers and do not require neutralization when tuned at frequencies below 50 MHz. The cascode amplifier shown in Fig. 11.9 is more compact and simpler than a neutralized stage, and either gate may be used for at least 10 dB of automatic gain control (AGC). Successful use of the cascode above the broadcast frequencies requires selecting FETs that have a low gate-to-drain capacitance and providing careful shielding and separation between the tuned input circuit and the rest of the amplifier. A cascode in RF applications is generally used as a single-stage, tuned, RF amplifier or as a remote preamplifier. A power gain of 30 dB may be easily obtained at 30 MHz, and with careful design 10 dB may be attained at 100 MHz without neutralization.

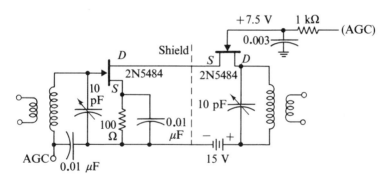

Figure 11.9. FET-FET cascode amplifier—1 to 100 MHz

11.11 JFET UHF AMPLIFIERS

Common-source JFET amplifiers are sometimes found that use tuned gate and tuned drain circuits up to frequencies as high as 200 MHz. These amplifiers generally have both the gate and drain tapped down on the tuned circuits to keep the gate-to-drain voltage gain less than 5 per stage, 15 dB. With neutralization the gain may be 3-to-6 dB higher. However, the superior advantages of the MOS-FET devices in RF applications make the tuned JFET common-source amplifier relatively obsolete.

Where low impedances and few components are particularly desired, the common-gate JFET amplifier may be used at frequencies between 100 and 600 MHz. The input impedance of the JFET with the gate grounded is

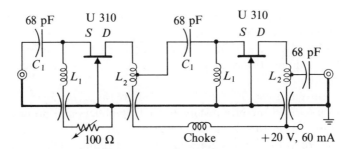

Figure 11.10. Wide-band JFET amplifier—10 to 600 MHz

approximately $1/g_m$, which is 100 Ω when g_m is 10,000 μmho. Thus, as shown in Fig. 11.10, a low-impedance line may be capacitor-coupled or direct-coupled to the FET source, and the inductance L_1 may be an RF choke. Power gain is obtained by making L_2 an inductance that resonates with the gate-to-drain capacitance. The load or a following stage is coupled to a tap on L_2 that matches the load impedance. At low frequencies a 15-dB gain per stage may be expected, and 6 or 8 dB may be obtained at frequencies as high as 600 MHz. At such high frequencies the attainable power gain is low and all components must be carefully matched to the JFET parameters. The common-gate amplifier has the advantages of requiring only one tuned circuit per stage, simplified input matching, and no Miller-effect increase of stray capacitance.

12 Video Amplifiers

A video amplifier has a frequency response that extends from a low audio frequency to tens of megahertz and is used to amplify pulse-type signals that must be reproduced without changing the signal waveform. Video amplifiers are found in TV systems, radars, satellite relays, and oscilloscopes. A broadband amplifier is generally a tuned RF amplifier that is used to transmit modulated signals that carry video or similar information bandwidths.

The design of high-frequency circuits is determined to a great extent by the increasing importance of capacitance as the signal frequency is increased. Thus, we generally find that the circuit impedances decrease progressively as the frequency is increased. In a similar manner the current gain of a transistor decreases with increasing frequency, and a wide-band frequency response is obtained only by accepting reduced gain in an exchange of gain for bandwidth. Furthermore, capacitance is reduced by making high-frequency semiconductors smaller and rated for low power or designed to be used with a heat sink.

Certain characteristics of the CB and CC amplifiers make these amplifiers useful at moderately high frequencies without the effort and understanding required to use the same devices in a CE stage at the same frequency. However, a CE stage offers greater power gain and more flexibility in the choice of input and output impedances.

With most of the amplifiers shown in this chapter the impedance level may be increased by reducing the upper cutoff frequency. Generally, if the upper cutoff frequency is 5 MHz, the resistors in the circuit may be increased by a factor of 10 with the cutoff frequency decreasing to 0.5 MHz. If the amplifier is for use entirely within the medium-frequency (MF) band, the coupling and bypass capacitors may be reduced to increase the low-frequency cutoff. The smaller capacitors reduce low-frequency noise and pickup, and may simplify power supply filtering.

210 Video Amplifiers   Ch. 12

12.1 VIDEO CB AMPLIFIER

A transistor operated with the base as the common terminal has a unit current gain, and for almost any RF transistor the current gain is constant up to at least 100 MHz. However, a CB stage has power gain only when the output impedance exceeds the input impedance, but the ease with which a low-impedance source may be coupled to a high-impedance load makes the CB stage attractive for coupling a low-impedance, high-frequency cable into an amplifier operating with a high-impedance load. Because the collector signal is in phase with the input signal, there is no Miller effect. The collector capacitance is across the load and is the principal cause of high-frequency cutoff. A CB stage has exceptionally linear and stable operating characteristics.

The single-stage amplifier shown in Fig. 12.1 may be used to couple a video cable into a circuit with an impedance of 1-to-3 kΩ. The base is grounded, so that both positive and negative power supplies are required, and they have equal load currents. The collector Q-point voltage is adjusted to 4 V with the emitter bias resistor, and the Q-point is a little less than one-half the supply voltage because the lower voltage allows a slightly higher peak output signal when the amplifier is coupled to an external load. The emitter current is about 2.7 mA, which makes the transistor input impedance 10 Ω. The 10-Ω transistor impedance with a 40-Ω series resistor terminates the cable in 50 Ω without reflecting an impedance that changes with frequency, as when the input is tuned.

A CB stage produces even harmonic distortion unless the equivalent source impedance is at least 5 times the transistor input resistance. In the amplifier shown in Fig. 12.1 the cable impedance and the series resistor make

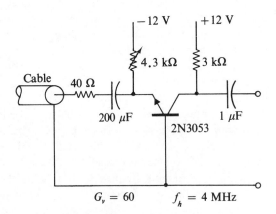

Figure 12.1. Video CB amplifier

the equivalent source impedance 90 Ω. The transistor input impedance varies inversely with the bias current, and the emitter current may be increased but should not be reduced.

The no-load voltage gain of the amplifier is 60, or nearly 36 dB. The 2N3053 transistor has an output capacitance of about 15 pF, which places the high-frequency cutoff at approximately 4 MHz. The 2N3053 has a minimum gain-bandwidth product of 100 MHz, which shows that the output and load capacities determine the high-frequency cutoff. The frequency response of the amplifier may be increased to 25 or 30 MHz by using a 1-kΩ load and a type 2N4258 VHF transistor.

Operating a CB transistor with the base grounded has the disadvantage that high-current positive and negative power supplies are needed to bias the transistor. When a few more components are added, the base may be operated 2 V off ground, as shown in Fig. 12.2, so that only a single polarity power supply is required. Although the amplifiers in Figs. 12.1 and 12.2 are similar, the second amplifier has about twice the emitter current and requires a +30-V supply. The higher emitter current is used to decrease the transistor input impedance and make the second amplifier linear to about 3 times the peak signal level of the first amplifier.

The base capacitor in Fig. 12.2 is not required in low-frequency applications. Removing the capacitor lowers the gain less than 3 dB, depending on the transistor current gain, but the high-frequency cutoff is down 6 dB at 1 MHz and decreases linearly with frequency. A 100-μH inductance may be used in series with the collector resistor to compensate a part of the loss at the upper cutoff frequency. The optimum inductance is best determined by trial while observing the transient response of the amplifier. An approximate value

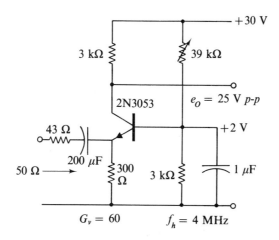

Figure 12.2. Video CB amplifier with base bias

for the inductance may be found by inserting an inductance that increases the gain 2-to-3 dB at the -3 dB frequency.

12.2 VIDEO CC AMPLIFIER

The CC amplifier is a broad-band amplifier used to couple high-impedance sources into low-impedance loads. The transistor has a current gain that decreases linearly with frequency from a value β_0 at medium frequencies down to 1 at the frequency f_T. Thus, the CC amplifier has a current gain S at the frequency f_T/S. The 2N3053 transistor that is shown in the CC stage of Fig. 12.3 has a current-gain bandwidth product $f_T = 100$ MHz (minimum). If we wish to operate a CC stage up to 5 MHz, we must limit the stage current gain to no more than $S = 100$ MHz/5 MHz $= 20$.

The collector capacitance C_{OB} is across the input of a CC stage and is a second cause of high-frequency cutoff. If we wish to use the 2N3053 transistor in a CC stage at 5 MHz and $C_{OB} = 15$ pF, we must use a parallel equivalent input and source impedance that is no more than 2 kΩ. Because each high-frequency cutoff contributes 3 dB loss at 5 MHz, a desire to have no more than -3 dB at 5 MHz suggests that we use a current gain of 10 and an equivalent impedance of 1 kΩ.

The CC stage shown in Fig. 12.3 may be used to couple a 2-kΩ signal source into a 100 Ω line and gives a power gain of approximately 5 dB up to 5 MHz. The impedance of the emitter as a signal source for the line increases with frequency and is approximately 50 Ω at 5 MHz. The resistor between the emitter and the line may be used to make the amplifier output-impedance match the line. Also, there may be transient ringing or instability if the line reflects a large capacitance component. The series resistor is helpful in

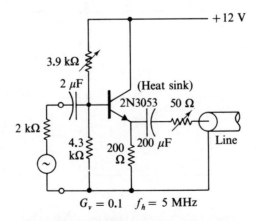

Figure 12.3. Video CC output stage

reducing the effect of capacitance, and the line should be terminated to minimize the reactance and standing waves.

12.3 VIDEO LINE-TO-LINE AMPLIFIERS

A video line-to-line amplifier may be constructed by direct-coupling the CC stage of Fig. 12.3 to the CB stage of Fig. 12.2. Direct coupling, as shown in Fig. 12.4, eliminates the bias resistors, and the emitter resistor may be lowered to 100 Ω without changing the ac current gain. The capacitance of both stages is in parallel with the 1-kΩ interstage resistor, and 30 pF produces a high-frequency cutoff near 50 MHz. Thus, the current-gain-bandwidth product of the second stage limits the response to 5 MHz, the measured voltage gain is 20, and the maximum output signal is nearly 10 V p-p. The 2-stage amplifier may be useful, but a single CE stage that uses a slightly better high-frequency transistor gives the same voltage gain and greater bandwidth. The CE amplifier has the advantage that the power gain of a single stage is essentially the sum of the power gains of a CB and a CC stage.

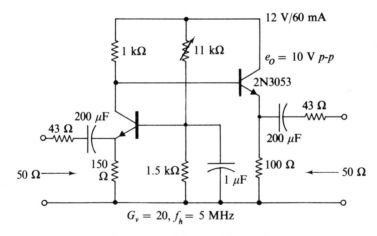

Figure 12.4. Video CB-CC amplifier

The single-stage CE amplifier shown in Fig. 12.5(a) uses a 2N2219 RF transistor that has $f_T > 300$ MHz when the collector current is 60 mA. The video stage has a partially bypassed emitter resistor that permits increasing the bias resistors and the RF signal efficiency. The 5-Ω emitter resistor is either left without a bypass, or a small capacitor may be used to compensate the frequency response at the upper cutoff frequency.

The 2N2219 transistor has relatively low input and feedback capacitances that produce a high-frequency cutoff at 50 MHz or above. The transistor f_T determines the video cutoff frequency, which is at approximately 25 MHz.

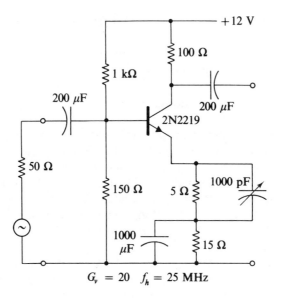

(a) Line-to-line stage

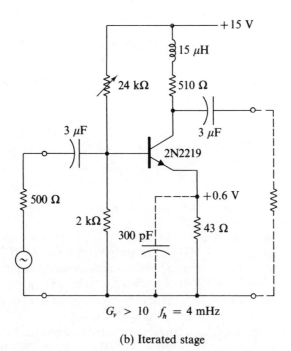

(b) Iterated stage

Figure 12.5. Single-stage video CE amplifiers

Because the capacitance cutoff is twice the f_T cutoff, the stage may be used with a 200-Ω load resistance without lowering the upper cutoff frequency below 20 MHz. This stage with a voltage gain of 40 requires careful shielding, bypassing, and short leads to prevent instability problems.

12.4 VIDEO ITERATED STAGE

The video line-to-line amplifier may be easily changed for use as a high impedance iterated stage, as shown in Fig. 12.5(b). The amplifier source and load impedances are increased by a factor of 10, and the emitter bypass capacitor is omitted to improve the low-frequency transient response. The high-frequency compensation is made with an inductance in series with the collector resistor. A capacitor across the emitter resistor lowers the input impedance at high frequencies, where high impedance is most difficult to obtain. The amplifier may be used on a higher supply voltage as long as the collector current does not exceed 100 mA and a heat sink is used to protect the transistor from overheating.

12.5 SINGLE-STAGE, CRT VIDEO DRIVER

A single-stage CE amplifier designed to drive a cathode ray tube with a 70-V p-p output signal at video frequencies is shown in Fig. 12.6. The amplifier provides adequate cathode drive for a small CRT and is driven to full output by a video IF detector with 3-V output. The amplifier has a 2-kΩ collector resistor and a 100-Ω emitter resistor at full gain. Therefore, the voltage gain is 20 and may be reduced to 6 by the variable cathode resistor. The video amplifier is biased with the three resistors used in an emitter-feedback stage. The bias voltage is obtained from a low-voltage dc supply that is used in the

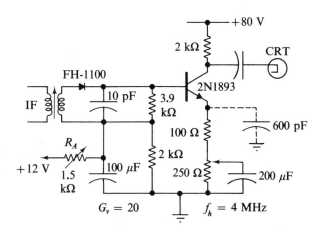

Figure 12.6. Single-stage CRT driver

IF amplifier. The low-voltage source makes the resistor R_A relatively low, but the bias resistors do not load the detector because the diode load is in series with the bias resistors and the transistor base. The transistor requires a small heat sink and careful shielding. The collector load capacitance should not exceed 25 pF for a 4-MHz bandwidth.

12.6 DIRECT-COUPLED, HIGH-FREQUENCY AMPLIFIERS

There are several important advantages in using direct-coupled pairs of transistors in medium frequency (MF) amplifiers and, where possible, in video amplifiers. The direct-coupled pair amplifiers have the advantage that fewer components are required, which means that the circuit has less distributed and stray capacitance. The dc feedback can be arranged to eliminate the need for at least one of the emitter capacitors, and therefore there is less low-frequency transient distortion. The transistors may be coupled as a cascode that almost eliminates the Miller feedback, and a pair may be made to share a high-voltage supply without either transistor being subjected to the full supply voltage. An *npn* stage followed by a *pnp* stage with the emitter connected to the collector supply has a high overall voltage gain but a low interstage impedance that minimizes the Miller feedback, as with a CE-CE cascode. With overall collector-to-emitter feedback the complementary pair has a high input impedance and does not need an emitter capacitor in either stage.

A transistor pair that is used at frequencies below 100 kHz may have feedback over both transistors without producing too much difficulty with instability. At medium frequencies, 0.3–3.0 MHz, feedback over 2 stages is usually difficult, except in an IC where the transistors and the circuit components are exceptionally small. The direct-coupled, 2-stage amplifiers that are designed for video applications may be easily changed to give higher gain or to have a higher input and load impedance for applications at frequencies below 0.5 MHz. In some cases these amplifiers may be used to advantage with 2-stage overall feedback. Similarly, the direct-coupled, 2-stage amplifiers that are designed for low-frequency applications may often be used with minor changes at frequencies as high as 0.5 MHz.

12.7 HIGH IMPEDANCE VIDEO AMPLIFIERS

Video amplifiers with a relatively high input impedance are used to amplify the low-voltage rectified output of an IF amplifier. A CE stage may have an input impedance of nearly 10 kΩ over the video band, provided the CE stage input capacitance is less than 10 pF. A CE-CB, direct-coupled cascode amplifier minimizes the collector-to-base feedback capacitance of the input stage. With emitter feedback the CE stage has a high input impedance up to a frequency of approximately the transistor f_T divided by the stage S-factor.

The cascode connection uses the CE stage for current gain and the CB stage for voltage gain. Because the current in the emitter resistor approximately equals the current in the collector resistor, the cascode has feedback that makes the voltage gain approximately equal to the R_L/R_E ratio unless the source impedance exceeds 10 or 20 times R_E.

The video amplifier shown in Fig. 12.7 has an input impedance of 2.5 kΩ and a voltage gain of 15 up to 5 MHz, except that some frequency compensation may be required to obtain a flat frequency response adequate for TV applications. For wide-band applications up to 100 kHz the amplifier may be used with all resistors increased by a factor of 10. The cascode has a low sensitivity to V_{CC} changes, and the CB stage-biasing is not critical. A cascode amplifier is often used as the preamplifier for an oscilloscope. With UHF transistors and a lower impedance level the cascode may be used to above 100 MHz. However, above 10 MHz an integrated circuit offers the advantages of exceedingly short lead lengths and the high f_T values of small transistors. The manufacturers offer several wide-band amplifiers operable to above 100 MHz with a choice of compensation circuits to shape the frequency or transient response.

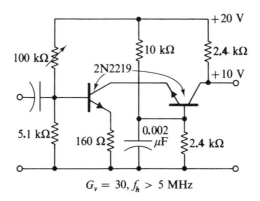

Figure 12.7. High-impedance video amplifier

12.8 CE-CE MEDIUM-FREQUENCY AMPLIFIER

The 2-stage CE-CE amplifier shown in Fig. 12.8 has useful characteristics for medium-frequency applications and up to 6 MHz. The amplifier has an emitter feedback stage followed by a collector-feedback stage, and there is dc feedback over both stages to control the Q-points. The amplifier has a voltage gain of 100, 40 dB, and a broad-band response that extends to 6 MHz when driven by a 50-Ω signal source. The high-frequency cutoff is caused mainly by the Miller effect in the first stage, which means that a higher source impedance

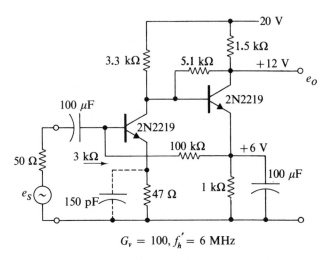

Figure 12.8. CE-CE medium-frequency amplifier

may be used in exchange for reduced bandwidth. Increasing the source impedance to 500 Ω and 5000 Ω, in turn, reduces the bandwidth to 0.5 MHz and approximately 50 kHz. The gain e_o/e_s is 100 for source impedances up to 1 kΩ. The amplifier has adequate dc feedback to make the amplifier operable over a wide temperature range.

12.9 HIGH-VOLTAGE VIDEO AMPLIFIERS (REF. 2)

A high-level video amplifier suitable for operation on an 80- to 100-V dc supply is shown in Fig. 12.9. The amplifier is a form of the cascode discussed in Sec. 12.7, except that a Zener diode is used to fix the base voltage of the output

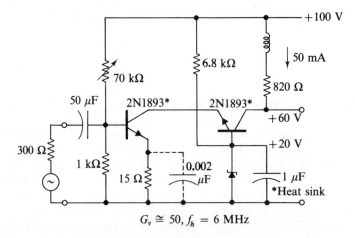

Figure 12.9. High-voltage video amplifier

stage. The Zener diode limits the collector-to-base voltage of the output transistor to 80 V and provides a fixed bias voltage at low audio frequencies without contributing the high-frequency disadvantages of an electrolytic bypass capacitor. The Zener diode is bypassed to reduce the impedance at high frequencies or sometimes to reduce the broad-band diode noise. The emitter capacitor in the input stage provides a high-frequency boost. The bias resistor is adjusted to make the dc output voltage 60 V.

The ac voltage gain of this CE-CB amplifier is approximately 50 up to the frequency $f_T/20$, where f_T is the gain-bandwidth rating of the input transistor at 50 mA collector current. The input impedance is high compared with the source impedance, except near the high-cutoff frequency, 5 MHz. Near the high-frequency edge of the band compensating inductors may be used in series with the collector resistor or the input signal source. The bandwidth may be increased by decreasing the source impedance and increasing the emitter resistor.

Since the development of high-voltage silicon transistors, the video output stage of a TV receiver may use a high-gain CE stage operating on a 150 V supply. The circuit of a video amplifier typical of those found in an all-transistor receiver is shown in Fig. 12.10. This amplifier uses a 250-V power transistor in the output stage, and the input stage has a low-power transistor operated CC for impedance step-down. The amplifier has a 3-kΩ input impedance and is used to couple a video IF detector to the cathode of a picture tube. The overall voltage gain is about 50, the current gain is approximately 10, and the bandwidth is 4 MHz. The intermediate gain control adjusts the picture contrast by changing the magnitude of the output signal. Adjusted for a maximum contrast, the peak-to-peak output signal is 90 V, which is adequate for a 19-in picture tube.

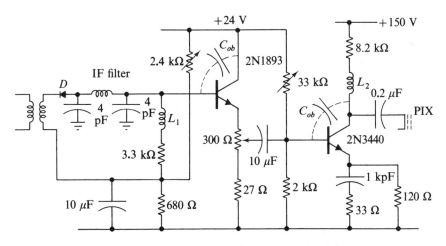

Figure 12.10. TV video amplifier (10 Hz to 4 MHz)

With a voltage gain in the output stage of 50 and $C_{OB} = 10$ pF, the Miller effect input capacitance is at least 300 pF. This relatively high input capacitance can be tolerated only because the CC stage reduces the equivalent ac source impedance to 50 Ω or less, depending on the gain setting. Thus the input capacitance produces a 3-dB loss at 10 MHz or higher. The ac S-factor is so low that the transistor $f_T = 15$ MHz determines the current-gain cutoff frequency. Together the Miller effect and the gain-bandwidth factor reduce the high frequency cutoff to approximately 6 MHz. Since the frequency response is degraded in both stages, the net bandwidth of the two-stage amplifier is about 4 MHz. Careful shielding and layout are required in a wideband amplifier because the response of the amplifier depends to a considerable extent on the capacitance of components and the stray capacitance between any two parts of the circuit.

A video amplifier usually has compensating inductances that are used to shape the transient response. After the amplifier is assembled, the compensating inductances are adjusted by spacing the coil windings until the desired transient response is obtained. The inductance resonates with the load capacitance and raises the gain about 3 dB at the 3-dB cutoff frequency. Because a video amplifier is sensitive to the physical layout, to the lead capacitance, and to compensating adjustments, the success of a particular circuit depends in a large measure on the skill and experience used in the performance tests and adjustments.

12.10 INTEGRATED-CIRCUIT VIDEO AMPLIFIERS (REF. 3)

A high-gain, multistage video amplifier may be designed by cascading low-gain stages that use discrete components and local feedback. Direct-coupled pairs require fewer components, particularly coupling capacitors, and have a higher loop gain, which makes the feedback more effective. For bandwidths up to 10 MHz the discrete construction offers high power output, low noise, and the high signal voltage required for driving a TV or radar picture tube. Above 10 MHz a designer is forced to use low-power transistors, and at these frequencies the capacitance and inductance of leads make it difficult to use feedback over more than a single stage.

Above 10 MHz an integrated circuit offers the advantages of a circuit with exceedingly small lead lengths and with high f_T values in the transistors. One such monolithic wide-band video amplifier (Sylvania SA 20) has a voltage gain of 10 with a 3-dB cutoff at 100 MHz. This amplifier uses three direct-coupled stages in a *CE-CE-CC* connection with overall feedback from the output to the first emitter. The amplifier is intended to use nearly 40-dB feedback, which improves the linearity and maintains a high input impedance. The circuit of the video amplifier, with minor omissions for simplification, is shown in Fig. 12.11.

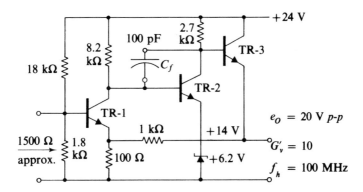

Figure 12.11. Integrated-circuit video amplifier

With the first emitter bypassed to ground, the amplifier has an open-loop gain $G_v = 1000$, which may be shaped in the high-frequency cutoff region by local feedback in the second stage. The collector-to-base feedback capacitor C_f may be adjusted to shape either the closed-loop gain or the transient response. In this way the frequency response may be maximally flat or caused to peak, as desired. With a larger value of the capacitor to shape the transient response, the amplifier has pulse rise and fall times of less than 10 nsec without overshoot.

12.11 MOS-FET VIDEO AMPLIFIER

A low-power video amplifier that uses a single-gate MOS-FET is illustrated in Fig. 12.12. This amplifier offers a simple circuit that does not require compensation, but the attainable gain is relatively low for reasons given in the MOS-FET chapter. The voltage gain with resistance coupling is usually less than 10. With shielded components the upper cutoff frequency is at 10 MHz, and compensation should not be necessary unless the amplifier has two

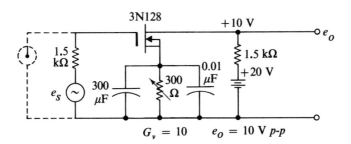

Figure 12.12. MOS-FET single-stage video amplifier

or more stages. The maximum output voltage is 10 V *p-p*. The amplifier may be used with supply voltages down to 10 V by making a small adjustment of the bias resistor. The voltage gain is reduced to 4 by removing the source bypass capacitor.

12.12 COMPENSATION TECHNIQUES (REFS. 2, 9)

The amplifiers used in cable or radio transmission and reception degrade the response characteristics of TV and sound systems. The response may be improved by circuit elements in the receiver that compensate for a frequency loss in another part of the transmission system. For an explanation of low-frequency compensation consider the two-stage amplifier shown in Fig. 12.13 and assume that the input capacitor introduces a loss and phase shift that are to be offset by compensation. The compensating components are the parallel-connected C_2 and R_5 that are in series with the base resistor R_2. Assume that the collector resistor and coupling capacitor may be neglected so that the load seen by the first-stage transistor TR-1 is the compensating network. If the resistance R_5 is large compared with R_2, the load seen by TR-1 is R_2 in series with C_2, and compensation is achieved by making the product $R_2 C_2$ equal to $R_1 C_1$. With this adjustment the low-frequency loss and phase shift in C_1 are exactly offset by the gain increase in the first stage. If R_5 can be made very large compared with R_2, the compensation removes both the loss and the phase shift produced by the input capacitor.

In a practical situation R_5 cannot be made very large compared with R_2, and, for example, R_5 may equal R_2. In this case the capacitor C_2 is selected by trial to achieve at least an approximate compensation, as shown in Fig. 12.14.

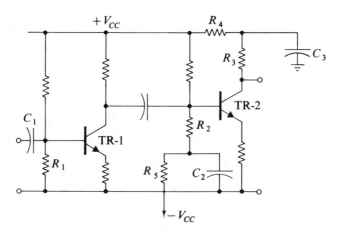

Figure 12.13. Amplifier with low-frequency compensation

Sec. 12.12 Compensation Techniques

In a similar manner the compensating components R_4 and C_3 may be used in series with the load resistor R_3 of the second transistor TR-2. Compensation by adjusting the product R_3C_3 may be used to offset a loss and phase shift that are external to the amplifier or in the load on TR-2.

The simplest means of increasing the high-frequency response or of increasing the bandwidth of an RC amplifier is by the addition of a small inductor in series with the base resistor or the collector load resistor, as shown in Fig. 12.15. This arrangement, called **shunt peaking**, decreases the transient rise time but may cause an overshoot, as indicated in Fig. 12.16. The fastest transient response without overshoot is obtained by increasing the high-frequency sine-wave gain about 2 dB at the 3-dB cutoff. By keeping the response at high frequencies, always slightly below the response at the lower frequencies, there is no overshoot. If the frequency response is only 1 dB greater at the high frequencies than at lower frequencies, as shown in Fig. 12.17, the rise time is slightly shorter but the resulting overshoot and transient ringing are generally undesirable.

The high-frequency performance of a video amplifier can be improved further when two capacitors can be separated by a series inductor, as shown in Fig. 12.18. With series peaking the capacitors act separately and the transient rise time may be reduced to about two-thirds the time with shunt peaking alone. Design values for similar circuits may be found in the vacuum-tube literature, but a trial value of the inductance may be obtained by making the inductive reactance equal to the reactance of the average value of the two capacitors at the desired cutoff frequency. Series peaking is most effective when the two capacitors are approximately equal.

The response characteristics of a video amplifier are greatly influenced by temperature, by the supply voltage, and by changes of stray capacitance caused by a relocation of lead wires. The high-frequency response and

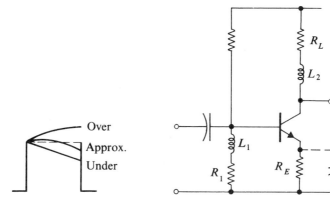

Figure 12.14. Wave forms with low-frequency compensation

Figure 12.15. Amplifier with high-frequency compensation

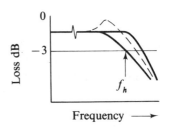

Figure 12.16. Pulse rise-time and overshoot

Figure 12.17. Video high-frequency response

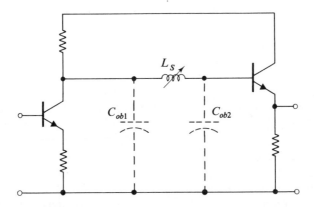

Figure 12.18. Amplifier with series compensation

compensation are particularly sensitive to the change of a circuit component. The effect of measurement probes and tools should be considered during measurements and adjustments.

Where speech and music are concerned, the audible sounds are faithfully reproduced when a transmission system has a flat frequency response throughout the audio spectrum. The response characteristics at the lower and upper edges of the band are relatively unimportant as long as the response falls off without resonant peaks. If peaks occur a transient sound excites ringing, and the peaks are usually eliminated by removing the positive feedback that is their cause. Compensation in an audio system is usually effected by the base and treble tone controls in the audio amplifier.

13 Diode Circuits and Applications

The semiconductor diode is perhaps the most versatile of all electronic devices. When we think of diodes, we usually think of their most common use as ac rectifiers or as detectors for converting high-frequency signals to an audio signal or a voltage reading. The improvement of diodes and the development of devices with a great variety of characteristics have given the diode a surprising versatility and many new applications.

Diodes may be used for steering or switching signals in response to the signal polarity or as a voltage-controlled gate that steers the signal, as in a synchronous detector. They are used for wave shaping, clipping, and noise control. At radio frequencies diodes are used for frequency multiplication, mixing, modulation, and detecting. They are used also as variable capacitors, voltage-controlled attenuators, and signal or noise sources.

Tunnel diodes are used as microwave amplifiers, oscillators, and detectors. Varactor diodes are used as voltage-variable capacitors for tuning high-frequeney circuits and for frequency multiplication. The PIN diode is used as a voltage-variable resistor in high-frequency attenuators and switches. At microwave and millimeter frequencies the point contact diode is indispensable as a detector and mixer.

The circuits in this chapter are mainly concerned with the applications of diodes as switches, wave shapers, modulators, and transient suppressors. The microwave applications are not included because they require different techniques and components than are available to most readers of transistor literature. The circuits for radio-frequency applications include a detector, RF voltmeter, modulator, mixer, and varactor. Examples of diode switches include standby-power switches, polarity and meter protectors, spark suppressors, AF and RF circuit switches, attenuators, clippers, limiters, clamps, and gates.

13.1 DIODE SWITCHES

A silicon diode is for practical purposes an open circuit when it is reverse-biased; when forward-biased, the diode's resistance is in the tens or hundreds of ohms and the voltage drop is 0.6 V. For these reasons a diode may be used as a switch that is controlled by a bias current. One-half volt is sufficient to reverse-bias or forward-bias a small diode, and the switching power required may be only a few milliwatts. Surprisingly, the power loss in a diode at rated current is not more than twice the loss in a typical power switch after half the normal operating life, but a diode is free from wear, contamination, noise, and vibration problems. The power loss in a switch with new silver contacts is only one-fifth the loss in a diode, but the reliability of a diode exceeds that of any mechanical switch.

The generator for charging an automobile battery must have a voltage-sensitive cutout to keep the battery from discharging through the generator when the generator voltage is less than the battery voltage. The relay cutouts used in automobile charging systems often cause a burned-out generator when the contacts stick or are closed accidentally. An inexpensive 50-A diode, connected as shown in Fig. 13.1, may be used to protect the generator from reverse currents, and the diode is simpler and more reliable than a cutout. Moreover, the diode protects the generator if the generator polarity is reversed, as sometimes happens.

A diode may be used to lower the power input to ac-operated heaters, soldering irons, incandescent lamps, and similar devices. As illustrated in Fig. 13.2, a diode opens the power line every other half cycle, and the power input to a resistance load is reduced to one-half the full wave input. In many applications a diode may be used in place of a voltage-variable transformer and is the simplest way of reducing the brilliance of decorative lamps or of extending the life of a lamp used in continuous service. However, in 120-V line service the diode should be carefully enclosed and insulated to prevent shock and fire hazards.

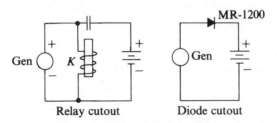

Figure 13.1. Generator cutout circuits

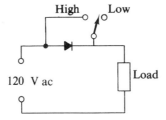

Figure 13.2. Diode power reducer

With a half-wave input to a resistance load the power input is reduced to one-half, the rms voltage is 71 per cent of the line voltage, and the dc voltage is 45 per cent of the line voltage. When the line voltage is 120 V, these voltages are 85 V rms and 54 V dc, respectively. The fact that the half-wave signal has a dc component and a high harmonic content is unimportant with resistance loads. However, with half-wave loads a small transformer may overheat and emit considerable audio-frequency noise. A diode cannot be used on the primary side of a transformer or with a highly reactive load.

Because a diode conducts only half time with a resistance load and the forward-voltage drop is 1.2 V at rated current, the power rating of the diode need be only $\frac{1}{2}(1.2/120)$ times the power rating of the 120-V load. A 1-A diode is suitable if the load does not exceed 100 W, and the power loss and the heat that must be dissipated are negligible. Lamps, motors, and loads that may change with the voltage input ordinarily take about three-fourths the full-wave power when supplied through a diode.

13.2 DIODE SWITCHES IN DC POWER SUPPLIES

Diodes are useful as automatic switches for protecting amplifiers and power inverters from damage when a power supply is connected accidentally with the polarity reversed. Most low-power circuits may be protected by connecting a diode in series with the power supply, as shown in Fig. 13.3(a). If the voltage loss in the diode is objectionable, protection may be assured by using a relay with a low-power diode in series with the coil, as shown in Fig. 13.3(b). In either protection circuit a polarity warning lamp may be connected

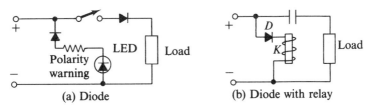

Figure 13.3. Polarity protection circuits

on the switch side of the power supply, as shown in Fig. 13.3(a). The high reliability of LED lamps makes them especially desirable as warning devices.

Crowbar circuits are used in power supplies to open a fuse when the supply voltage or current exceeds a certain maximum, or when the supply is connected with the wrong polarity. A diode crowbar may be used to protect a meter or sensitive instrument from reverse polarity, as shown in Fig. 13.4(a). An SCR may be used to protect a load from an over-voltage, as shown in Fig. 13.4(b), and a Triac may be used to open a fuse with either an over-voltage of correct polarity or with power applied with a reversed polarity. A transistor may be used to open a fuse when a load current exceeds a given maximum by using the circuit shown in Fig. 13.4(c). The series resistor is selected to provide a voltage drop of 0.6 V at the maximum current, and the collector resistor is selected to limit the collector current to about twice the fuse rating. The series diode is required to protect the transistor if the polarity of the supply is reversible.

Fig. 13.5 shows how a transistor may be used with a remotely located, low-power switch to connect a load to a power supply. For loads up to 1 A a magnetic reed switch may be used if rated to carry 3 per cent of the load current. For applications requiring a low-noise power switch, a large capacitor

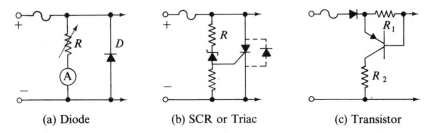

(a) Diode   (b) SCR or Triac   (c) Transistor

Figure 13.4. Crowbar circuits

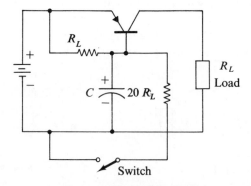

Figure 13.5. Low-noise switch

may be connected across the high-resistance switch circuit. Depending on the characteristics of the load and the size of the base capacitor, noise produced by the switch is reduced by a factor of 10 to 1000.

A circuit that uses a diode to connect a standby power supply is shown in Fig. 13.6. The standby battery drives the load through the diode D_1 whenever the regular supply voltage drops more than 0.6 V below the standby voltage. The simplicity and reliability of a diode for this purpose cannot be surpassed, and the 0.6-V change of the supply voltage may be corrected by a voltage regulator between the voltage supply and the load. The transistor shown in the circuit suggests one way of connecting a pilot lamp to show when the standby power is in use. However, the transistor should be rated to carry a base current equal to the load current, and the diode D_1 may not be necessary. The diode D_2 is used to protect the standby battery from a short circuit in the main power supply.

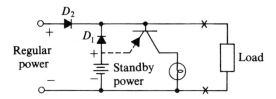

Figure 13.6. Standby power switch

13.3 SPARK AND NOISE SUPPRESSORS

An important application of diodes prevents arcing of switch contacts and suppresses the high-frequency transients produced when an electromagnet or a relay is released. A diode connected across a dc motor or an electric clutch, as shown in Fig. 13.7, protects the switch from arcs and prevents voltage breakdown in the magnet coil. Especially where high reliability is desired, circuits with small switches and relay contacts should be protected from the switching arcs that produce cumulative contact erosion. Magnetic reed switches are particularly susceptible to sticking and should be protected, even though a spark is not observed. Thus, a routine use of diodes across relays may prevent contact problems and increase reliability.

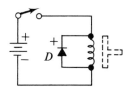

Figure 13.7. Diode spark suppressor

Because a diode suppresses transients picked up by sensitive circuits, an inexpensive low-voltage diode installed across all relays is both time-saving and good design practice. Opening a relay circuit produces a voltage across the coil that is approximately 100 times the operating voltage. A diode, connected as shown in Fig. 13.8(a), provides a conducting path for the coil current so that the stored magnetic energy is dissipated slowly in the internal resistance. The diode limits the terminal voltage to about one-half volt and must carry a peak current equal to the operating current of the relay.

Because the diode tends to maintain current in the coil after the switch is opened, a relay with a simple diode suppressor may operate too slowly. Faster operation may be obtained by connecting a resistor in series with the diode, as shown in Fig. 13.8(b). The value of the resistor R may approximately equal the coil resistance, and a capacitor across the coil may help reduce the RF interference. For some applications shorter release times may be obtained by shunting the coil with a resistor that is ten times the coil resistance and by using a diode and resistor to keep the transient voltage off the control circuit, as shown in Fig. 13.8(c).

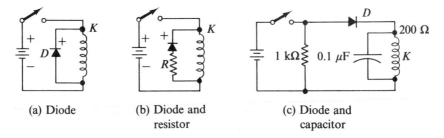

(a) Diode   (b) Diode and resistor   (c) Diode and capacitor

Figure 13.8. Transient-pickup suppressors

13.4 DIODE METER PROTECTION

Sensitive meters may be protected from minor overloads by a pair of diodes connected back-to-back, as shown in Fig. 13.9. A sensitive meter usually reads full scale when the terminal voltage is between 0.1 and 0.25 V, and the silicon diodes present a high resistance and negligible error up to full-scale deflection of the meter. If a high voltage is accidentally applied to the meter, the forward-

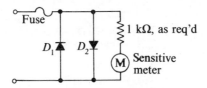

Figure 13.9. Diode meter protector

biased diode limits the terminal voltage to about 0.6 V, which is a low multiple of the full-scale voltage, even for input currents up to several amperes.

A meter may be protected from a sustained overload by the diodes and fuse shown in Fig. 13.9. A sensitive meter cannot be protected by a fuse alone, but the shunt diodes increase the overload current exponentially and open the fuse in time to protect both the meter and the diodes. In effect the diodes operate as fast and reliable voltage-sensitive switches. A resistor in series with the meter may be used to limit the overload deflection. If the meter is used as a bridge balance detector, the resistor may be chosen to give the meter an extended range with the diodes reducing the full-scale sensitivity. The internal resistance of meters with a full-scale sensitivity of 100 μA, or better, usually exceeds 2 kΩ, and a suitable resistor has a value that reduces the full-scale voltage approximately 0.3 V.

13.5 PREVENTING TRANSISTOR BREAKDOWN

Power transistors and sensitive RF transistors are sometimes protected by connecting a diode across the base-emitter junction, as shown in Fig. 13.10(a). When the transistor is transformer-coupled to a driver stage, large signals may reverse-bias the base-emitter diode and produce destructive voltages or annoying transients. Small transformers have enough resistance in the winding to protect the transistor, but a suppressor diode should be used whenever the transformer has a large core and a low winding resistance. Diodes may be needed across both transistors in a Darlington stage.

Sometimes the diode D is replaced by a low-voltage Zener diode to limit both the forward and reverse voltages. In limiting the forward drive, the transistor may be protected from most collector-current overloads. RF transistors have a high internal base resistance and are easily protected from reverse voltages by a parallel-connected, high-frequency diode. These transistors tend to be self-protecting with a forward bias, but they may be protected by connecting 2 high-speed switching diodes in series and across the base-emitter junction.

The output transistors in a power amplifier are sometimes protected by a diode connected between the emitter and collector. In a class-B stage, as in Fig. 13.10(b), the diodes protect the emitter junction from a reverse bias that may exist when the load has a high reactance or a short circuit. Switching amplifiers and inverters may use Zener diodes, as in Fig. 13.10(c), to prevent reversal of the collector-emitter voltage and to limit transient switching spikes that increase the peak collector-emitter voltage and impair the transistor reliability or cause breakdown failures. Damper diodes are used in TV horizontal-deflection amplifiers, as shown in Fig. 13.10(d). These diodes may have a 10-A, 400-V rating and are connected to prevent reversal of the capacitor voltage and yoke current in the resonant LC load. These diodes are

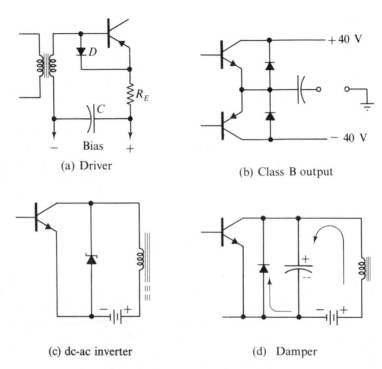

Figure 13.10. Diodes in power amplifiers

designed to damp oscillations and appear in some circuits as if used to prevent reversal of the collector-emitter voltage.

13.6 DIODE SWITCHES IN LOW-LEVEL CIRCUITS (REF. 3)

Diodes may be used to switch low-level signals with the circuit illustrated in Fig. 13.11(a). Depending on the polarity of the control voltage, the signal may be switched to one or the other channel, and the control current may be transmitted from a remote location via a single-wire, high-resistance circuit. With a control voltage of only 1 V, 150 μA, the loss in the diode switch is negligible, and the ON channel transmits undistorted signals up to 100 mV peak input voltage. Increasing the control voltage to 12 V permits the transmission of signal peaks up to 1 V. In audio-frequency applications the loss in the open diode is high, and crossover into the open channel is attenuated at least 60 dB up to 20 kHz. With the control voltage OFF both channels are open for signals up to 300 mV peak.

Figure 13.11(b) shows how a 3-step diode attenuator may be controlled using a single wire and a 3-position switch. The attenuator loss is determined

Sec. 13.6 Diode Switches in Low-Level Circuits 233

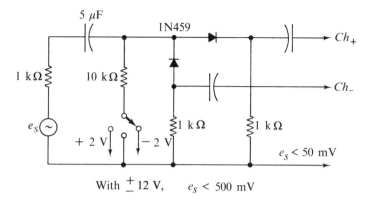

(a) 2-channel switch

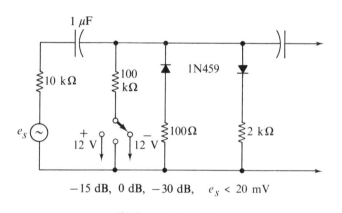

(b) 2-step attenuator

Figure 13.11. Diode AF switches

by the resistors in series with the diodes. With a 30-dB loss the peak input signal may be as high as 30 mV, and with 15-dB loss the peak signal may be 300 mV. With the control voltage OFF the attenuation is negligible and the channel is distortion-free for peak signals up to 150 mV.

Two examples of adjustable diode attenuators for audio applications are shown in Fig. 13.12. The loss in the single-diode attenuator varies inversely with the applied voltage up to at least 24 V. The loss is 20 dB with 4 V, and the peak input signal should be less than 20 mV. The loss in the 2-diode attenuator varies inversely with the square of the applied voltage. When the control voltage is 1.4 V, the loss is 20 dB. With 12 V, the loss is 40 dB. A resistor may be used in series with one of the diodes to limit the maximum loss. The maximum undistorted input signal is the same as for a single diode attenuator.

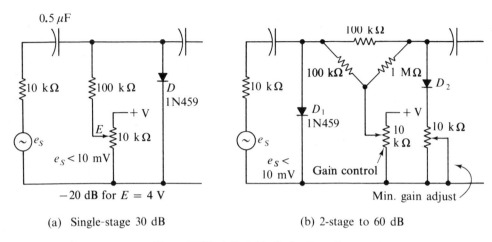

Figure 13.12. Adjustable diode attenuators

Diode switches are used in RF circuits when the capacitance of a manual switch is high enough to affect the circuit adversely. As illustrated in Fig. 13.13(a), the crystals in a transmitter may be selected by a remotely located switch that applies a few milliamperes dc to forward-bias a diode in the crystal ground return. The diode and the RF choke are placed near the crystal to minimize capacitance and RF feedback in the control circuit. Any high-frequency diode may be used, and PIN diodes are available that are especially designed for RF switching. In ac switching circuits the bias current must be approximately twice the peak ac signal current.

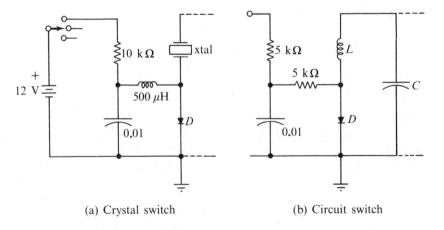

Figure 13.13. Diode RF switches

13.7 DIODE STEERING CIRCUITS

Diode steering circuits are found in communication and indicator applications where a signal is steered to one or another load, depending on the polarity of a dc carrier. The full-wave diode bridge circuit shown in Fig. 13.14 uses diodes as polarity-sensitive switches to ensure that an input pulse is always connected to an amplifier with the same polarity. The diode bridge may be used where a circuit is to be triggered regardless of the incoming polarity, or where a meter is to be operated without concern for the polarity of the input voltage.

The rectifier-type voltmeter shown in Fig. 13.15 illustrates the use of a diode bridge to steer an ac signal to a dc load so that the signal is inverted every other half cycle. The voltage to be measured is applied through the range resistor R to opposite corners of the bridge. The dc meter reads the average of the rectified current, but the meter is usually calibrated to read the rms value of a sine wave. As a result, such instruments do not read correctly when used with nonsinusoidal waveforms. Because each diode requires about one-half of one volt for switching the diode, steering and rectifier circuits are not effective at low signal levels. The lowest range on a rectifier-type voltmeter is usually about 2 V, and this range requires a specially calibrated scale. A filter capacitor is not usually required because the meter responds too slowly to follow the ac components of the rectified signal. However, with input frequencies below 5 Hz an electrolytic capacitor may be used to increase the response time and the accuracy. With high frequencies, a small capacitor may be needed to bypass the meter reactance. With a capacitor connected across the meter the response is proportional to the peak voltage of the input signal.

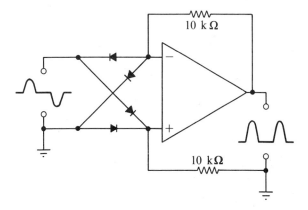

Figure 13.14. Pulse steering bridge

For measuring low-level ac signals the sensitivity of a rectifier-type ac meter may be increased by using the voltage-doubling rectifier circuit shown in Fig. 13.16. The meter reading is doubled, provided the signal source has an internal impedance that is low compared with the impedance looking into the ac terminals. The capacitors increase the meter-response time and give a 40 per cent higher meter deflection than when a capacitor is used across the full-wave rectifier load.

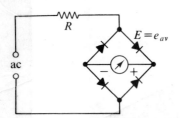
Figure 13.15. Rectifier-type voltmeter

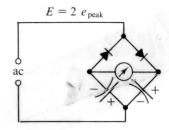
Figure 13.16. Voltage-doubling rectifier

13.8 DIODE MODULATORS AND CHOPPERS

A modulator is a circuit for changing the frequency of a signal for transmission over a communication channel or by radio. Modulators generally use a diode or a nonlinear amplifier that multiplies one signal by the instantaneous amplitude of another, the carrier. The high-frequency modulated signal is transmitted and then demodulated by a diode detector or a synchronous detector. A chopper is a form of modulator often used in instruments to increase the frequency of low-frequency signals and thereby facilitate their amplification in an ac amplifier.

Low-level signals may be switched, modulated, or chopped in a ring modulator that uses diodes driven by a separate ac power source. A diode circuit for modulating or chopping low-frequency signals is shown in Fig. 13.17(a). Depending on the polarity of the switching signal e_s, the diodes are biased to conduct the input e_I through D_1 and D_2, or inverted through D_3 and D_4. In this way the switching signal converts a low-frequency input to a modulated high-frequency signal that may be amplified in an ac amplifier or transmitted in a high-frequency system, as in a radio. However, the modulated signal does not include a carrier at the switching frequency, and demodulation of a suppressed carrier wave cannot be affected by simple rectification because the filtered output signal follows the wave envelope and has double the frequency of the information signal. Simple rectification may be used if the waveform of the demodulated signal is unimportant, as in a control system.

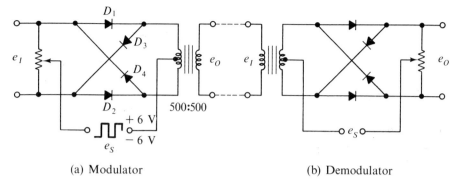

(a) Modulator (b) Demodulator

Figure 13.17. Diode modulator and demodulator

When the waveform of the low-frequency signal is required, as in the transmission of speech, the modulated signal may be demodulated in a synchronous detector by using a switching signal with the same frequency as that used in the modulator. The simplicity of the diode modulators makes them particularly useful in a complex communication system, and IC demodulators are available that lock in on the carrier frequency when the signal frequency and waveform are needed.

A difficulty with diode modulators is that the diodes and circuit components must be closely matched. Even when the circuit is carefully adjusted, low-level output signals are concealed by a signal produced by the unbalanced diode voltage drops. As the diode voltage varies with time and temperature, the resulting output appears as a low-frequency drift or noise. This difficulty is greatly reduced by using FETs in place of the diodes.

The circuit of a synchronous detector that uses FETs in the modulator is shown in Fig. 13.18. The bridge has 2 n-channel and 2 p-channel FETs that are switched ON alternately by the switching signal. The FETs eliminate problems caused by the diode voltage drops because the channel conductance is controlled by the gate without superimposing an appreciable current in the bridge. With a 10-V peak gate signal the residual unbalance is less than 10 mV rms, and the input signals are usable from 10 mV to 10 V.

A synchronous demodulator may be used as a high-Q narrow-band system for detecting signals deeply buried in noise. If the output is observed with a long-period dc meter, the meter responds only to signals that have a frequency close to the switching frequency. A typical 50 μA meter used as the output indicator responds to a 3-Hz bandwidth. If the switching frequency is 6 kHz, the meter responds to signals in a 3-Hz bandwidth between 3-dB frequencies, and the system has an effective Q of 2000. With higher reference frequencies and a longer period meter the Q of the system may be increased several orders of magnitude. By synchronous, or phase-locked, detection, signals deeply buried in noise are retrieved and measured with accuracy.

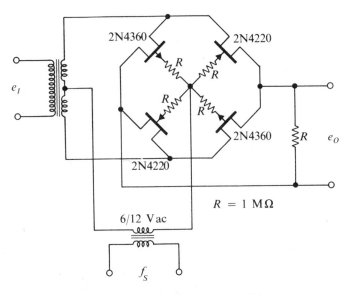

Figure 13.18. Synchronous FET modulator

13.9 DIODE AM MODULATOR

A single diode may be used as a simple, inexpensive AM modulator, as shown in Fig. 13.19. The carrier and the modulating signal are impressed on the diode through summing resistors, and the modulated signal is observed as a voltage across a tuned circuit that is in series with the diode. With a carrier voltage between 5 and 50 V the carrier is modulated up to 100 per cent with a low-frequency signal amplitude that is 15 per cent of the carrier amplitude.

The diode modulator may be used up to radio frequencies with silicon signal diodes and up to the microwave frequencies with hot carrier diodes. The Q of the tuned circuit is relatively unimportant as long as the -3-dB bandwidth is adequate for the side-band frequencies. The inductor and

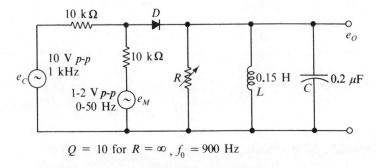

$Q = 10$ for $R = \infty$, $f_0 = 900$ Hz

Figure 13.19. Diode AM modulator

capacitor should have a reactance at the resonant frequency of one-tenth to one-half the resistance of a series resistor.

The circuit is adjusted with the carrier applied and the modulating input grounded. The resonant circuit is tuned for a maximum carrier output and then detuned to a lower frequency where the carrier is reduced to half maximum. A modulating signal that is 12 to 15 per cent of the carrier voltage should give nearly 100 per cent modulation with low distortion.

The Q of the tuned circuit may be reduced for a wider bandwidth by shunting the tuned circuit with a resistor. For a reasonable efficiency with low Qs the series resistors should approximately equal the shunt resistor. In the circuit shown in Fig. 13.19 all 3 resistors may be 3 kΩ and $Q = 1$. With a high-frequency carrier, the resistors should be reduced and may be as low as 50 Ω with a 50 MHz carrier.

13.10 DIODE DETECTORS (REFS. 2, 3)

Detectors use the nonlinear characteristics of diodes to convert amplitude changes of a high-frequency signal to a low-frequency signal that represents the information originally modulated on the high-frequency signal. The circuit for a diode detector is essentially that of any single-diode rectifier with the output driving a low-frequency amplifier. The circuit of a detector used in a broadcast-band radio is usually similar to that shown in Fig. 13.20(a). The diode may be almost any germanium diode. If the detector is used above 50 MHz, the diode may be a point-contact silicon or a Schottky *hot-carrier* diode. The detector used in a TV video system may have a selected point-contact germanium diode that is followed by an IF filter, as shown in Fig. 13.20(b).

A square-law detector is used for measuring low-level, high-frequency signals and operates in the region where the output dc is proportional to the square of the input signal. The output of a square-law detector is measured with a sensitive microammeter, or is amplified in a dc amplifier. The circuit of the square-law detector is shown in Fig. 13.21. The diode may be a point-

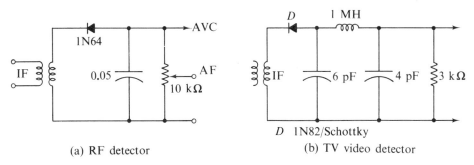

(a) RF detector

(b) TV video detector

Figure 13.20. Diode RF detectors

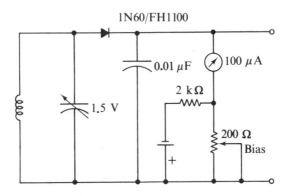

Figure 13.21. Square-law detector

contact germanium device with a low reverse-current rating. However, in high-frequency applications the Schottky diodes are more uniform, equally sensitive, and less subject to damage by an overload. Both types may be used for power measurements from 0.001 μW to 1 μW, and both require a small forward bias for a maximum sensitivity.

13.11 UHF DIODE VOLTMETER

The application of diodes for voltage measurements at high frequencies is illustrated by the circuit in Fig. 13.22. This circuit uses a Schottky diode to charge the 0.01 disc capacitor. The meter is generally calibrated to read rms voltage, but the deflection is actually proportional to the peak-to-peak signal voltage. The variable resistor at the left of the meter is necessary to provide a circuit for dc current and provides a convenient means for calibrating the meter reading. The input capacitor protects the diode from dc voltages, provided the capacitor is not too large.

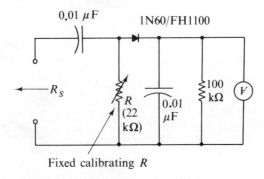

Figure 13.22. High-frequency ac voltmeter

The voltmeter calibration is independent of frequency until the diode capacitance shunts the nonlinear junction and reduces the ac voltage across the diode. Thus, the cutoff frequency varies with the impedance R_S and with the input voltage. When the input signal is sinusoidal and a few volts rms, the rectification efficiency is reduced to 50 per cent when the reactance of the diode capacitance equals the source impedance R_S. For example, the capacitance of the Schottky diode is given as 1.2 pF, so the calculated cutoff frequency of the voltmeter is 14 MHz. The observed cutoff frequency is about 25 MHz, and with a point-contact germanium diode the cutoff frequency is 40 MHz.

The cutoff frequency of the voltmeter may be increased approximately 1 decade for each factor-of-10 reduction of the source impedance. When R_S is 1 kΩ the cutoff frequency with a germanium diode is about 400 MHz. With lower impedances the rectifier provides useful indications at microwave frequencies. Decreasing the source impedance forces the diode to rectify at a lower impedance level and reduces the effect of the capacitance, but increases the amount of RF power required to produce a given signal. Probes for high-frequency TV servicing use similar circuits and with germanium diodes give accuracies of ± 2 dB (26 per cent) up to 250 MHz.

The range of input voltages that is handled best by a crystal voltmeter is between 0.1 and 2 V. When the applied voltage exceeds 20 V rms, an attenuator must be inserted to reduce the voltage actually applied to the crystal. Capacitance attenuators are preferred at UHF and GHz frequencies.

Because appreciable power is required to operate high-frequency voltmeters, the diode that converts the ac signal to dc tends to clip the signal on one or both peaks and loads the circuit being tested. The load on the circuit is determined in part by the dc load; hence, the dc meter should have a high sensitivity.

13.12 DIODE MIXERS

A mixer is a circuit for converting an incoming high-frequency signal to a lower frequency signal. Diode mixers are often used at UHF and microwave frequencies where the construction of an amplifier is either impractical or uneconomical. As illustrated in Fig. 13.23, frequency conversion is accomplished by driving a diode simultaneously with the high-frequency input signal and the local oscillator (LO). The circuit of the diode mixer is similar to that of the AM modulator shown in Fig. 13.19, except that the input signals are introduced by a tuned circuit in place of the resistors in the AM modulator. The series-resonant circuit L_1 and C_1 is tuned at the input frequency, and the parallel tuned circuit L_2 and C_2 is tuned at the IF frequency. In a TV tuner a 470-MHz UHF input signal and a 515-MHz LO signal are mixed to produce a 45-MHz signal that can be amplified in the IF amplifier of the TV receiver.

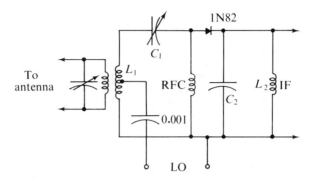

Figure 13.23. Diode mixer (RF)

When selected for low capacitance, the point-contact diodes are preferred for UHF and microwave applications. Germanium diodes are inexpensive and able to rectify low-voltage signals but have a high reverse leakage and are temperature sensitive. Silicon diodes are less sensitive but operate better at high temperatures. The Schottky diodes are more uniform and more predictable than point-contact devices, but the latter still excell in applications at frequencies above about 1 GHz. With further improvement the Schottky diodes may supplant the point-contact devices.

The Schottky diode is a highly efficient mixer. It has less conversion loss and thereby produces lower mixer noise than for any other diode. Its lower impedance level provides better impedance matching, and the high efficiency at high signal levels reduces adjacent channel interference and distortion problems.

13.13 VARACTOR DIODE TUNING (REF. 8)

The capacitance of a reversed-biased diode may be reduced by increasing the applied voltage. Silicon tuning diodes provide a capacitance variation of 3 to 1, and in a few low-frequency types a capacitance variation of 10 to 1. The voltage-variable capacitors offer advantages in automatic frequency control and for amplifier tuning from 1 MHz through microwave frequencies. Varactor diodes are available with nominal capacitance values from 5 to 200 pF with 2- or 4-V reverse bias. The diodes are used with reverse voltages between 1 and 20 V, and the capacitance varies inversely with the square root of the voltage.

The advantages of the diodes for tuning include simplified circuits and construction, a higher reactance range, faster response times (vital in frequency sweeping circuits), and the ability to control a multiplicity of devices simultaneously.

Sec. 13.14   Diode Clipping Circuits   **243**

Voltage-variable capacitors, VVC, are commonly used as the voltage-variable tuning element in tuned amplifiers, for automatic frequency control of oscillators, and for frequency modulating an oscillator. In these applications the amplitude of the ac signal across the varactor is relatively small, so the capacitance is only a function of the dc, or low-frequency, control voltage. Tuning varactors are designed to have high Q values and large capacitance variations with voltage.

A circuit used for tuning an L-C resonant circuit is illustrated in Fig. 13.24. The tuned circuit is formed of L_1, which is tuned by C_1 and the diode capacitance C. The diode voltage is varied by the potentiometer R_1. The resistor R_2 isolates the diode from other diodes that may be connected to the potentiometer, and the resistor has a large value to prevent loading of the tuned circuit. There is a negligible current in R_2. The capacitor C_2 is used to block dc from the tuned circuit and may be a small capacitor if the capacitance change in the diode is more than needed for frequency control.

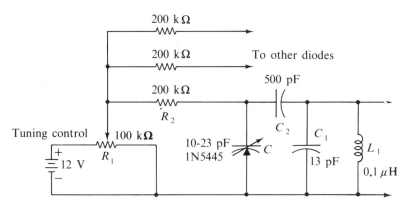

Figure 13.24. Varactor tuning 88 to 108 MHz

13.14 DIODE CLIPPING CIRCUITS

In electronic circuits extensive use is made of various kinds of nonsinusoidal waves, such as square and sawtooth waves, gated and chopped waves, and pulses. Many waveforms may be produced by using diodes to modify a sine wave by clipping, clamping, or gating.

Clipping flattens a portion of a wave by limiting the peak amplitude to an arbitrary level that is lower than the amplitude of the original signal. Clippers are classified as ***peak clippers***, ***base clippers***, or ***slicers***, depending on the way they operate on the wave.

A ***peak limiter*** operates by preventing either the positive or the negative, or both, amplitudes of an input wave from exceeding a value set by the clipper. Examples of positive peak clippers are shown in Fig. 13.25. The common

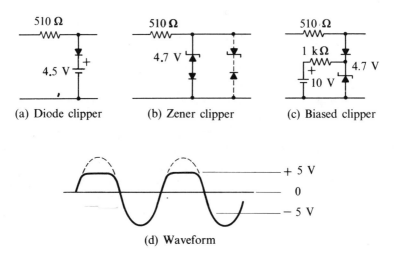

Figure 13.25. Diode clipping circuits

diode clipper shown in Fig. 13.25(a) uses a battery to determine the positive peak output signal. If, as shown, the battery has 4.5 V and the forward voltage drop of the diode is about 0.5 V, then the diode is effectively open for input signals less than $+4.5$ V or for negative signals, but the diode closes to prevent a positive-going output from exceeding $+5$ V. A Zener diode limiter that produces essentially the same result is illustrated in Fig. 13.25(b). Here, because two diodes are operating in series, the peaks are not so sharply limited as with a diode limiter, but the Zener limiter does not require a battery. Another form of clipper, illustrated in Fig. 13.25(c), has the advantage of the diode clipper—i.e., more abrupt limiting—and uses a Zener diode to set the limiting voltage. A current for forward-biasing the Zener diode may be supplied from any convenient supply (10 V in the example), and the clipping level is easily adjusted by changing the Zener diode.

All three clippers illustrated in Fig. 13.25 produce essentially the same output waveform as shown in Fig. 13-25(d). Because only one diode is involved, the clippers shown in Figs. 13.25(a) and (c) produce more abrupt clipping than does a circuit having several diodes in series. Furthermore, Zener diodes do not have a sharp breakdown in the very low voltage range. Therefore, a diode clipper should be used for clipping peaks in the 1 V to 4 V range.

A Zener diode used without the series diode makes a positive and negative peak clipper that limits the peaks of one polarity at the voltage rating of the Zener diode and limits the peaks of the opposite polarity to about 0.5 V. This clipper, as shown in Fig. 13.26(a) and 13.26(b), is used in triggering applications where a limited trigger signal of fixed polarity must be assured.

Both positive and negative peaks are clipped when a second Zener diode and a plain diode are connected, both in reverse, as indicated by the dotted connection in Fig. 13.25(b). By similar modifications of the circuits shown in Fig. 13.25(a) or 13.25(c), these also become double peak clippers. However, as shown in Fig. 13.26(c), only two back-to-back Zener diodes are required to produce a double peak clipper. When the input voltage is positive, the Zener diode Z_1 is forward biased and conducts, but diode Z_2 is open for any voltage below the breakdown value. When the input voltage exceeds the breakdown voltage of Z_2 plus about 0.5 V in Z_1, the output signal is clipped. With negative input voltages the diodes exchange roles and the negative peaks are determined by the breakdown value of Z_1. The waveform of a double peak clipper is shown in Fig. 13.26(d).

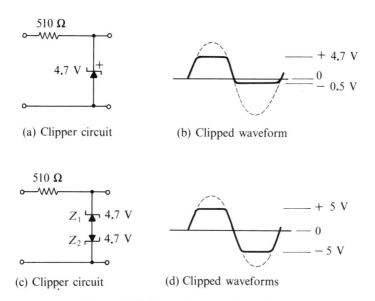

Figure 13.26. Zener clippers and waveforms

13.15 BASE CLIPPER (NOISE SUPPRESSOR)

In many circuits, particularly for counting and switching, a circuit must be made immune to small noise signals and permitted to operate only when the input signal exceeds a particular value. A *noise suppressor*, also called a *base clipper*, is shown in Fig. 13.27(a). With input signals less than $+4.5$ V or negative, the diode is forward-biased, and the output terminal is held at the 4.5 V base line, as shown in Fig. 13.27(b). Positive signals having amplitudes exceeding the battery voltage open the diode and are passed without attenuation. Noise signals less than 4.5 V cannot open the diode and are attenuated. If the series resistor is 10 kΩ, the noise is attenuated by at least 40 db.

A noise suppression circuit using the Zener diode, as shown in Fig. 13.27(c), transmits the positive peaks but has the advantage of leaving the reference base line at 0 V. Signals too low to effect Zener diode breakdown are attenuated. The Zener circuit is useful in direct coupling an amplifier to a trigger circuit, provided the Zener diode rating is high enough to block the Q-point voltage of the amplifier with an additional allowance to cover the noise signals. For some applications the Zener diode is preferable to a capacitor because the latter stores charge and, in discharging, makes the base line vary. With the Zener circuit, the positive-going signal is easily limited by making the supply voltage only a few volts above the Zener diode rating.

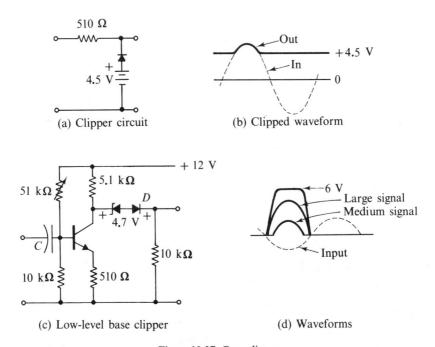

Figure 13.27. Base clippers

13.16 LOW-LEVEL LIMITER AND NOISE SUPPRESSOR

For low-level signal applications a pair of back-to-back silicon diodes make a simple and effective peak limiter. Using ordinary low-power silicon diodes, the circuit in Fig. 13.28(a) limits the output signal to about 0.4 to 0.6 peak-to-peak volts, depending on the input level. The limited signal has rounded shoulders, but the low-level signals are transmitted with negligible distortion.

When the diodes are connected as a series element as in Fig. 13.28(c), the circuit makes an effective noise suppressor that attenuates signals less than 0.1 V rms by about 40-dB and passes signals exceeding 1 V rms with less than a 3-dB loss. With large input signals the diode switching transients are practically negligible.

By substituting gold-doped or Schottky diodes in Fig. 13.28(a) and 13.28(c), and by increasing the resistor to 100 kΩ, the clipping level and the noise suppression level of the diode circuits may be lowered about 5 times. Even lower switching levels can be reached by using the back-to-back diodes as elements in the feedback path of an operational amplifier.

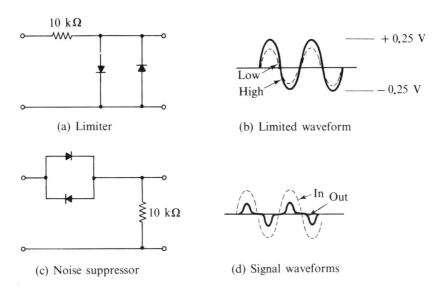

Figure 13.28. Diode limiter and noise suppressor

13.17 IGNITION NOISE SUPPRESSORS

The diode noise suppressors shown in Fig. 13.29 may be used to clip the pulse-type noise spikes that are produced in a radio receiver by automobile ignition. The clippers are designed to be connected across a point in the receiver where the desired signal is less than 1 V p-p. The clipper acts to short circuit spikes that exceed the voltage drop of a forward-biased diode.

The noise suppressor shown in Fig. 13.29(a) may be connected across an IF transformer where the signal frequencies are relatively high, and the suppressor is usually most effective in the last IF stage. The noise suppressor shown in Fig. 13.29(b) may be used across the audio gain control provided an AGC keeps the signal peaks below 1 V. The best place to connect a noise

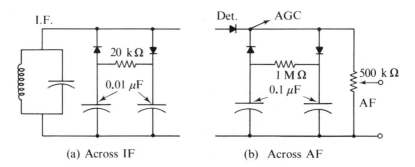

(a) Across IF   (b) Across AF

Figure 13.29. Impulse noise limiter

suppressor should be determined by a trial under service conditions. Ignition noise suppressors are usually more effective when the engine is operating at cruising speeds than when idling.

13.18 DIODE CLAMP

A clamping circuit generally uses a diode as a switch to hold one extreme of a wave at fixed potential or ground. When, as in a television amplifier, the dc component of the picture signal must be reinserted, the clamp is called a *dc restorer*. As shown in Fig. 13.30(a), a clamping circuit is formed by combining a diode and a long time constant RC coupling circuit. The diode rectifies—i.e., clips—a small part of the signal peaks and charges the capacitor until the output signal moves over and is clamped to the negative peaks. The drive circuit should have a low resistance to enable easy charging of the capacitor. The shunt resistor should have a high resistance so that the capacitor does not require frequent recharging.

In effect, the diode connects the output terminal to ground whenever the input is negative, and the long time constant maintains the clamped state without requiring too much distortion of the signal peaks. By inserting a battery between the diode and ground [at X in Fig. 13.30(a)], the peaks can be

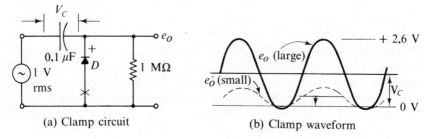

(a) Clamp circuit   (b) Clamp waveform

Figure 13.30. Diode clamp

clamped to a voltage off ground. By reversing the diode polarity, the positive peaks of the signal are clamped.

Almost any resistance-coupled amplifier tends to clamp large signal peaks by signal rectification. In a similar manner signal rectification in the collector or drain circuit may produce a dc component that shifts the operating Q-point. Clamping circuits are used whenever the bias point of a circuit must be made to follow the amplitude of an input signal. Often the grid or gate diode of an amplifier is used as the clamp diode, and the fact that a circuit is expected to clamp the input signal may not be obvious.

13.19 DIODE GATE

The circuit in Fig. 13.31(a) shows one form of a diode gate. The signal channel is from A to B, and the gate control signal is applied at G. The resistors in the channel ensure that the channel impedance is high. The resistors are not required if the source and load have equivalent or higher resistance. The gate diode D_1 is closed when the control at G is $+12$ V, and any signals that are negative or positive up to $+12$ V are inhibited. Depending on the impedance of the gating source, the channel loss is about 40 dB. When the gate signal is OFF, the diode is open for any positive-going pulse or half-wave, as shown in Fig. 13.31(b). Full-wave signals are transmitted by applying -12 V at the gate G.

The diode gate may be operated by switching the gate from $+12$ V to -12 V, or the gate may be closed by returning the 500 Ω resistor through $+12$ V to ground. The gate is then opened by applying a -24 V pulse at G. Diode gates have the disadvantage that the turn-ON gate signal enters the signal channel as an unwanted transient. Transistor gates have the advantage that the ON voltage from the collector to ground is several times smaller than the ON voltage of a diode gate.

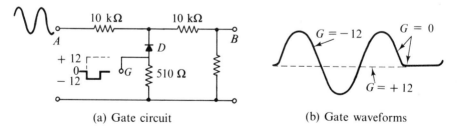

(a) Gate circuit   (b) Gate waveforms

Figure 13.31. Diode gates

14 Oscillators and Inverters

An oscillator is an active circuit that produces an ac output signal by converting dc power to ac power. Oscillators are widely used as convenient sources of sine-wave, square-wave, ramp, triangular, or pulse signals for testing, control, and frequency conversion. Sinusoidal oscillators are used in signalling over audio-frequency circuits and for the transmission of information or control signals, as in telemetry. Square-wave and pulse signals are used for counting, timing, and control, as in a digital computer. Ramp signals are used for converting analog to digital signals, as in digital voltmeters.

Sine waves are produced in linear circuits that have positive feedback. Square-wave, saw-tooth, and pulse signals are produced by multivibrators. Pulse signals are produced in blocking oscillators or multivibrators. An inverter is a multivibrator-like oscillator that is designed to generate ac power that may be converted back to dc, as in a converter, for use at a voltage level different from the input voltage.

14.1 SINUSOIDAL OSCILLATORS

The circuit requirements that make a good sine-wave oscillator are easily described by referring to the Franklin oscillator shown in Fig. 14.1. A Franklin oscillator has a tuned LC circuit that determines the oscillator frequency and a 2-stage amplifier that provides positive feedback and just enough power gain to offset the power lost in the tuned circuit. Each stage of the amplifier inverts the signal, so the feedback returned to the input is in phase with and augments an assumed input signal. However, the feedback

Sec. 14.1 Sinusoidal Oscillators 251

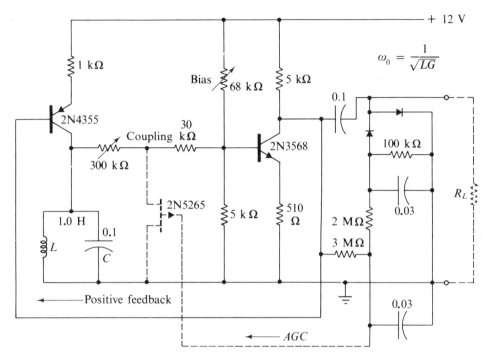

Figure 14.1. Franklin oscillator with AGC, 500 Hz

signal is in phase at the resonant frequency of the tuned circuit only if the spurious phase shifts in the circuit are small. The circuit shown in Fig. 14.1 has no capacitors in the feedback loop, and at audio frequencies we may neglect the component and transistor capacities.

The power lost in the tuned circuit is offset when the voltage gain around the feedback loop is 1, but a gain slightly more than 1 is required to ensure easy starting and a reasonably stable signal amplitude. On the other hand, some means is usually required to reduce the gain and prevent the signal from becoming too large and distorted. In the oscillator shown in Fig. 14.1 the loop gain is controlled by sampling the loop signal and converting it to dc, which is used to vary the drain-to-source resistance of the FET. The dc is polarized to reduce the drain resistance when the output signal is too large.

A sine-wave oscillator is biased to operate with linear class-A Q-points. In the Franklin oscillator the coupling resistor is so large that the Q-point of the second stage is not affected by the Q-point of the first stage. A second requirement for a quality oscillator is that the frequency should be independent of the supply voltage and the external load. Frequency stability is obtained by using a high-Q tuned circuit with a rapidly changing phase characteristic at the operating frequency. Thus, a small variation of the loop phase shift is offset by only a small change of the operating frequency. Moreover, the frequency-determining network should have a stable phase characteristic and be independent of frequency. Most tuned circuit oscillators that use a single transistor have reactance characteristics that make the oscillator frequency considerably different from the calculated L-C resonant frequency. The frequencies shown in the oscillator circuits are the measured and not the L-C frequency.

When an oscillator is operated with a fixed-loop gain, the signal starts at a small amplitude and builds up until the signal amplitude is limited by overload in some part of the amplifier. When low distortion is required, some means is used to reduce the loop gain before the signal amplitude is high enough to cause excess waveform distortion. The oscillator shown in Fig. 14.1 uses an FET as a voltage-controlled variable resistor connected across the second-stage input. When the signal amplitude is about 2-V peak, the rectified voltage is filtered and applied to make the FET gate voltage less positive, which decreases the drain-to-source resistance and reduces the loop gain. The automatic gain control (AGC) has the additional advantage that the output amplitude is relatively independent of frequency.

A requirement for a stable oscillator is that the external load should not affect the adjustments of the oscillator. In the example under discussion we can assume that this requirement is met by readjusting the coupling resistor if the load is changed, but for better frequency and amplitude stability the load may be separated by adding a buffer stage. The Franklin oscillator was used as a laboratory signal source before the development of the phase-shift and Wien bridge oscillators.

14.2 RESONANT CIRCUIT OSCILLATORS
(REFS. 2, 3, 4, 9)

Resonant circuit, sine-wave oscillators are commonly used in radio receivers and transmitters. The resonant circuit for determining the operating frequency is usually a simple coil and capacitor or an equivalent electromechanical resonator. With care in the design, such oscillators can be made to have a good power efficiency, a reasonable waveform, and very good frequency stability. The best known resonant circuit oscillators are named after the radio pioneers Hartley, Colpitts, and Clapp. Positive feedback exists in these oscillators because the base and collector are connected to opposite ends of the tuned circuit. The emitter is connected to an intermediate point of the tuned circuit, and any one of the three connections may be grounded.

The well-known Hartley oscillator, shown in Fig. 14.2, uses a tapped coil and a single tuning capacitor that couples the collector and base circuits. The emitter and the tap on the coil are at ground potential. The collector and the base circuits share a single resonant circuit, and the amount of the base drive is determined by the position of the tap on the coil. The coupling between the two parts of the coil is not important. A good waveform is obtained if the Q of the coil exceeds about 20 and the output loading is relatively light. An output signal for low-impedance loads is obtained by coupling a third coil, and, for high-impedance loads, by capacitor-coupling to the collector.

The *Hartley oscillator* in Fig. 14.2 is designed to operate in the supersonic frequency range from about 8 kHz to 64 kHz. Higher or lower frequencies may be obtained by changing the reactance components so that the impedance values at the desired frequency are the same as those illustrated in the figure. For example, the Hartley oscillator operates at a 10 times higher frequency by reducing all capacitors by a factor of 10 and by reducing the inductance of the coil by a factor of 10. The Q of the coil should be at least 10 at the lowest operating frequency for good stability and waveform.

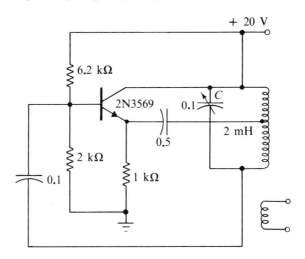

Figure 14.2. Hartly oscillator, 8 kHz to 64 kHz

A *Colpitts oscillator*, shown in Fig. 14.3, is similar to the Hartley oscillator except that the emitter is tapped to a point on the capacitor side of the resonant circuit. Usually the capacitor at the base end of the tuned circuit is fixed, and the oscillator frequency is changed by varying the capacitor on the collector end. Because the fixed base capacitor tends to reduce the base drive at high frequencies, the waveform of the Colpitts oscillator tends to improve at high frequencies. For such reasons, the choice between the Hartley or the Colpitts oscillator depends on minor differences between their performance

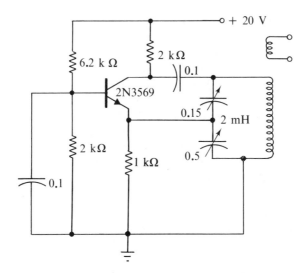

Figure 14.3. Colpitts oscillator, 8 kHz to 64 kHz

characteristics or on the ease with which one or the other oscillator can be adjusted to meet the requirements of a design.

The circuits of the Hartley and Colpitts oscillators are proportioned to exhibit the similarities between them. Each of the oscillators has emitter and base resistors that fix the operating Q-point in essentially the same way bias resistors are used in an amplifier. Each oscillator has three points coupled to the transistor and each has the collector and the base at opposite ends of the tuned circuit. Both oscillators can be tuned up to an 8 times higher frequency by reducing the tuning capacitor from the value indicated. However, the effect of the resistors in these oscillator circuits is to make the actual operating frequency as much as twice the frequency calculated from the equivalent L and C values of the resonant circuit.

The **Clapp oscillator** is used where a high degree of frequency stability is required in a variable-frequency oscillator. The Clapp oscillator in Fig. 14.4 is similar to the Colpitts oscillator except that the inductance is replaced by a series resonant circuit. By making the collector and base coupling capacitors large—i.e., low reactances—the transistor is effectively removed from the frequency determining elements which are the series resonant LC circuit. The Clapp oscillator operates as a variable frequency oscillator over only a limited frequency range. Hence, except for fixed frequency applications or in frequency standards where complicated range switches are acceptable, it is not often used. In the circuit shown a 2 to 1 increase in frequency is produced by a 4 to 1 reduction of the tuning capacitor. A greater frequency change requires a change of the coil and the coupling capacitors.

Sec. 14.3   Tuned Emitter and Tuned Collector Oscillators   255

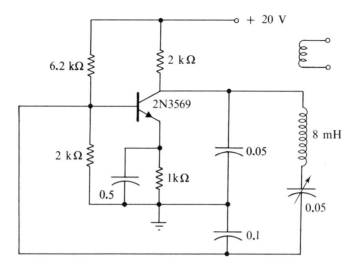

Figure 14.4. Clapp oscillator, 12 kHz to 25 kHz

14.3 TUNED EMITTER AND TUNED COLLECTOR OSCILLATORS

One circuit used in broadcast receivers is the tuned emitter oscillator shown in Fig. 14.5. The circuit found in a receiver may have additional circuit elements, either to permit using the oscillator for a dual purpose or to provide coupling. In the tuned emitter oscillator a resonant circuit is coupled to the emitter, the base is bypassed to ground, and regenerative feedback is obtained

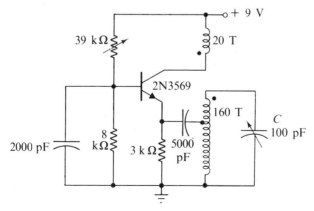

Figure 14.5. Tuned-emitter oscillator: typical broadcast, 0.5 to 1.5 MHz

by closely coupling a coil in the collector circuit to the tuned circuit coil. The two coils are phased as indicated by the dots on the coils in the figure. Output power is usually obtained by coupling a third coil. In this oscillator the transistor must supply the power demand of the load and enough power to the emitter to maintain the oscillations. Because the load is closely coupled to the frequency determining circuit, this type of oscillator has poor frequency stability and a poor waveform; however, a receiver load is small and constant, and high Q-factors are easily obtained at radio frequencies. The resistors shown in the circuit are selected to bias the transistor and to stabilize the operating point in much the same way that resistors are selected for a linear amplifier.

A capacitor feedback oscillator is shown in Fig. 14.6. Inductances are used in both the base and the collector, but they are not coupled, and the feedback signal is provided by the capacitor connected between the collector and base. Usually it is sufficient to tune either the collector or the base circuit, and the

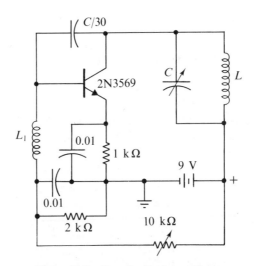

Figure 14.6. Tuned-collector oscillator

correct phase turnover occurs at a frequency slightly below the resonant frequency. If both circuits are tuned, the oscillator operates more readily, and the internal capacitance of the transistor generally provides enough feedback. For these reasons the capacitor feedback oscillator is often used at high frequencies where the internal capacities of the transistor naturally become a part of the circuit. Similarly, almost any amplifier having a tuned base circuit and a tuned collector circuit tends to oscillate unless the stage voltage gain is limited enough to ensure stable operation.

14.4 TUNED-CIRCUIT POWER OSCILLATORS

A tuned-circuit oscillator required to deliver considerable power may be designed to operate class B or class C, especially at radio frequencies. For a fixed size transistor, class-B or class-C biasing gives a higher power efficiency and permits higher output powers. Single-sided—i.e., *single-ended*—class-C operation is used in radio frequency circuits because tuned circuits have a high Q at these frequencies, and the flywheel effect ensures an acceptable waveform. Push-pull class-C operation may be used where a better waveform or higher output powers are required.

A common difficulty in the design of class-C oscillators comes from the need to bias the oscillator approximately class A for starting and to provide some method for increasing the bias as the oscillations build up. A variable bias is usually obtained by rectifying a part of the oscillator signal and filtering the bias, using a large capacitor. In a simple oscillator the base-emitter diode may serve as the rectifier with the base coupling capacitor acting as the bias filter capacitor. Because the bias changes with the amplitude of the oscillator signal, the capacitor charge must also change. If the capacitor is too large, the charge changes too slowly, and the oscillator operates intermittently as a blocking oscillator. If the capacitor is too small, the oscillator will not start easily, or its waveform will show considerable second harmonic distortion.

Any oscillator must be over-driven enough to ensure starting under the most unfavorable operating conditions of supply voltage, temperature, etc. If an oscillator is required to have an excellent waveform also, the transistors are usually operated class A and the signal amplitude is limited by an automatic gain control (AGC). In a typical AGC circuit a fraction of the output signal is rectified, filtered, and used to control the resistance of an FET or transistor that acts to limit the loop gain of the oscillator. The AGC action not only ensures a good waveform but also holds the output amplitude independent of the operating frequency and improves the frequency stability.

Additional information concerning tuned oscillators is found in the descriptions of typical radio frequency circuits given in the transistor handbooks. The problems of oscillator design are much the same, whether transistor or tube driven, and considerable information concerning oscillator problems and oscillator design is found in the vacuum tube literature.

14.5 CRYSTAL OSCILLATORS (REFS. 3, 4.)

A crystal oscillator known as a *Pierce oscillator* is illustrated in Fig. 14.7(a). A quartz crystal is a very high Q resonant element that behaves as a series

258 Oscillators and Inverters Ch. 14

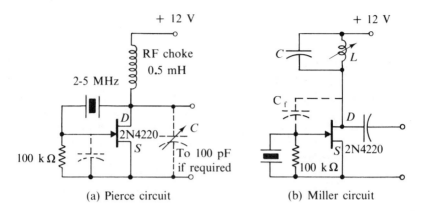

Figure 14.7. FET crystal oscillators

resonant circuit in parallel with a capacitor. Because the series resonant frequency and the parallel resonant frequency of the crystal are close together (within 1 per cent), the reactance of a crystal changes rapidly for a small change in frequency near resonance. For this reason and because a crystal is time and temperature stable, a quartz crystal is capable of holding an oscillator frequency to better than ± 0.01 per cent of the nominal frequency.

The use of an FET for the active element in a crystal oscillator has two advantages: first, the high impedance gate does not adversely load the crystal and lower the Q; second, the low drain voltage protects the crystal from damage by over-excitation or high-voltage breakdown. The Pierce oscillator requires a capacitive reactance across both the gate and the drain, and the crystal provides an inductive reactance at the operating frequency. This reactance π (pi) configuration is necessary to establish the 180° phase turnover. If the choke is inductive at the crystal frequency, a shunt capacitance across the drain may be needed, but the drain circuit is not tuned in the ordinary sense. Because capacities exist naturally across the gate and drain, the Pierce oscillator is generally preferred when the crystal is operated at its fundamental frequency.

The Miller oscillator shown in Fig. 14.7(b) has the crystal connected across the gate. The oscillator operates as a capacitance feedback oscillator, and both the gate and drain must have an inductive reactance. An inductive drain reactance is obtained by tuning the drain circuit at a frequency above the crystal frequency. The Miller oscillator has the advantage that the drain circuit can be tuned to an odd harmonic frequency either to operate the crystal at an overtone frequency or to select an odd harmonic of the crystal frequency.

14.6 PHASE-SHIFT OSCILLATORS (REF. 1)

Audio frequency oscillators are usually a form of a resistance-capacitance feedback oscillator. A sine-wave RC oscillator is generally a low distortion class-A amplifier, followed by an RC network that shifts the signal phase until the in-phase condition is satisfied. The RC network introduces a considerable power loss that must be made up in the amplifier, but the RC circuits avoid the use of inductances, which are unsatisfactory at audio frequencies.

The RC oscillators can be classified according to whether the amplifier has an odd or even number of stages. If the collector of a single-stage amplifier is returned to the base through a simple ladder network, the ladder must provide the 180° phase shift. The phase-shift oscillator of Fig. 14.8 accomplishes the phase turnover by providing a 60° phase shift in each section of the three identical RC sections. The resistance of the last section is provided in part by the input resistance of the transistor. With three identical sections and an external collector resistor equal to the other resistors, the 180° phase shift is obtained when the input-to-output current ratio is 56. To bring the loop current gain up to 1, the transistor must have a minimum β of 56. When the collector resistor value is between two and three times the value of the resistors in the ladder, a minimum current gain of 46 will suffice. Lower current gains can be used by adding a fourth section in the RC ladder and by decreasing the impedance level of each section progressing toward the base. With these changes and with the base section resistor shorted, the required current gain may be as low as 30.

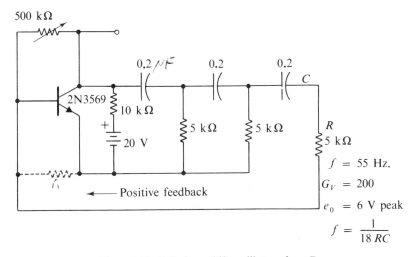

Figure 14.8. RC phase-shift oscillator, shunt R

For the RC oscillator shown in Fig. 14.8 the wave form is a sine wave, and the frequency is most stable when the circuit current gain is just sufficient to maintain the oscillations. A circuit current gain of 1 is established by adjusting the supply voltage or by shunting the base with a resistor that lowers the current gain. Best results are secured when the transistor current gain is not above the required minimum.

Phase-shift oscillators are constructed by using field-effect or insulated-gate devices. A three-section ladder of like sections requires a voltage gain of nearly 30, which is more than is obtainable from many FET devices. In the circuit illustrated in Fig. 14.9 the ladder has each section progressively at a three times higher impedance level. The 180° phase turnover is obtained in this ladder when the overall voltage loss is about 18. With the output of the ladder returned to the FET gate, oscillations are sustained if the FET has a voltage gain exceeding 18. By increasing the impedance step-up to 10 times, the loading of each section on the previous section is almost negligible and the circuit oscillates if the voltage gain exceeds 12. However, because the FET output impedance is not zero, as is usually assumed, the circuit must have about 3 dB more voltage gain than the theoretical minimum.

The frequency commonly given for a phase-shift oscillator is derived by assuming impractical values for the active element load resistor, and the calculated frequency may be in error by a factor of two. In round numbers, the frequency of a three-section shunt R oscillator is given by the equation:

$$f \cong \frac{1}{18RC} \tag{14.1}$$

and the frequency of a three-section shunt C oscillator is given by the equation:

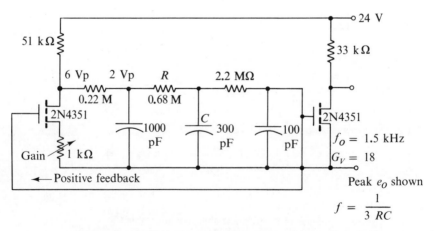

Figure 14.9. MOS-FET low-distortion phase oscillator, shunt C

$$f \cong \frac{1}{3RC} \tag{14.2}$$

where RC is the time constant of a typical section. With four sections, the frequency of the shunt R oscillator is a factor of 2 higher and the frequency of a shunt C oscillator is a factor of 2 lower. A required frequency is easily obtained by adjusting one or two of the resistors.

Phase-shift oscillators usually produce 5 to 10 per cent distortion. The oscillator of Fig. 14.9 is unusual because it will supply a very low distortion signal. The second FET, on the right side, is driven by the oscillator gate signal, which is quite free of harmonics because the ladder is a low-pass filter. The enhancement-mode MOS transistors eliminate the need for a series blocking capacitor and allow high resistances in the ladder.

In general, phase-shift oscillators are best adapted for essentially fixed frequency operation. The large capacitors required in transistor circuits have to be changed in steps. Small frequency changes can be made by adjusting one or more of the resistors, but changing the resistors changes the required current gain, and the waveform will be distorted at the extreme frequency adjustments. The important advantages of the phase-shift oscillators are their simplicity, frequency stability, and reasonably good waveform.

14.7 STEPPED LOW-FREQUENCY, PHASE-SHIFT OSCILLATOR

An advantage of the MOS-FET phase-shift oscillator is that the frequency may be varied by changing the series resistors, and within limits the resistors may be changed without changing the loop gain. This advantage greatly facilitates the construction of variable low-frequency oscillators. The frequency may be changed either in steps or a single variable resistor may be used in series with any of the resistors to produce up to a 10 per cent continuous frequency change.

The stepped oscillator shown in Fig. 14.10 has three capacitors connected to three separate sections of a 4-pole, 11-position switch. The taps of the switches connect to a double string of series-connected resistors. Each resistor in the series has a standard RTMA value that is approximately 1.5 times the resistor on its left. In effect, the switch moves the capacitors along the contacts between resistors and increases the frequency by a factor of 1.5 for each step that the switch moves to the left. With an 11-pole switch (which is readily available) a 10-to-1 frequency change is obtained in 6 steps.

The resistor and capacitor values shown in Fig. 14.10 cover a frequency range from 1-to-10 Hz. Increasing or decreasing the capacitors by a factor of 10 moves the range of frequencies down or up by a factor of 10. In this way the oscillator may be designed to operate at frequencies from 0.01 Hz to 10 kHz.

262 Oscillators and Inverters Ch. 14

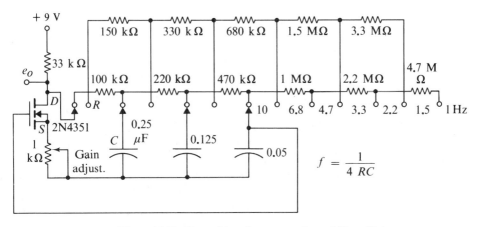

Figure 14.10. Stepped low-frequency, phase-shift oscillator

A variable source feedback resistor may be used to adjust the loop gain and minimize the distortion. With a 9-V supply the drain output signal is 1-to-2 V peak-to-peak. For a higher output and less distortion the oscillator may be constructed with a direct-coupled output stage, as shown in Fig. 14.9.

14.8 TWIN-*T*, R-C OSCILLATORS

The twin-*T* oscillator in Fig. 14.11 is an interesting example of a simple transistor oscillator. A twin-*T* or a bridge-*T* null filter behaves like a high-*Q* series-resonant circuit across the feedback path. At the frequency $f = 1/2\pi RC$ the twin-*T* has a sharp null, or balance. By decreasing the shunt resistor slightly, the twin-*T* transmits a small in-phase signal that is rapidly changing in phase at the null frequency. A collector feedback stage offers the advantage of a high loop gain that makes the oscillator frequency almost independent of the external circuit and of the transistor parameters.

The collector feedback oscillator shown in Fig. 14.11 is remarkable for its high-stability of frequency and low-distortion waveform. The oscillator uses the collector-to-base bias resistor for one side of a twin-*T* null network, and in the other side a variable resistor is used to adjust the output waveform and the loop gain. The second transistor, a CC buffer stage, is required only when the oscillator needs isolation from a variable or low-impedance load.

The oscillator operates on a wide range of the supply voltage, and the frequency is relatively independent of normal changes of this voltage. With a fixed waveform adjustment the frequency stability is better than 0.1 per cent. The frequency is 1/5 RC, a little higher than the exact null frequency. The oscillator is useful for frequency comparisons and as a timing oscillator and may be used in testing or checking audio and high-fidelity systems. The twin-*T* oscillator has about 5 per cent distortion, and better waveform may be

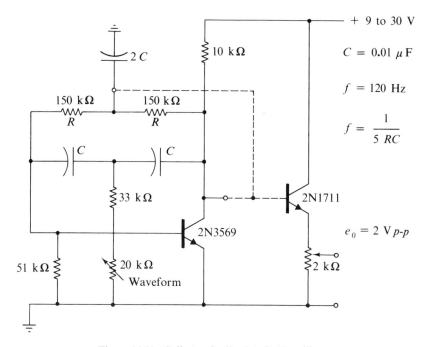

Figure 14.11. Collector-feedback twin-T oscillator

obtained with reduced output by connecting the output stage to the shunt capacitor 2C.

14.9 WIEN BRIDGE OSCILLATOR

A more constant operating frequency can be obtained in RC oscillators by employing a feedback network that shifts the phase rapidly for a given small change in the signal frequency. Simple RC circuits that shift the phase rapidly are high-loss networks like the twin-T and the Wien bridge. A high loss in the feedback network makes a two-stage amplifier necessary. A two-stage amplifier provides an output signal that is in phase with the input. Hence, the feedback network should have a 0° net phase shift at the operating frequency. A Wien bridge and a twin-T meet the conditions required of the feedback network, and, when the circuit loss is high, both circuits provide a rapid rate of phase shift with frequency. Either circuit makes an RC oscillator with excellent frequency stability. The twin-T oscillator may be changed to a 2-stage high-loss oscillator with the in-phase feedback obtained by adjusting the shunt resistor R slightly larger than the null value.

The Wien bridge oscillator is used as an audio frequency test oscillator and has been extensively described in the vacuum tube literature. When the FET became available, a number of oscillator circuits were published that were essentially a tube-type circuit with FETs replacing the vacuum tubes. The Wien bridge oscillator shown in Fig. 14.12 uses a high-gain integrated circuit differential amplifier with the + and − inputs connected to the output of the bridge. The output of the amplifier is connected to the bridge input. Since bias current for one side of the amplifier flows through the high resistance bridge arm R, the amplifier must be a type requiring low bias currents.

The left side of the bridge is made up of the series RC arm and the parallel RC arm. The right side of the bridge is made up of a fixed resistor and a voltage-controlled FET resistor. The voltage that controls the FET resistor is obtained by rectifying and filtering part of the output signal peaks. With no output signal the FET resistance is low, the bridge is off balance, and the signal input to the differential amplifier is largest on the + (in-phase) side of the amplifier. When the output signal builds up to the desired level, the AGC signal biases the FET toward pinchoff and reduces the differential input signal.

The frequency of the Wien bridge oscillator is the frequency at which the phase angle of the arm Z_1 is the same as the phase angle of the arm Z_2, namely,

$$f = \frac{1}{2\pi RC} \tag{14.3}$$

Because the differential amplifier has a high gain, the bridge is almost balanced at this frequency, and the signal at the input to the amplifier is in phase with the signal input to the bridge. The AGC circuit maintains just enough unbalance in the bridge to keep the loop gain slightly more than 1.

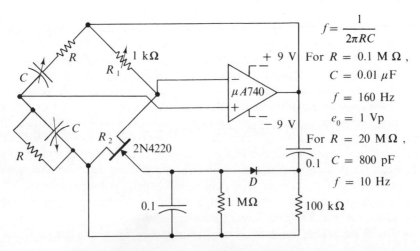

Figure 14.12. Wien bridge oscillator with FET gain control

The frequency is varied over a 10-to-1 frequency range by varying the ganged variable air capacitors C. The frequency decade is changed by a 10-to-1 change of the resistors R. The resistor R_1 is used only to make an initial adjustment of the bridge operating point and is about twice the resistance of the FET.

14.10 NONSINUSOIDAL OSCILLATORS

Oscillators that produce a nonsinusoidal waveform have many important applications. Usually they are a form of *relaxation oscillator*—an oscillator having an excess of loop gain so that the active element is driven well into cutoff. For a part of the cycle, energy is stored rapidly in one of the reactive elements of the circuit, and at a later part of the cycle this energy is discharged more slowly. These circuits are characterized by a tightly closed high-gain loop, one or more energy storage elements, and an amplifier that operates as an ON-OFF switch. Examples of these oscillators are known as multivibrators, blocking oscillators, and relaxation oscillators.

14.11 MULTIVIBRATORS (REF. 7)

A multivibrator, Fig. 14.13, is a 2-stage amplifier that is connected so that there is an excess of positive ac feedback. The collector-to-base resistors hold the static Q-points in the active region and supply a part of the positive feedback. The capacitors that cross-couple the collectors to the bases increase the loop gain and force the transistors to switch alternately ON and OFF between temporarily stable states. The astable, or free-running, multivibrator generates a square wave that is used in computers as a clock-timing frequency,

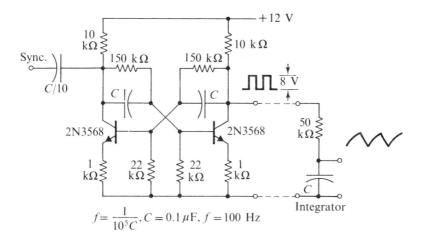

Figure 14.13. Multivibrator, synchronized

266 Oscillators and Inverters Ch. 14

and the square wave is easily converted to a triangular wave by driving an RC integrator or converted to a pulse by an RC differentiator.

The switching period of the multivibrator shown in Fig. 14.13 is determined mainly by the resistors and capacitors in the circuit but is partly dependent on the transistors and the supply voltage. With fixed values of the resistors and equal feedback capacitors the frequency varies inversely with C, $f = k/C$, and the value of the constant k is best determined experimentally. The frequency of the multivibrator in Fig. 14.13 is 100 Hz when $C = 0.1\ \mu F$ and may be increased to 50 kHz by changing C to 100 pF. Higher frequencies require RF transistors and high-frequency techniques to minimize circuit capacitance. The multivibrator is a low-current type that is easily synchronized to a standard frequency by capacitor-coupling a synchronizing pulse to a collector.

The multivibrator shown in Fig. 14.14 uses a collector feedback amplifier that may be operated with supply voltages from 1.5- to 40-V. This simplified multivibrator makes a very satisfactory signal source for applications requiring a dependable constant-frequency square wave. With a 20-V supply and the 8-V regulator the frequency stability is approximately 0.05 per cent per volt depending on the regulation of the Zener diode. The stability with temperature is approximately -0.3 per cent per degree centigrade. The spacing between alternate pulses may be equalized by connecting a high-resistance bias resistor as shown in the figure.

The transistor Q-points should fall in the active region if the capacitors are removed. When the circuit resistances are symmetrical, the Q-points may

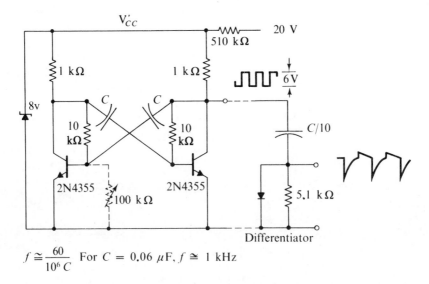

$f \cong \dfrac{60}{10^6\,C}$ For $C = 0.06\ \mu F$, $f \cong 1$ kHz

Figure 14.14. Multivibrator, simplified, frequency-stable

be observed by connecting both bases or collectors. This Q-point check provides an easy way of examining the Q-point and locating design problems. Sometimes, to ensure easy starting, a diode is connected in series with one of the feedback resistors so that the circuit is asymmetrical during the starting instant, and, when operating, the symmetry is restored. Figure 14.14 shows an RC differentiator that may be coupled to a collector to convert the output square wave to a pulse with a duration of approximately 100 μs. An RC integrator may be used, as shown in Fig. 14.13, to produce a triangular output signal. The waveform of the square wave may be improved by a clipper or by transmission through an inverter.

14.12 RINGING CONVERTER OR FLYBACK OSCILLATOR

Low-power, high-voltage, dc-to-dc converters often use a single transistor in a circuit known as a *ringing converter* or *flyback oscillator*. The circuit has the advantages of unusual simplicity and of producing a high-voltage output with efficiencies of 50 to 75 per cent. Flyback oscillators are used in TVs to convert 150-V dc to more than 20 kV, and the components and circuits for a 20-kV supply may be obtained from a radio supply firm. Low-voltage converters are easily designed and built to convert 6- or 12-V battery power to the voltages needed for timing and photo-flash lamps.

Ringing converters operate by storing energy in a transformer core when the transistor is ON, and the energy is transferred to the load when the transistor is turned OFF. By making the energy-release time shorter then 10 per cent of the storage time, the output voltage may be several times larger than the input voltage multiplied by the transformer-turns ratio.

The circuit of a ringing converter is shown in Fig. 14.15, and the transformer connections are indicated by polarity dots. During the conduction period the transistor is driven ON as a switch by regenerative feedback of the base winding, and the collector current rises linearly until the core saturates. When the base drive is suddenly decreased by core saturation, the transformer voltage reverses and drives the transistor into cutoff. The sudden change of collector current produces an "inductive kick," and the load capacitor is charged by the relatively high and reversed secondary voltage.

The size and shape of the ferrite core is relatively unimportant as long as there is adequate window area for the windings. For a 10-to-20 kHz switching frequency the primary needs 10-to-20 turns per dc volt input. The primary-to-secondary turns ratio depends on the required load voltage and the inductive boost. The transformer shown in Fig. 14.15 has a 20-to-1 turns ratio that makes the secondary voltage 240 V. Thus, if we assume the no-load voltage is 2000 V, the no-load boost is a factor of 8, and the boost that can be attained in practice is the least predictable factor in a given design. The attainable boost is most easily found and controlled by experiment. The

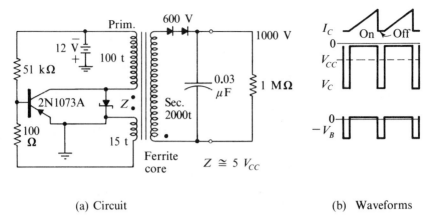

(a) Circuit (b) Waveforms

Figure 14.15. Ringing or flyback oscillator

loaded output voltage shows that the collector breakdown voltage must be at least 50 V, and the Zener diode rating should be about 60 V.

By operating at a high audio-frequency a flyback oscillator may use a relatively small transformer. Consequently, the core from a TV transformer may be easily rewound for low-voltage applications. A high efficiency is obtained by using a square-loop ferrite material and a bifilar winding to minimize leakage inductance. The voltage boost produced by the inductive kick is maximized by winding high-Q coils and by using a high-frequency switching transistor. The transistor ON time is determined by the supply voltage, the number of turns on the primary winding, and the core. The OFF time is determined by the secondary capacitor and the load.

The transistor in a ringing converter must withstand somewhat more than the inductive voltage rise of the primary winding and is ordinarily protected by a Zener diode if the load is removable, as in an ignition system. This problem of absorbing the no-load energy tends to limit the power capacity of these converters. The efficiency of a ringing converter varies with the load resistance. It the load reduces the output voltage to one-half the no-load voltage, the efficiency is between 50 and 70 per cent.

14.13 BLOCKING OSCILLATORS

A blocking oscillator is similar to a flyback oscillator except that the transistor is ON for short periods and is OFF for longer times. In this oscillator the output is a pulse that is shaped by the transformer and circuit components. Because the transistor is OFF for a long period, a high-energy pulse may be delivered to the output without exceeding the power capability of a small transistor. Blocking oscillators are used as free-running or synchronized

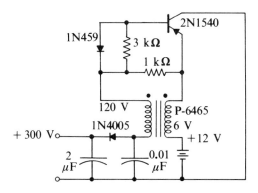

Figure 14.16. Blocking-oscillator converter

oscillators, as sources of steep-wavefront pulses, and as simple dc-to-ac inverters.

The blocking oscillator shown in Fig. 14.16 is designed to supply power in the 200- to 500-V range with a 12-V battery as the power supply. This converter uses a 2-winding, 60-Hz, filament transformer, and the base drive is obtained by connecting the base to the low-voltage end of the high-voltage winding. The resistors and diode in the base circuit reduce the base drive when the transistor is OFF and increase the drive when the transistor is turned ON. The no-load output voltage is 400 V, and with a 33-kΩ load the voltage is reduced to 200 V, the power output is 1.2 W, and the efficiency is 70 per cent. For intermittent timing and flash-lamp applications an inverter of this type recovers so quickly that the output voltage is almost the no-load voltage. An inverter with a separate base winding has better switching and may be designed to give a higher inductive boost than is obtained with the 2-winding transformer.

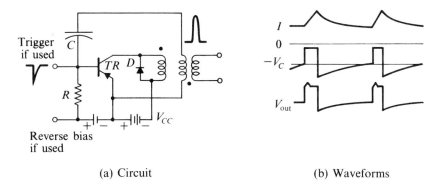

(a) Circuit   (b) Waveforms

Figure 14.17. Blocking oscillator

270 Oscillators and Inverters   Ch. 14

The circuit of a blocking oscillator is shown in Fig. 14.17, and typical waveforms are given in Fig. 14.17(b). The transformer coupling the collector to the base has a step-down ratio of about 5-to-1, and the series capacitor is charged rapidly and discharged slowly. The output pulse is generated by the rapid turn-ON of the collector current, and the rise-time of the pulse is limited only by the leakage inductance of the transformer. The top of the pulse is flattened by collector-current saturation, and the pulse terminates when the transformer core is saturated. The time-constant RC of the base resistor and capacitor determines the quiescent time between pulses. Notice that the diode D protects the collector junction from the high voltages produced by the transformer at the end of the pulse.

Because the pulse depends in a complicated way on the circuit components, pulse transformers are designed to give a particular pulse shape at a particular operating frequency, and the intended circuit is usually supplied with the transformer. Transformers are readily available that produce pulse-rise times of 0.01 μs to several μs, and the pulse duration is usually 20 times the rise-time. Blocking-oscillator converter transformers are made that supply 1 kV with a 20-μA to 50-μA load current. Pulse transformers are made for SCR trigger circuits, but they are easily built by winding 20 turns or more for each winding on a one-half or one inch length of ferrite rod cut from a loopstick antenna. A triggered blocking oscillator may be used as the pulse source, and transformers designed for SCR applications are offered in electronic supply catalogs.

14.14 VARIABLE-FREQUENCY PULSE OSCILLATOR (REF. 7)

A pulse with a rise-time less than a microsecond is sometimes needed for operating counters and switching circuits. The variable-frequency unijunction oscillator shown in Fig. 14.18 makes an easily assembled and inexpensive pulse source. The circuit uses a programmable unijunction transistor as the

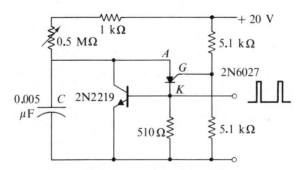

Figure 14.18. Variable-frequency pulse oscillator

pulse oscillator and a transistor to discharge the timing capacitor and shorten the pulse. With the components shown the output pulse is 0.5 V with a rise time less than 0.5 μs and a pulse duration of less than 10 μs. The pulse-repetition frequency may be varied from 300 Hz to 100 kHz with a single value of the timing capacitor and the 0.5 MΩ variable resistor.

14.15 INVERTERS (REFS. 3, 4, 5)

An *inverter* is a multivibrator or switching amplifier that is used to change dc to ac power. Inverters are either self-excited or are driven by a separate square-wave oscillator to increase the efficiency and make the output frequency independent of the load. In these circuits a pair of transistors operates in a push-pull transformer output configuration and a part of the output is returned to each input base. The bases are driven hard enough to switch the collectors alternately ON and OFF. The result is that one side of the output transformer is effectively connected to the battery through the emitter resistor until the magnetizing current builds up and saturates the core. At this point the ON transistor is switched OFF, the feedback voltage reverses, and the opposite transistor is turned ON for the second half of the cycle. The half cycle is determined by the time the collector current requires to saturate the core. Hence, the operating frequency is determined by the core and winding design and is proportional to the applied voltage. Power conversion efficiencies of 80–95 per cent are obtained at switching frequencies in the audio range upward to at least 100 kHz.

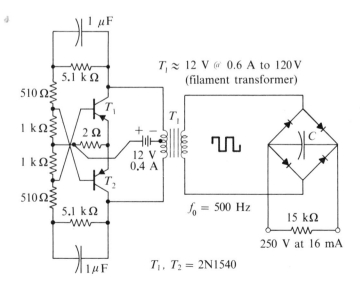

Figure 14.19. Low-power inverter, 4 W, 500 Hz

The inverter shown in Fig. 14.19 is interesting because the transformer is a small 60-Hz filament transformer and the circuit operates with an efficiency of 75 per cent. The output waveform is a square wave, and, to obtain 250 V dc, the load must be relatively small, 4 W. Because the secondary voltage is a square wave, only a small capacitor C is required to smooth out the switching transients, and the dc voltage is the same with or without the capacitor. The maximum load should make the input power approximately equal to the VA rating of the transformer.

A low-power inverter may use a 2-winding transformer with a multivibrator feedback network, as shown in Fig. 14.20. This inverter uses a readily available 24-V CT transformer and may be used to drive 120-V ac equipment from a 12-V automobile battery. For operation at 60 Hz the transformer has a laminated iron core and must be operated near the rated currents and voltages. The square-wave output is satisfactory for driving series-ac motors, electric shavers, soldering irons, lamps, and similar loads up to 50 W. For radio, phonograph, and audio equipment an approximately sine-wave output may be required, and for these loads the inverter is equipped with a tuning capacitor and hash filter, as shown in Fig. 14.20. The capacitor C is selected with a particular ac load, and the optimum value usually gives an approximately trapezoidal waveform.

Inverters for 120-V ac loads exceeding 50 W may be purchased in kit form or already assembled. These inverters use a transformer especially designed for inverter use, and the base drive is supplied by a feedback winding. The 3-winding transformer makes the inverter less sensitive to load changes, increases the overall efficiency to 70 or 80 per cent, and permits the use of reactive loads or a capacitor for improved waveform. Inverters suitable for operating a 150-W, 120-V TV on a 12- or 24-V battery are available, but inverters for higher load powers are for special applications, because the input current at 12 V is impractically high for an automobile battery. When

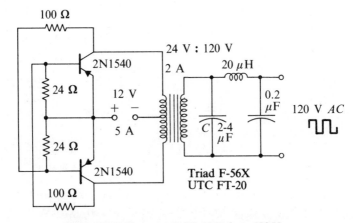

Figure 14.20. Inverter, 50 W, 12 V dc to 120 V ac

the load power exceeds 200 W, an inverter usually has a low-power, self-excited inverter driving a switching amplifier. The switching transients are easily controlled in the low-power inverter while the switching amplifier uses a linear output transformer that is not driven into saturation.

The transistors in an inverter or converter should be selected with concern for the second-breakdown requirements. The transistors must carry the input current with zero collector voltage when ON and twice the supply voltage when OFF. Second-breakdown is rarely a problem with pure resistance loads, but the transformer and a relatively small load capacitor may cause breakdown problems. The transistors are usually in the safe operating area (SOA) if the point designated by the collector input current and the supply voltage falls below the 1-ms curve on the SOA chart.

High-voltage switching spikes are produced in square-wave switching circuits, in power amplifiers, and especially in dc-to-ac inverters. These transient spikes may have such a short duration that they are not seen on a low-frequency oscilloscope, or the spikes are seen at a small fraction of their true peak value. Transient spikes may be reduced by connecting a Zener diode across both sides of the output transformer, and the diodes should have a breakdown rating that is 20 to 50 per cent above the peak voltage. In a 12-V inverter the diodes may need to be rated as high as 20 V because the input voltage supplied by an automobile battery is sometimes as high as 18 V.

An important reason for limiting the transient spikes in an inverter is that the transistors may be destroyed unexpectedly after passing satisfactory performance tests. The switching spikes are reduced by using carefully constructed transformers with bifilar windings to minimize leakage inductance. A small capacitor across the primary or secondary of the transformer usually helps to reduce the spikes and pickup in the load and nearby equipment. The optimum value of the capacitor is best found experimentally.

14.16 CONVERTERS (REFS. 3, 7)

A converter is a circuit for changing power at one dc voltage to another voltage, as in changing a 12-V supply to 40 V for a power amplifier. A converter is usually an inverter operating at 400 Hz to 2 kHz with the ac output converted to dc by a full-wave rectifier and smoothing filter. The high operating frequency permits the use of relatively small transformers and filter components, and with a square-wave output the filter is needed mainly to remove high-frequency hash and spikes. A 70-to-85 per cent overall efficiency is usually obtained with 10 per cent of the input power lost in the transformer, 5 per cent in the circuit resistors, and 5 per cent may be dissipated as switching power in the transistors.

The converter shown in Fig. 14.21 illustrates the design of a low-voltage circuit that may be built with a simple hand-wound coil on a ferrite core. The core may have a rectangular or toroidal shape and should have a cross-

274 Oscillators and Inverters Ch. 14

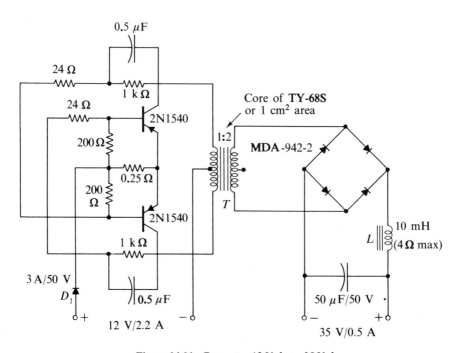

Figure 14.21. Converter 12 V dc to 35 V dc

sectional area of at least 1 cm². A larger core may be used without requiring a change of the winding specifications.

The coils are wound with four strands of 22- or 24-gauge wire and may be hand-wound as a single 4-strand wire. The primary wires may be cut to a length of 1 meter with 2 or 3 cm used at each end for the external leads. The secondary uses 2 wires, each 2 meters long. Two strands are used in series for both the primary and secondary. If a higher output voltage is required, the strands used for the output winding may be made longer with the extra wire wound at the start of the winding. If the output need not be insulated from the input supply, additional output power may be obtained by connecting the output strands in parallel and in series with the input supply as a voltage boost. With the series connection, twice the power may be obtained at the same output voltage. The 16-Ω winding of a small transistor audio transformer may be used as a 10-mH filter inductor.

The converter operates at a frequency of approximately 10 kHz and has 70 per cent efficiency at full load. The transistors may be any switching power device with a 1-ms breakdown rating of 2.5 A at 30 V. The transistors require a small heat sink that may be a flat sheet with a surface area of at least 20 cm² for each transistor. The entire package may need a volume of only 300 cc.

14.17 HIGH-VOLTAGE CONVERTERS

An LC oscillator operating at 10 kHz or higher frequency may be used as a converter for outputs of less than 1 W. The oscillator has the advantage of producing negligible RF interference and the disadvantage of only 40-to-50 per cent efficiency. The high-voltage winding of a TV flyback transformer may be used for the secondary winding, and the primary winding may be tapped or rewound for an oscillator operating on a 12-to-20 V supply. Because the transformer is operating with a sine-wave input, the voltage step-up is a factor of 5-to-10 times lower than when the transformer is used with a flyback oscillator.

Transformers needed for 200-to-400 V square-wave converters are readily available with suggested circuits and recommended components. The transformers usually operate at approximately 1 kHz, but where weight and space are important, they may be designed to operate up to 100 kHz.

Figure 14.22 shows the circuit for a 30-W converter that uses a commercially available toroidal transformer. The converter uses germanium transistors and supplies 30 W at 75 per cent efficiency. With light loads the inverter may be operated without the dc bias resistor R_2, and for reliable starting under heavy loads the resistor should be adjusted to make the base-emitter dc voltage between 0.6 and 1.0 V. Silicon switching transistors may be used with the bias adjusted to a value between 0.8 and 1.2 V. With germanium transistors a finned heat sink is required with a base area of approximately 100 cm², and with silicon transistors the base area may be only 50 cm². If a

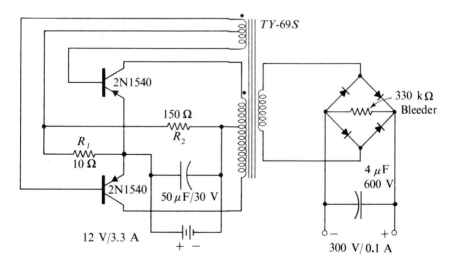

Figure 14.22. Converter, 30 W, dc to dc

flat heat sink is used, the total radiating area should be twice the base area of the finned radiator.

The input voltage of an inverter may be changed within limits imposed by the transistors. The output voltage and the frequency change with the input, and in this respect an inverter has the voltage-transforming characteristics of a transformer. Most inverters may be operated with a supply voltage much less than the rated input, provided the load power is reduced and the lower operating frequency is acceptable. With 6-V or lower supply voltages germanium transistors are generally preferred and offer higher efficiencies than silicon transistors.

14.18 UHF OSCILLATORS

Transistor oscillators for service above the transistor f_T rating generally use the transistor in the CB configuration. Since the ac collector voltage is in phase with the emitter voltage, a CB stage tends to be unstable with sufficient collector-to-base feedback capacitance or with an impedance between the base and ground. With tuned emitter and collector circuits the regenerative feedback is generally high enough to make a transistor useful as an oscillator up to about twice the f_T rating. At these relatively high frequencies the stray inductance in the emitter and collector leads may either aid or reduce the regenerative feedback so the physical layout of the components usually has a marked effect on the oscillator performance. Thus, when the parasitic lead inductances and stray capacities become of first order importance, the exact form of a microwave circuit may not be clearly defined.

The UHF CB oscillator shown in Fig. 14.23(a) has a tuned-collector LC load, and a high-impedance external load may be capacitor-coupled to the collector. Regenerative feedback is obtained by collector-to-base capacitance and by inductance in the transistor base lead. The tunable UHF oscillator may be used in a UHF mixer for a TV. The oscillator shown in Fig. 14.23(b) is a similar CB oscillator for use in driving a line-connected load at a fixed frequency. The inductance for the collector load is produced by a capacitor connected one-quarter wavelength away from the collector, and the line is continued to a matching load. Capacitance for the tuned circuit is supplied by the internal transistor capacitance. At lower frequencies a small capacitance may be connected between the collector and emitter, and the emitter lead may be lengthened to increase the positive feedback.

A high-frequency voltage-controlled oscillator (VCO) is illustrated by the circuit shown in Fig. 14.24. The transistor is connected for CB operation with a $+12$ V bias supply and a -12 V collector supply. The emitter and collector circuits are tuned by the varactors C_1 and C_2. Since the capacitance of a varactor may be changed by a factor of 4, the oscillator frequency may be

Sec. 14.18   UHF Oscillators   277

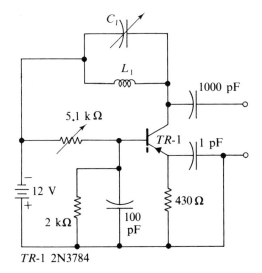

(a) High-impedance load

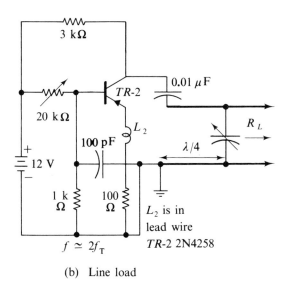

(b) Line load

Figure 14.23. UHF CB oscillators, 0.5 to 1 GHz

changed by a factor of 2. Similarly, the oscillator may be frequency-modulated by a signal that varies the voltage applied to one or both varactors.

The inductance and capacitance values shown in the oscillator circuit

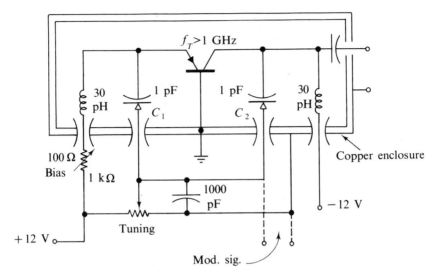

Figure 14.24. VCO or modulated UHF oscillator

are approximately correct for operating frequencies from 500 to 1000 MHz provided the transistor is biased for an f_T value of at least 1 GHz. At these frequencies the transistor capacitance usually provides adequate feedback, in which case the external capacitor C_f may be omitted. Frequencies a factor of 2 higher may be obtained by interposing an output filter tuned to select the second harmonic. For lower frequencies the CB oscillator operates as a Colpitts oscillator by increasing the collector-to-base feedback capacitance C_f and retuning the emitter and collector circuits. Tuning the emitter circuit below the operating frequency makes the emitter-to-base impedance a capacitance and tuning the collector circuit above resonance makes the collector impedance an inductance. Thus, the oscillator has the form of a Colpitts oscillator.

14.19 PARASITIC OSCILLATIONS AND FEEDBACK INSTABILITY (REF. 3)

Anyone working with high-gain circuits knows too well the difficulties encountered in preventing spurious oscillations. At high frequencies a short length of wire has considerable inductive reactance, and the small capacities between components may cause considerable coupling between circuits. Together, these reactances are the source of most parasitic oscillations in high-gain circuits, especially at high frequencies. At low frequencies unwanted oscillations are more likely to be caused by mutual coupling in the power supply filters. A helpful solution of spurious feedback problems is to isolate the circuit responsible for the oscillation and then find which specific elements are responsible for the circuit's meeting the Barkhausen conditions. The prevention of oscillations, or ringing, in negative feedback amplifiers is an interesting and related problem.

The amount of feedback that can be used in a high-gain amplifier depends on how well the loop gain and phase shift can be controlled to avoid the Barkhausen conditions, particularly at high frequencies. The amount of feedback is generally controlled by a resistor connected from output to input. If the resistor is too small, the amplifier usually oscillates at a frequency well above the frequency range for which the amplifier is intended. High-frequency oscillations of this kind may be stopped by increasing the feedback resistor or decreasing the amplifier gain. Either of these changes should reduce the loop gain to less than 1 at the frequency of oscillation. However, the feedback need not be reduced if the phase shift at the critical frequency can be reduced to make the feedback negative (degenerative) instead of positive (regenerative).

The phase shift in the feedback loop is determined by the rate at which the amplitude-frequency response falls off at high frequencies. Since the phase shift with a single high cutoff is only 90°, the loop phase shift may be reduced by making one cutoff occur at a much lower frequency than all other cutoffs. This technique makes an OP amp relatively stable with considerable feedback.

Another technique reduces the cutoff rate at frequencies just above the critical frequency. For this purpose a small capacitor is used in parallel with the feedback resistor, and the optimum value may be found by trial. Because an uncomplicated circuit usually has less phase shift, a feedback amplifier tends to be more stable when simply designed and carefully built to avoid unnecessary and stray capacitance. A ferrite bead surrounding the transistor base lead is sometimes helpful in reducing parasitic or high-frequency oscillations.

15 Special-Purpose Circuits—AF and AGC

A selection of interesting special-purpose circuits is described in this chapter. These circuits include audio and automatic gain-control amplifiers, which deserve a more complete description than may be found easily in the published literature. The first of these circuits is a simple hearing-aid amplifier that can be built and enjoyed by many persons who do not want to buy a hearing aid. A second group of circuits for phonograph applications includes tone controls, an audio mixer, and two high-fidelity amplifiers. Phono preamplifiers are described in Chapter 7.

There are many applications for automatic gain-control (AGC) amplifiers but few circuits are available of simple, well-designed AGC amplifiers. An AGC amplifier is easily built when the amplifier gain may be allowed to change relatively slowly, as in a radio receiver. In instrument amplifiers where the gain is required to adjust rapidly without distorting the signal the design may be difficult and force a compromise among several objectives. Examples of voltage and photo-controlled attenuators and several complete AGC amplifiers are described.

15.1 AN INEXPENSIVE HEARING AID OR INTERCOM

Many persons with hearing difficulties can be helped by an inexpensive audio system. Some people find even the most expensive hearing aids unsatisfactory because of their difficulty in discriminating between the voice of another person and background noise, especially in a room where several people are talking simultaneously. When a hearing aid is built with a separate microphone, as in a public address system, a person speaking with one who is hard-of-hearing may hold the microphone close to his (the speaker's) lips and make his speech louder than the background noise. The only difficulties with this kind of hearing aid are that the separate microphone may embarrass the hard-of-hearing person, and the microphone cable is an additional maintenance problem. Nevertheless, an amplifier with a separate microphone is helpful to persons with a partial loss of hearing who are reluctant to purchase an expensive hearing aid or are troubled mainly by background noise.

The hearing aid is designed to use either a low-impedance or a high-impedance, cassette-type dynamic microphone. The amplifier is rated at 0.3 W output and operates class B for a low standby power and a long battery life. The earphone may be a low-impedance headset or any good quality 10-Ω transistor earphone. The heavier and more expensive earphones are more sensitive and better able to utilize the maximum output. For portability the battery may be a 9-V transistor radio battery, which should give 40 to 50 hours of continuous service. The NEDA 1603 and the smaller 1605 are pocket-sized 9-V batteries that are relatively inexpensive and offer a longer service life in everyday use where the battery may occasionally be left ON.

The amplifier shown in Fig. 15.1 has a collector-feedback stage for voltage gain followed by a three-stage complementary-symmetry power amplifier. A high-impedance microphone may be connected to the input base for a 10-kΩ input impedance, or a loud speaker used as a microphone may be connected across the emitter resistor, provided the base input terminal is grounded.

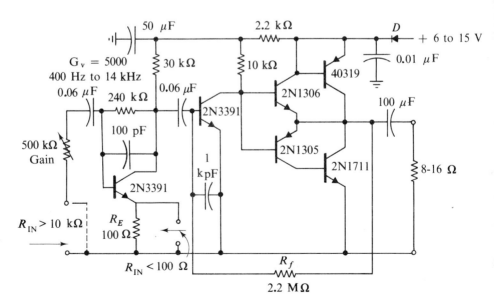

Figure 15.1. Hearing aid or intercom amplifier

With either input the overall voltage gain is 70 dB, and full output may be obtained with a 300 μV rms input signal. The gain may be reduced 33 dB by increasing the variable base resistor R_B to 0.5 MΩ. Because of the high gain, care must be exercised to shield and separate the input from the output. Instability caused by a resonant earphone load is prevented by connecting a 51 Ω resistor across the output.

The amplifier uses silicon transistors except for two germanium devices that may be almost any low-power npn and pnp transistors with a β of 50. By operating the germanium transistors with a low-value base current, the crossover distortion is negligible with supply voltages from 6 to 15 V, and the standby current is an exceptionally low 1.5 mA. The power stage Q-point is adjusted by selecting R_f to make the collector voltage one-half the supply voltage.

The frequency response is from 400 Hz to 14 kHz as is needed to offset the common high-frequency hearing loss. The amplifier makes an excellent intercom that gives a crisp, clear reproduction of speech sounds. With a 13-V supply the power output is 0.5 W with a 16-Ω load. The power amplifier may be used alone for 40-dB gain with a 20-kΩ input impedance.

The amplifier may also be used to help a second person hear a telephone conversation without making a connection to the telephone circuit. The amplifier input signal is obtained from the stray magnetic field at one side of the telephone subset or at the receiver end of the handset. The pickup coil may be purchased, or a coil with 10-to-50 Ω dc resistance may be removed from a small transformer. If hand wound, the coil should have 150 or more turns, using any convenient size magnet wire wound with a 15-cm inside diameter. The coil may be connected to the low-impedance input, or, for greater gain, may be coupled through a small, 50:10,000-Ω input transformer to the high-impedance input.

With a ferrite-core antenna and a variable-air capacitor the amplifier makes a broadcast receiver, and, with the input stage replaced by the phono preamplifier shown in Fig. 7.9 there is adequate gain for a small record player.

15.2 TONE CONTROLS (REF. 4)

Tone controls are an important but somewhat neglected part of a high-fidelity audio system. Two controls are usually provided that permit frequency adjustments similar to those shown in Fig. 15.2. The base control may be used to increase or decrease the slope of the frequency response as much as 15 dB per decade between 20 Hz and 1 kHz, while the treble control may increase or decrease the slope of the response above 1 kHz. Together the controls may be used to cut or boost both the low-frequency and the high-frequency responses relative to that at 1 kHz, or the controls may be used to provide a uniformly

Sec. 15.2 Tone Controls 283

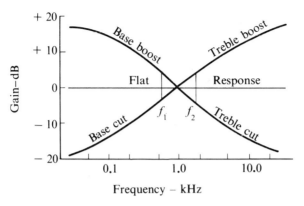

Figure 15.2. Tone control frequency response

increasing or decreasing adjustment of the response throughout the audio spectrum. A tone control is usually expected to have a flat frequency response when the potentiometers are set in mid position.

The base and treble tone controls are generally constructed as a back-to-back pair of resistor-capacitor networks that operate almost independently of each other. Figure 15.3 shows base and treble controls that are driven by a single capacitor-coupled signal source. The output of both controls connects at point A to a capacitor-coupled load. Since the tone controls have only resistors and capacitors, both controls are designed to reduce the signal 20 to 23 dB when the potentiometers are in the mid position where the frequency response is flat from 20 Hz to 20 kHz. The 20-dB boost is obtained by moving the controls to the upper end, where the loss is reduced 20 dB at low frequencies when C_1 and C_2 may be neglected, and at high frequencies when C_3 may be neglected as a series loss. The base and treble frequency cuts are obtained by moving the controls to the lower end, where the low-frequency loss is approximately R_1/R_2. The capacitors C_1 and C_2 shunt the potentiometer R_1 and determine the frequency at which the base control is bypassed and ineffective. The capacitors C_3 and C_4 determine the frequency at which the treble control becomes operative. The crossover frequency of the controls may be changed by increasing or decreasing the capacitors, but all 4 capacitors should be changed up or down together unless a special uneven frequency characteristic is desired near the center frequency when the controlled frequencies either overlap or fail to meet.

The need to minimize the loss in the network complicates the design of the loss-type tone control and makes the base and treble adjustments interdependent. The resistor R_5 in Fig. 15.3 is needed to make the 3-dB, high-cut frequency the same as the 3-dB high-boost frequency, and the resistor R_4 is used to make the frequency response relatively constant above 20 kHz, instead of

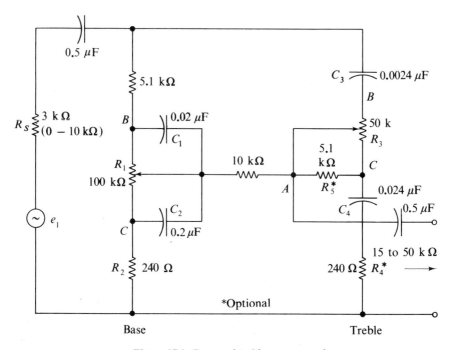

Figure 15.3. Base and treble tone controls

decreasing continuously with increasing frequency. Resistor R_4 may usually be eliminated by a short, and resistor R_5 may be omitted if the 3-dB, high-cut frequency is permitted to be at 1 kHz instead of 2.5 kHz.

The impedance of the signal source is preferably made less than 3 kΩ, but a 5-kΩ source may be used by offsetting the resulting low-frequency loss with the base boost and accepting a slightly uneven response for the nominally flat characteristic. The output load resistance should be 50 kΩ or more. If the load impedance is appreciably less than 50 kΩ, all impedances in the tone controls and the signal source may be reduced by a common factor.

A disadvantage of the passive-element tone control is that the potentiometers should have an *audio taper* in order that the flat response may be obtained with the controls in the mid position. In the mid position the control should have a low resistance on the cut side of the moving contact and a high resistance on the boost side. Thus, the controls should increase logarithmically with a clockwise rotation. Because the loss-type controls may have as much as 40-dB loss, an amplifier is required to restore the signal level and the controls cannot be placed where the signal is too close to the noise level.

The base control may be used alone by removing the treble control. If the treble control is used alone, the base control must be replaced by a 100-kΩ resistor connected from point A to the input and by a 10-kΩ resistor connected between A and ground. The 10:1 voltage divider is needed to provide the low-frequency signals that are removed by the capacitor C_3.

15.3 FEEDBACK TONE CONTROL (REF. 4)

The simplest and most satisfactory tone control is the feedback type shown in Fig. 15.4. A control with the potentiometers in the feedback loop requires relatively few components, simplifies adjustments of the frequency response, and uses linear potentiometers. Moreover, the tone-control network may be used with an amplifier input impedance as low as 1 kΩ, and the feedback connection may be moved from X to Y to provide 12-dB overall gain, or, to lower the amplifier output impedance, X and Y may be connected. The feedback tone control has the frequency response characteristics shown by the curves in Fig. 15.2.

On the base side of the control the resistors R_2 and R_3 fix the amount of low-frequency boost and cut. Resistor R_5 with a 0.24-MΩ resistance may be connected across either side of the potentiometer to make the flat frequency

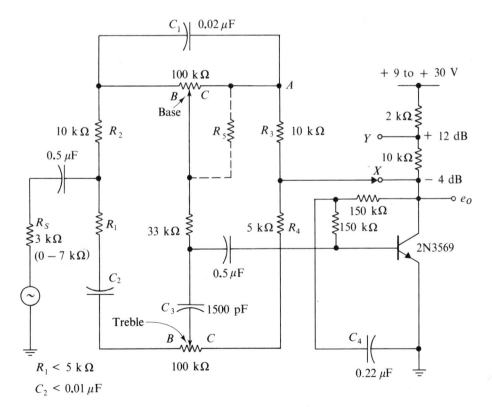

Figure 15.4. Feedback tone control

response occur at the exact center of the control, or the resistor may be omitted for simplicity. The capacitor C_1 may be changed to adjust the frequency f_1 at which the base control has a 3-dB cut or boost. Similarly, the capacitor C_3 in the treble control may be adjusted to change the frequency f_2 of the 3-dB high-frequency cut and boost. A smooth transition is ensured at the center crossover frequency by making C_1 10 to 15 times C_3.

The capacitor C_2 on the treble side of the control may be omitted or adjusted to make the flat frequency response occur when the control is at the exact center position. The resistor R_1 should be omitted when the source impedance exceeds 2 kΩ.

Because the ideal tone-control frequency response depends to a large degree on personal taste, most tone controls offer only an approximation of the characteristics shown in Fig. 15.2. A satisfactory, but simplified, tone control may be built as shown in Fig. 15.5. With C_4 removed, a slightly higher loop gain is obtained by connecting the 300-kΩ bias resistor between the transistor base and point A on the base control.

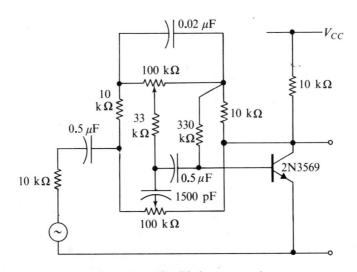

Figure 15.5. Simplified tone control

15.4 LOW-DISTORTION AF MIXER

The mixing stage shown in Fig. 15.6 is designed to add signals from several sources while preventing cross-feed from one source to another. There is the further advantage that the signal input from any one channel is not changed by connecting or disconnecting any of the other channels. The circuit takes advantage of the low input impedance and linear response of a CB stage.

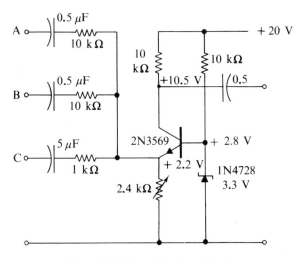

Figure 15.6. Low-distortion AF mixer

This summing amplifier has a wide-band frequency response, excellent waveform up to 5-V rms output, and a voltage gain of 1 in channels A and B.

Each channel is shown with a separate capacitor, which should have an impedance at the lowest frequency of interest that is one-tenth or less of the channel impedance. Any number of inputs may be added so long as the signal sum does not overload the collector circuit, which tolerates up to nearly 14 V, peak-to-peak.

In some applications the resistor shown between the emitter and the channel input may be omitted. Because the source impedance is in series with the resistor, the resistor may be omitted unless the source is adversely affected by a low impedance or is significantly less than 10 kΩ. Because the series resistor is the load seen by a channel, the input signal from a high-impedance channel is reduced more than from a low-impedance channel, and it may be desirable to increase the resistor values to offset gain differences. Channel C, which has a gain of 10 from the input terminal, shows how the channel gain may be increased by changing the series resistor and capacitor, provided, of course, the source impedance is small compared with 1 kΩ.

The amplifier emitter current is fixed by the emitter resistor and by the Zener diode connected to the base. The simplest way to change the Q-point is by changing the emitter resistor. The Zener diode is operated on the knee of the curve, where the diode voltage is about 15 per cent larger than the nominal voltage, and a small adjustment of the Q-point may be obtained by changing the base resistor. For mixing at lower signal levels the CB stage may be replaced by a collector feedback stage with the advantage that fewer components are needed and the bias is self-adjusting.

15.5 A 7-W HIGH-FIDELITY AMPLIFIER (REF. 7)

The circuit of the 7-W high-fidelity amplifier shown in Fig. 15.7 is designed to give satisfactory performance with a minimum number of components. The first three stages are a preamplifier and tone control, and the last six transistors are a quasi-complementary-symmetry power amplifier. A stereo phonograph requires two complete amplifiers and may be rated to deliver 14 W with a sine-wave signal.

The preamplifier has a low-noise FET input stage and a direct-coupled CE transistor stage with a feedback network to produce the RIAA frequency response. A 22-V Zener diode and an RC smoothing filter are used in the collector supply to protect the input stage from voltage changes and noise. The tone control uses the simplified circuit described in Sec. 15.3, except that the collector load resistor is reduced to 4 kΩ and has a tap to reduce the amount of feedback. With the feedback connected at point Y the tone-control stage has nearly 12 dB more gain than with the tap at X. This choice of the feedback connection permits an adjustment of the amplifier gain or permits additional feedback in the power amplifier. On the other hand, the tone control has a slightly greater range of adjustment if the feedback is connected to X, the collector.

The amplifier gain control R_3 is between the tone control and the power amplifier where there is no danger of distortion at low-gain settings of the control. An approximate Fletcher-Munson frequency compensation is provided at low-gain settings by the 0.5 μF capacitor connected in the ground side of the gain control. The input stage of the power amplifier has a CE configuration to minimize interaction and a tendency for instability in the power amplifier. Bias for the CE stage is obtained from the 22-V regulated supply.

The power section of the amplifier has a CE-CE pair amplifier for gain, followed by the complementary-symmetry drivers. The output stage uses 2N1540 germanium power transistors that may be replaced by 2N1542 transistors if higher reliability is required. However, amplifiers using 2N2869 or 2N1540 devices have withstood short circuits and are reliable. The fuse in the PA supply is used to protect the power transformer and serve as the emitter-feedback resistor. The amplifier may be used with 4-, 8-, or 16-Ω loads. If a stereo pair is used, the gain and tone-control potentiometers should be dual controls, and the gains may be balanced by shunting the feedback resistors R_5 with a 30-kΩ or larger resistor.

Although the amplifier circuit shows values for all resistors, a few adjustments are usually required when the circuit is assembled. The most important adjustments are of the bias resistors R_1, R_2, and R_4. The bias of the input stage is adjusted to make the collector voltage $+19$ V or slightly lower. The

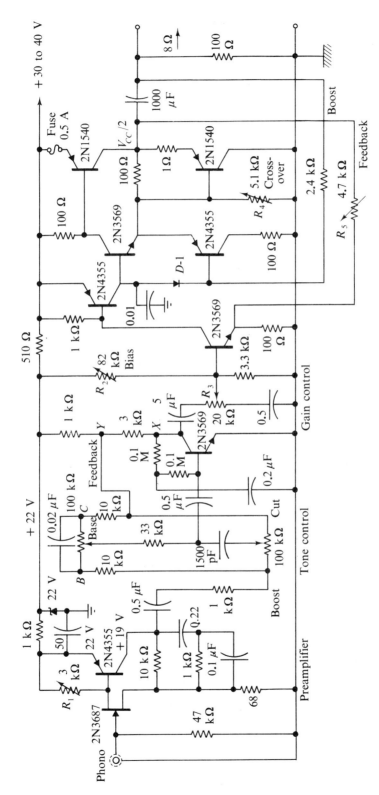

Figure 15.7. A 7-W high-fidelity power amplifier

power amplifier bias is adjusted with R_2 to make the output-stage collector voltage one-half the supply voltage with no input signal. The crossover distortion is adjusted by R_4, which has the effect of moving the distortion to a high signal level where the residual distortion is relatively unimportant. The resistance of R_4 should be increased if the output transistors show any tendency to feel hot, rather than warm, or to show a marked Q-point drift when operated in a high ambient temperature.

A small change of the overall gain may be obtained by adjusting the feedback resistor R_5, and this change generally requires an accompanying adjustment of the bias resistor R_2 to correct the dc output voltage. Although the power section has 30-to-40 dB feedback, the amplifier is stable for all ordinary loads and does not require a feedback gain step. If a square-wave signal source is available, the transient response may be adjusted by shunting the feedback resistor R_5 with a 100-to-500 pF capacitor.

The power stages are protected from a short-circuited load mainly by the 100-Ω resistor in the 2N4355 collector. The bias for the complementary driver transistors is obtained by connecting the 2.4-kΩ base resistor to the load. This connection, a simplification of the usual bootstrap, effectively removes the bias resistor as a load on the 2N4355 driver and increases both the peak signal output and the amount of feedback. Diode D-1 is any low-current silicon diode. The amplifier uses the 34-V, 25-VA power supply shown in Fig. 6.7.

15.6 POWER AMPLIFIER WITH IC DRIVER
(REF. 6)

The amplifier shown in Fig. 15.8 combines the advantages of an IC OP amp with those of a 2-stage complementary-symmetry amplifier. The amplifier has a voltage gain of 30 and produces 30-W output with a 500-mV input signal. The frequency response is constant up to 20 kHz, and there is 50-dB feedback in the audio-frequency range below 100 Hz. A disadvantage of the 741 driver is that its voltage gain decreases with frequency, and there is hardly more than significant feedback above 10 kHz. At frequencies where there is adequate feedback the distortion may be expected to be less than 0.2 per cent with 20-W output.

The amplifier requires a power supply with the ground connection at the center tap and has the advantage that the load may be connected without a coupling capacitor, which is expensive, a source of low-frequency instability, and a coupling loss. The 5000-pF capacitor connected between the load and the output of the 741 driver supplies feedback above 0.5 MHz to prevent instability in the output stages. Feedback over the entire amplifier is supplied with the 100-kΩ resistor that has a 10-pF shunt capacitor to reduce the gain above 200 kHz. The input to the amplifier is connected to a 3.3 kΩ resistor at the + side of the 741 amplifier. Usually the resistor on the + side may be as high as 10 kΩ if a higher input impedance is desired.

Sec. 15.7 Audio Crossover Network **291**

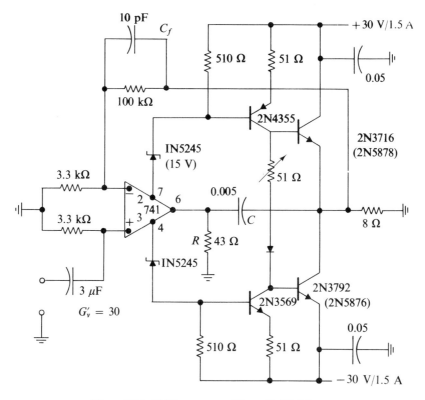

Figure 15.8. 30-W power amplifier with IC driver

The Zener diodes protect the 741 amplifier from more than the rated + and −15-V supply voltage. If the supply voltage exceeds 30 V on a side, the diode voltages should be raised enough to offset the voltage increase. For 30-W output with an 8-Ω load the power supply must be capable of supplying 4-A peak and 1.5-A average load current. The filter capacitors on each side should be rated from 3000 to 4000 μF at 50 V, and two 0.05 μF ceramic capacitors are connected to the collectors of the output transistors to bypass high frequencies to ground.

15.7 AUDIO CROSSOVER NETWORK

The audio-frequency crossover network shown in Fig. 15.9 is easily constructed and adjusted to provide a smooth transition from a low-frequency loud speaker to a smaller high-frequency speaker. The crossover network consists of L and C in the figure; R_A and R_B represent the nominal speaker load impedances, and a resistor network is included for testing the crossover.

The inductor L is wound on a 2- or 3-cm diameter, nonmetallic core using 50 meters of 20- or 22-gauge magnet wire. The coil may be scramble-wound

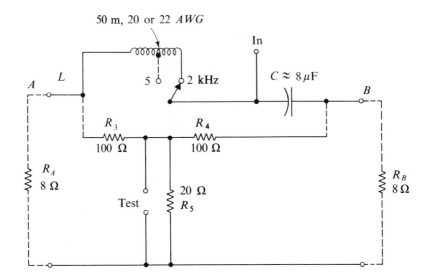

Note: Adjust C to make test voltage constant

Figure 15.9. LC cross-over network

in any shape as long as the turns are closely confined with a finished length that does not greatly exceed the outside diameter. The process of winding the coil may be simplified by cutting the wire into four equal lengths that are wound as a single 4-strand wire. After the winding is completed, the strands may be connected series-aiding with the connections serving as taps for adjusting the inductance.

The crossover is adjusted by connecting a low-impedance source or a power amplifier to the input and selecting a capacitor C that produces a constant voltage at the test point when the signal frequency is swept from low to high frequencies. The crossover frequency is changed by using a different tap on the coil and selecting a new capacitor value. If the loud speakers have unequal power sensitivities, a resistor may be connected in series with the more sensitive speaker, but small differences may be removed with the amplifier tone control.

15.8 AUTOMATIC GAIN-CONTROL AMPLIFIERS

An automatic gain-control (AGC) amplifier is used when the output of an amplifier is required to be substantially independent of the input. In an AGC amplifier a part of the output signal is rectified and used to control the amplifier gain. Radio receivers have an AGC to make both weak and strong stations equally loud and to offset signal fading. The AGC also greatly simplifies the tuning and adjustment of a high-gain receiver. Sound recording systems generally need an AGC amplifier to prevent overloading and distortion when the sound input is too loud because most recording media tolerate only a limited signal range. AGC amplifiers are used also for recording and observing transients with a wide dynamic range, such as those obtained in seismic exploration.

An AGC amplifier uses part of the output signal to control the amplifier gain. The control signal is rectified and smoothed in a filter to prevent feedback and instability. The output of the filter is proportional to the amplifier output and is used to reduce the signal input or the amplifier gain. If a close control of the output is required, the amplifier must have a high gain and the AGC system may be blocked for an appreciable time after a sudden increase of the input signal. If the output is permitted to change noticeably with the input, the filter time constants can be adjusted either to minimize distortion of the transient waveform or to effect faster control.

The time constants in the filter determine the stability of the feedback loop, the effect of gain changes on the signal, and the speed with which the gain recovers when the signal stops. The gain adjustment time with a small signal change is usually one-fourth to one-tenth the filter time constant, and, with a signal that overloads the amplifier, the recovery time is approximately the filter time constant. Thus, the performance of an AGC amplifier with a time-varying signal depends in a complicated way on both the linear and the overload characteristics of the circuit, and an AGC amplifier usually requires different circuit and time-constant adjustments for different kinds of service.

15.9 GAIN-CONTROL CIRCUITS

The gain of a transistor stage is not easily controlled by changing a bias current or voltage, and gain control is usually obtained by using a variable resistance element in a loss network. An FET may be used as the voltage-controlled resistor (VCR) in a high-impedance loss network, and a diode bridge or a photo resistor may be used in a low-impedance network. The gain of a transistor stage may be reduced up to a factor of 10 by a bias change, and this change is adequate in a multistage IF amplifier when three or more stages are controlled. If the gain of a single transistor stage is reduced more than 30 dB, gain control is difficult because the transistor tends to turn OFF, and the gain may drop 20 to 40 dB with each one-tenth volt increase of bias. Thus, a transistor is more easily operated as a switch than as a variable gain element, and AGC amplifiers needing a smoothly controlled 40-to-60 dB gain change generally use a voltage-controlled loss network.

Two VCR loss circuits that use FETs as variable resistors are shown in Fig. 15.10. Figure 15.10(a) uses a low-pinchoff voltage 2N2386 FET that gives up to 60-dB attenuation with a relatively uniform voltage control. As an

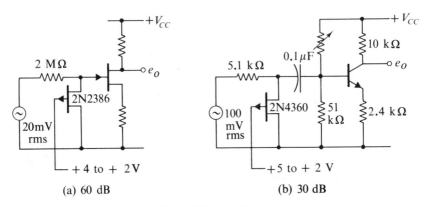

Figure 15.10. FET variable attenuators

attenuator the maximum input signal should not exceed 20 mV rms, but with AGC control to maintain a constant ac voltage across the FET the input may be as high as 10 to 20 V. A similar low-impedance FET attenuator uses a 2N4360 FET, as shown in Fig. 15.10(b). The low-impedance attenuator makes a satisfactory AGC, or voltage-controlled, audio amplifier with two stages in tandem for 50-to-60 dB control.

A silicon diode, Fig. 15.11(a), gives 40 to 60 dB of attenuation control with approximately 10-dB increased attenuation for each one-half volt change at low attenuations; a 1-to-2 volts change is required at higher attenuations. Since the control characteristic is more uniform than with an FET attenuator, an AGC loop that uses a diode has less tendency to pinch out strong transients. The diode attenuator has a simplicity that suggests its use to prevent overdriving amplifiers in which a 2-to-3 V change of the collector voltage is available for controlling gain.

The photo-resistor attenuator shown in Fig. 15.11(b) offers the smoothest available distortion-free gain control, and the lamp and photo resistor may be

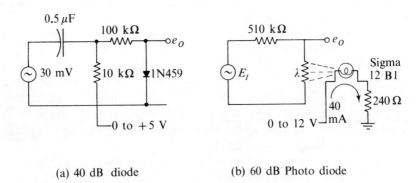

Figure 15.11. Diode and LDR attenuators

purchased in a light-tight assembly. The light sources are available as incandescent lamps with turn-ON and turn-OFF times of 10 to 100 ms or with neon lamps with similar operating times. The response times provide built-in time constants that may serve as the bias filter in some applications or may be too slow for others. The incandescent lamps have 50-to-500 Ω resistance and require up to 100 mW, which can be supplied only by using an additional power stage in the feedback loop.

The all-transistor AGC amplifier shown in Fig. 15.12 is similar to those used in an IC AGC amplifier. With an additional stage for gain and a diode rectifier the circuit may be built with a single IC transistor array to make a compact and useful AGC package.

The transistor-FET gain control shown in Fig. 15.13 has control characteristics similar to those of the all-transistor gain control. The gain may be changed by a factor of 300, and a choice of the control characteristic may be obtained by selecting the FET used to change the emitter impedance. The transistor-FET control has the advantage of presenting a 10-kΩ input impedance and of requiring both lower control voltage and power.

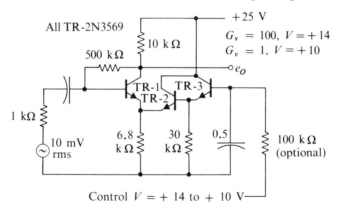

Figure 15.12. Transistor gain control, 40 dB

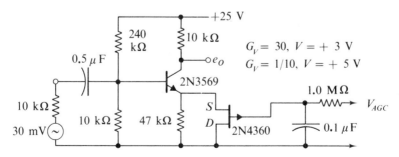

Figure 15.13. Transistor-FET gain-control stage

296 Special-Purpose Circuits—AF and AGC Ch. 15

When an attenuator is used in an AGC feedback loop, the loop gain should be high enough to hold the ac signal across the control element about 10 dB below the maximum input shown in the figures. Thus, with enough loop gain the maximum input signal may be higher than the input shown by the amount of loss in the attenuator. However, the gain controls shown in Figs. 15.12 and 15.13 are exceptions in which the input should be less than the maximum indicated, regardless of the loop gain.

15.10 AUDIO AGC AMPLIFIERS

An AGC amplifier that is designed to limit the amplitude of voice signals in audio systems is shown in Fig. 15.14. To preserve the characteristics of speech the AGC is designed to reduce loud input sounds quickly enough to prevent overloading, and the gain does not recover too rapidly, as, for instance, between words. Such gain controls are said to have a *fast attack* and a *slow release*. The amplifier uses a diode D_1 to control the gain of the input stage. A desirable feature of this control is that it produces a negligible shift of the dc Q-points; otherwise, gain changes are accompanied by an objectionable dc thump.

The AGC amplifier has an output-input characteristic that Fig. 15.15 shows is a 20-dB gain reduction with a 30-dB increase of the input signal. As the curve indicates, the amplifier should operate with a 10-mV average input where the output approximately doubles when the input increases by a factor of 10. The frequency response of the amplifier is constant from 50 Hz to

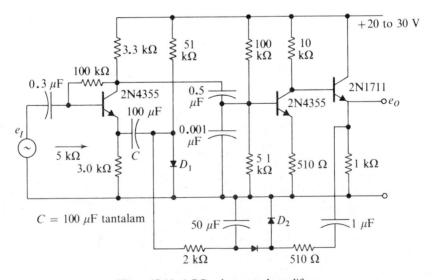

Figure 15.14. AGC voice-control amplifier

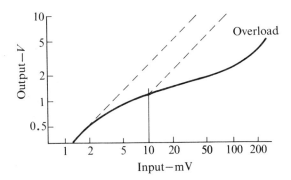

Figure 15.15. Amplifier AGC characteristics

30 kHz, and the signal level is controlled with signal frequencies above 100 Hz. The control characteristics are slightly degraded with signal source impedances exceeding 5 kΩ, but the performance is relatively independent of the supply voltage.

Low-frequency signals or noise below 100 Hz should be removed in a preamplifier to prevent distortion by lack of AGC control at low frequencies. For voice frequencies the harmonic distortion is relatively low at all signal levels and peaks are not clipped unless the input signal is excessive. Slightly lower distortion may be obtained by making the first stage a direct-coupled pair, as in Fig. 3.8, and by reducing the impedance level of the second stage by a factor of 5. These circuit changes reduce the input millivolts shown in Fig. 15.15 by a factor of 2 but do not change the output or the overload voltage.

15.11 INSTRUMENT AGC AMPLIFIERS

For instrument applications an AGC may be designed to produce a nearly constant output for a 60-to-80 dB range of input signals, as for recording time-decaying transients from a seismograph. This type of AGC is useful for levelling the output of an audio oscillator or for gain control in some audio systems.

The AGC amplifier shown in Fig. 15.16 is designed to level input signals between 15 mV and 15 V, and, with the 60-dB input change, the output increases from 150 mV to 300. Thus, the output increases 1 dB for each 10-dB increase of the input signal. The amplifier levels signals at frequencies between 50 Hz and 10 kHz, and the lower control frequency may be change by increasing or decreasing all capacitors in the circuit by a common factor. With the capacitors shown, the amplifier brings a 20-dB signal change under control in approximately 50 ms. The response time varies linearly with the capacitor values, and some adjustment of the response characteristics may be

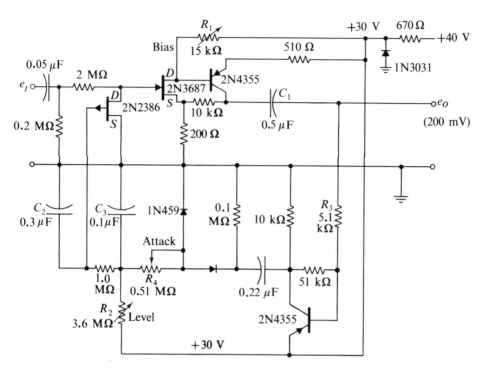

Figure 15.16. Automatic gain control amplifier

obtained by changing the filter capacitors and by making C_2 either 2, 3, or 4 times C_3. The attack time is shortened by reducing R_4 and increasing C_3.

The AGC amplifier is easily reproduced by constructing the circuit exactly as shown. The amplifier uses a voltage-controlled FET resistor across the input of an FET-transistor pair amplifier with a voltage gain of 25. The output of the amplifier is amplified in a collector feedback stage, rectified, and returned through a 2-stage filter to the gate of the FET resistor.

The 2N2386 that is used as the voltage-controlled variable resistor (VCVR) is readily available and inexpensive enough to permit selecting a device with the AGC characteristics desired. Approximately one-half the 2N2386 FETs are usable, and the most satisfactory devices usually have a 500-Ω, drain-to-source resistance R_{DSS} when measured with an ohmmeter. The FET should be a p-channel amplifier type with a 4-to-5 V pinchoff voltage.

The amplifier bias resistor is adjusted at assembly to make the dc collector voltage of the output transistor equal to one-half the supply voltage. With a signal input of about 0.1 V the resistor R_2 is adjusted to make the output 150 mV, more or less. An adjustment of the dynamic response may be obtained by changing the level with R_2 and the series resistor R_3.

The bias amplifier uses a collector-feedback stage that has the collector Q-point within a few volts of the supply voltage. The collector-feedback stage permits relatively large negative excursions and seems to produce a better dynamic response than is obtained with a CE driver stage. The AGC amplifier operates with supply voltages from 15 to 30 V, except that the dynamic response to a transient input signal changes with the supply voltage.

15.12 RADIO-FREQUENCY AGC AMPLIFIERS (REF. 4)

When a radio receiver is tuned from one station to another, the incoming RF signals vary from a few microvolts to several hundred millivolts. A receiver needs AGC to prevent distortion and to make the detected signal independent of the RF signal strength. AGC may be obtained by using the dc component of the detector output to control the gain of the IF amplifier, and sometimes the RF stage is included also. There are several ways the control voltage may be used to vary the gain of transistors.

High-frequency amplifiers are designed with the emitter resistor bypassed to eliminate ac feedback. With the feedback eliminated, the gain of a transistor stage is approximately proportional to the product of the transistor current gain and the dc emitter current. The current gain β varies somewhat more slowly than the transconductance at low emitter currents but falls off rapidly at high emitter currents. The net effect of a Q-point change can be represented as a dependence of the stage power gain on the emitter current, as shown in Fig. 15.17. By operating the RF and IF stages at the peak of the gain curve, the gain can be reduced either by decreasing or increasing the dc emitter

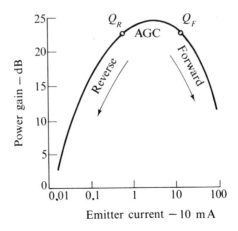

Figure 15.17. IF amplifier power gain vs. emitter current

current. Because the impedance matching changes with a change of Q-point, the AGC gain reduction is more than is indicated by the curve.

The most common form of an IF AGC amplifier, shown in Fig. 15.18, operates the variable gain stage with a constant collector voltage, and the AGC bias is polarized to reduce the power gain by decreasing the emitter current. A less frequently used system has the AGC bias polarized to increase the bias current. By supplying the collector voltage through a series resistor, the increasing emitter current is accompanied by a decrease in the collector voltage.

AGC systems that decrease the gain by increasing the emitter current are called *forward AGC*. Those that decrease the emitter current are called **reverse AGC**. In both AGC systems the Q-point change produces an impedance mismatch, and a considerable change in the power gain can be obtained with a small change of the AGC bias. The reverse AGC requires few components and few control stages. The forward AGC has the advantage of accepting higher signal levels as the gain is reduced. Since most systems of AGC tend to overload at high signal levels, the auxiliary diode D_1, shown in Fig. 15.18, is nearly always used as a variable resistor across the input. The collector resistors R_1 and R_2 are proportioned to back bias the diode at low signal levels and to forward bias the diode when the conventional AGC changes the collector current and the voltage drop in R_2. Because the auxiliary diode shunt

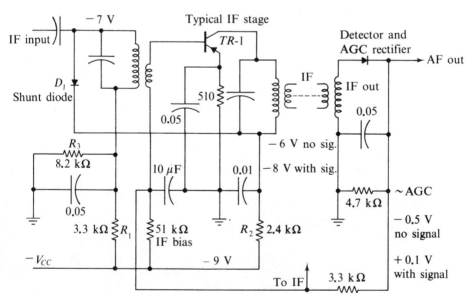

Note: R_3 or preceding collector make $V_D = -7$ V
TR_1 is any germanium IF *pnp* transistor (2N1524)

Figure 15.18. Reverse AGC circuit with shunt diode

eliminates the large signal problem, most radios now use reverse AGC. Occasionally an auxiliary dc amplifier stage is used in the AGC line as an AGC voltage amplifier or as a voltage-variable emitter resistor to change the AGC gain. The choice between the different types of AGC systems is determined by weighing the cost of a complex system against the limited performance characteristics of a simpler circuit. The only observable difference between a forward and a reverse AGC may be the polarity of the diode D_2 and that of the diode D_1, if the latter is used.

15.13 IC AMPLIFIERS WITH AGC (REF. 6)

Several IC amplifiers are available for use as IF amplifiers with built-in detectors and AGC control. The ICs are intended for use in IF and tuned RF applications at frequencies up to 2 MHz. Signals between 50 μV and 50 mV are closely controlled to produce 0.3-V *p-p* audio output. These amplifiers may be preceded by a single or double-tuned, ferrite-core antenna to operate as a TRF receiver, or they may be preceded by a mixer and a multisection IF filter for use as a superheterodyne receiver. Since all the coupling and bypass capacitors are connected externally, the amplifiers may be used in low-frequency instrument and audio applications where the transient response of the AGC is not an important factor.

The amplifier shown in Fig. 15.19 is intended for AGC applications with carrier frequencies down to 50 kHz and may be used in systems with a lower-frequency carrier by increasing the capacitors. The LM372 shown in the figure is intended for room temperature applications, 0°C to 70°C, and may be used with supply voltages between +6 and +15 V.

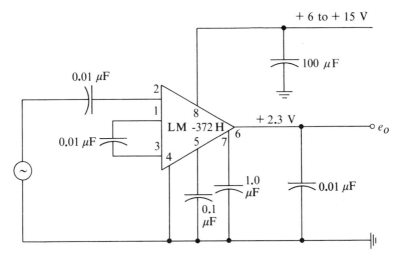

Figure 15.19. IC amplifier with AGC

16 Transistor Switching Circuits

Transistor switches offer small size, high reliability, low power loss, nano second switching speeds, and noiseless operation. They are used as high-speed commutators in power inverters, and as voltage-controlled switches they substitute as low-power replacements for thyristors. Many millions of transistor switches are used in computers and in communication switching systems.

This chapter includes examples of transistor switching circuits that may be used as switching amplifiers, relays, silicon-controlled rectifiers, voltage-controlled switches, or voltage-controlled diodes. Practical applications of these circuits include a burglar and fire alarm and a flashlight-operated switch. Other circuits in this chapter include inverters, flipflops, multivibrators, triggers, and gates. These switching circuits may be used to shape or delay pulses, count, divide by two, and to control switching. These circuits are essential elements of pulse systems and digital computers.

16.1 TRANSISTOR SWITCHES AND RELAYS

For switching applications the collector and emitter terminals of a transistor are the terminals of a switch. When the base is connected to the emitter or reverse-biased, the collector is like an open switch except for a leakage current that is usually negligible. With a forward-biasing base current the collector circuit is like a closed switch for collector currents that are less than β times the base current. For moderate currents the voltage drop in the transistor switch is often less than 0.15 V, the voltage drop in a good quality mechanical switch.

In many applications a transistor may be switched ON by applying a base current that is 5 to 10 per cent of the maximum expected collector current and may be switched OFF by removing the base current or by connecting the base to the emitter. In circuits where the transistor is switched OFF by removing the base current a resistor may be connected across the base-emitter junction, or the base resistor may be returned to a reverse voltage source to ensure turn-OFF. The optimum value of the resistor or of the reverse voltage is best determined by experiment.

One of the common applications of transistor switches is for the control of power relays, as shown in Fig. 16.1(a). A transistor driver is used to operate the relay where the amount of control power is limited or where the electronic circuit must be isolated from the contact circuit. The relay in the figure is operated by supplying about 0.5 W to the coil, and the contacts turn on another device (such as a small motor). The transistor operates as a simple switch—not as an amplifier—because the base current is much higher than that required for an amplifier. The advantage of using a transistor to operate the relay is that the required base control power is less than one-tenth the relay power and the switch contacts are not subjected to high-current inductive arcs. Observe that the transistor must be protected from high-voltage inductive spikes by shunting the relay coil with a diode.

The circuit in Fig. 16.1(b) illustrates the use of a high-current power transistor for switching nearly a kW (kilowatt) of power. This load power is 20 times the power dissipation rating of the transistor at 60°C, and the load current is 10 times the current in the operating switch. By adding a driver

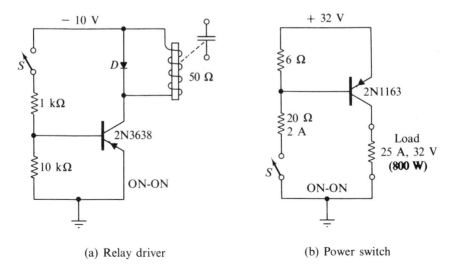

(a) Relay driver   (b) Power switch

Figure 16.1. Transistor switches

stage, the operating current can be lowered by a factor of 10. Higher load currents can be switched by paralleling transistors, if transistors with higher current ratings are not available.

16.2 SWITCHING AMPLIFIERS

Figures 16.2 and 16.3 show examples of 2-stage transistor switching amplifiers. In these circuits both transistors operate as switches, either full ON or full OFF, and the input switching currents are about one-tenth the current required for a single-stage transistor switch. Figure 16.2 requires a pair of *npn* transistors and Fig. 16.3 uses alternate *pnp* and *npn*. Otherwise, there is little difference between the two circuits. Both circuits have the emitters essentially connected to one side or the other of the power supply, and both have a 1-kΩ resistor in the first-stage collector circuit in order to limit the second-stage base current. By observing these practices, we may easily add one or more power stages and switch much higher power levels.

The two-stage switching amplifier in Fig. 16.4 is sometimes used as a way of reversing the switching action, so that closing the switch *S* turns OFF the load current. In this circuit the second-stage transistor is OFF when the first stage is ON.

Circuits similar to those in Figs. 16.2, 16.3, and 16.4 are often used as switching amplifiers or as squaring amplifiers, i.e., converting irregular waveforms to square waves. The purpose of such amplifiers is to produce a square wave or a series of variable-width square waves that are under the control of a sine wave or a series of pulses. The amplifiers shown produce a square wave signal across the load when driven by a 5-to-10 V rms signal.

The circuit in Fig. 16.5 illustrates the use of a switching amplifier to obtain equal and opposite switching signals in response to a 1-V input pulse. When

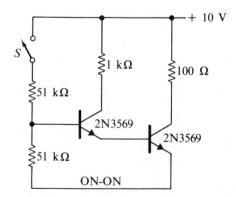

Figure 16.2. CC-CE switching amplifier (like transistors)

Sec. 16.2 Switching Amplifiers **305**

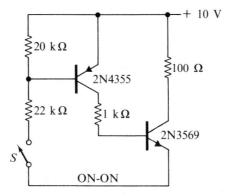

Figure 16.3. CE-CE switching amplifier (opposite transistors)

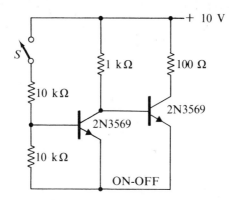

Figure 16.4. CE-CE switching amplifier (like transistors)

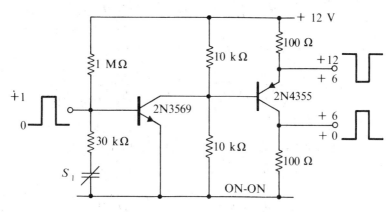

Figure 16.5. Switching phase-inverter

the collector of the output stage is connected back to the input base, regenerative feedback makes the switch lock-ON in response to a 1-V pulse. Switching circuits with regenerative feedback may be used as latch-up switches and as low-power replacements for thyristors.

16.3 ALL-PURPOSE LATCH-UP SWITCH (REF. 7)

The direct-coupled transistor pair shown in Fig. 16.6 has positive feedback that forces the transistors to remain OFF or ON, so that there are either a few microamperes through the load or nearly the full supply voltage across the load. With suitable values of the external resistors the pair may be used as a low-power thyristor, a Shockley diode, or a silicon controlled switch (SCS).

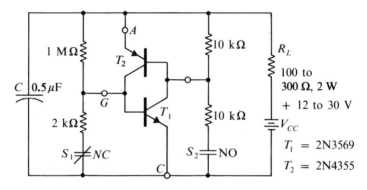

Figure 16.6. Transistor-pair lock-up switch

With the components shown in Fig. 16.6 the latch-up switch may be turned ON either by opening the normally-closed contacts (NC) or by closing the normally-open contacts (NO). The current through the load is less than 30 μA when the switch is OFF, and, when ON, the load is across the power supply, except for a 1-V voltage-drop in the switch. With a 100-Ω load, supply voltages from 3 to 15 V may be used, and with a 300-Ω load the switch may be used with supply voltages up to 30 V. The transistors should be high-β switching transistors rated for the expected load current. Those shown in Fig. 16.6 are rated for load currents up to 150 mA. The capacitor C is needed to prevent turn-ON by power supply transients, or the capacitor may be omitted to make the transistors turn-ON when the power line is pulsed.

The transistor switch is useful for operating loads directly or through a relay with switching currents less than 30 μA in the NC circuit and currents less than 1 mA in the NO circuit. The all-purpose switch may be used to convert a NO switch to an NC switch and vice versa. With power switching transistors the circuit may be used at higher currents as a substitute for a thyristor. With a capacitor-coupled input the switch may be turned ON by a pulse. An important advantage or disadvantage, as the case may be, is that the switch is turned OFF only by interrupting the power supply. However, as a latch-up switch the circuit is relatively uncomplicated.

16.4 TRANSISTOR SCR EQUIVALENT (REF. 7)

A positive feedback transistor pair may be used as a low-voltage, silicon controlled rectifier, as shown in Fig. 16.7. The anode of the pair is connected to the power source through the 100-Ω, 2-W load resistor. The control terminal, or gate, G is protected from over-currents by the 100-Ω series resistor. The shunt resistor and the capacitor C are to prevent false triggering by leakage currents and by capacitor-coupling to the anode A. The diode D is not required with a dc power supply, but with an ac supply the diode is needed because the emitter diodes have a relatively low reverse-breakdown voltage.

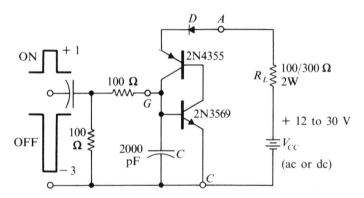

Figure 16.7. Transistor-pair SCR

The SCR may be turned ON by a positive gate signal greater than 0.5 V and may be turned OFF by a negative signal greater than 3 V. Because the gate turn-OFF signal is relatively large, SCRs are usually turned OFF by removing the anode supply. If the supply is ac, the SCR turns OFF each cycle, and with dc the SCR turns OFF when the gate signal is removed if the anode current is less than the minimum holding current. Thus, if the load resistor exceeds about 1 kΩ, the transistor pair turns OFF when the gate signal is removed. The voltage drop in the transistor SCR is 0.75 V, and with the diode the voltage drop is about 1.3 V.

16.5 TRANSISTOR BREAKDOWN PAIR

The transistor pair shown in Fig. 16.8 is the equivalent of a Shockley 4-layer breakdown diode. A Shockley diode is a 2-terminal, 4-layer diode that does not conduct until a specified breakdown voltage is reached, at which the diode

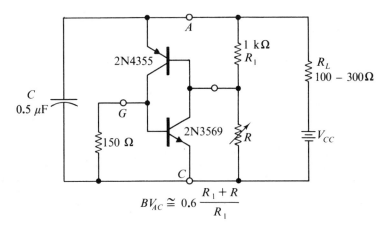

Figure 16.8. Transistor-pair break-down diode

turns ON. The breakdown voltage in the reverse direction is higher and not well defined. Shockley diodes are generally low-power devices that have switching voltages from 10 to 200 V and dissipation ratings from 0.5 to 5 W.

The transistor equivalent conducts when the voltage across R_1 exceeds 0.6 V. The resistors R and R_1 form a voltage divider across the supply voltage, and, by adjusting R, the diode may be made to turn ON at any voltage between 1 V and the voltage rating of the switching transistors. If R is 9 times R_1, the diode turn-ON voltage is 10 times 0.6 V, or 6 V. Replacing the resistor R by a Zener diode makes the breakdown pair turn ON at about 1 V above the Zener-diode voltage rating.

16.6 TEMPERATURE- OR LIGHT-OPERATED SWITCH

A relay may be operated by a light beam or by a temperature change using the transistor-pair lock-up switch shown in Fig. 16.9. The switch is activated when the resistor R increases to a value that makes the peak gate voltage approxi-

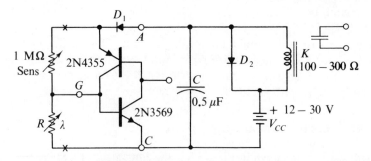

Figure 16.9. Photo-resistor switch

mately 1 V. When the resistor connections at x-x in the figure are interchanged, the relay turns ON when the resistance R changes from a high to a low value. If the temperature or light-sensitive element changes by a factor of 100, say from more than 1 MΩ to less than 10 kΩ, the circuit sensitivity may be adjusted with a 1-MΩ variable resistor.

With a dc supply and a dc relay the switch stays ON until the power is disconnected. With an ac relay and a low-voltage ac supply the relay follows the state of the light-sensing resistor, as in a door opener. However, in ac applications the ac supply should be connected through a diode D_1 to protect the transistors from breakdown when the anode voltage is reversed. The capacitor C is not used with an ac supply, and triggering by the rate effect may be prevented by connecting a small capacitor between the gate and cathode. The capacitor may be selected by trial, and 2000 pF is suggested as a suitable value.

16.7 A BURGLAR AND FIRE ALARM

A burglar alarm should be reliable and require negligible standby power. The alarm should be battery-operated to prevent failure when the ac supply is OFF, and the alarm should turn OFF without ringing too long. The alarm circuit shown in Fig. 16.10 is designed to meet these power requirements and is made reliable by a careful choice of the components and by using as few components as practicable.

A 25-cm Faraday bell may be purchased that operates on 12 dc and less than 150 mA. With such low power required the 12 V dc may be obtained by connecting a 9-V transistor battery in series with a $4\frac{1}{2}$-V D-cell booster. The transistor battery operates the bell for 30 to 60 minutes and runs down if

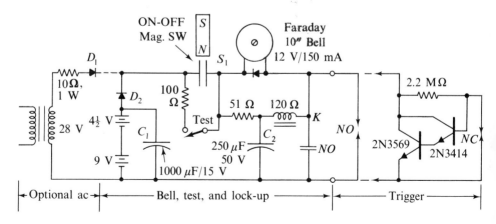

Figure 16.10. Burgler and fire alarm

the alarm cannot be turned OFF when the owner is away. The remainder of the circuit is mainly an activating switch S_1 in series with the bell and a magnetic-switch relay connected to latch-up if a trigger is tripped.

A magnetically-operated switch is recommended for S_1 because it is reliable and easily concealed. A normally-OFF switch may be turned ON or OFF with a small magnet used as a key to set or disable the alarm. Because the bell uses a low operating current, it may be located 30 m from the control package and connected with 24-gauge wire. The 1000-μF capacitor across the battery is needed because the transistor battery may be unable to supply the peak current required by the bell.

With the circuit shown in Fig. 16.10 the alarm may be operated, if necessary, on a 28-V ac supply such as is available in a home furnace. The half-wave rectifier and the 10-Ω series resistor reduce the dc voltage to about 13 V. Both the ac supply and the battery are connected through diodes that allow operating the alarm when either the ac is OFF or the battery is short-circuited. The resistor and capacitor in the relay supply ensure that the relay remains closed if the supply voltage drops momentarily, as with the half-wave rectified supply.

The test switch with the 1-kΩ series resistor may be used to charge the capacitor C_2 and operate the bell for a single stroke when a trigger switch is tested. Replacing the resistor with a 6-V, 0.1-A pilot lamp permits testing all but the bell with a visual indicator.

The trigger circuits of most burglar alarms use a normally-closed loop that may be disabled by a skillful intruder. A normally-opened trigger has the advantage that switches may be placed at a few strategic locations to present several opportunities for detection, whereas protecting against entry may require many circuits, and a single disabled switch offers no further protection. The trigger switches for either system may be made with 14-gauge, hard-drawn silver wire that can be purchased at a hobby shop or jewelry supply shop. The silver wire should be used to ensure contact reliability, and a pair of 10-cm lengths of wire are easily shaped to make contact on a 5-mm displacement of a door, a drawer, or a window.

A series trigger circuit is more reliable for a fire alarm. A Darlington inverter may be used to connect NC switches to the NO trigger circuit, as shown in Fig. 16.10. The Darlington pair is turned ON by 6 μA and with a current gain of 10,000 may be used to turn ON a NO trigger that requires as much as 60 mA. The 6-to-10 μA standby current that flows when the NC trigger is closed is not large enough to reduce the battery life below the normal shelf life. Switches for the fire alarm may be made with the silver wires and triggered by melting a nylon cord. An NC switch may be made of a length of soft solder placed close to a fire hazard, and heat-sensitive switches opened by a rapid rate of temperature rise may be purchased. A bedside switch may be installed to trigger the alarm and summon help in an emergency.

16.8 FLASHLIGHT-OPERATED TV SOUND SWITCH

The circuit shown in Fig. 16.11 is designed to permit turning the TV loud speaker sound OFF or ON by use of a flashlight. A reed relay K_1 is connected across the voice coil of the speaker, and the speaker is shorted when the light cell L_1 is illuminated by the light beam of a flashlight. After one minute the 10-μF capacitor discharges, and the speaker output resumes the normal sound level. When desired, the sound may be turned ON earlier by illuminating the cell L_2.

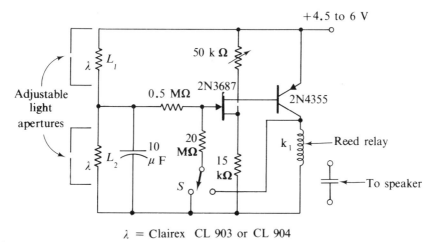

Figure 16.11. Low-power light-operated relay (TV ad-killer)

Important advantages of this circuit are that it may be operated on flashlight cells and the 100-μA standby current is so low that the batteries last more than a year without a battery switch. The circuit may be safely connected to the TV because the voice coil is usually insulated from the chassis or the power circuit, and a shorted voice coil does not damage the audio power circuits used in TVs. If desired, a small resistor may be inserted in series with the relay contacts, so that the sound level is only reduced instead of being turned OFF.

The low standby power and the use of a battery supply without a power switch make this circuit useful in many relay applications. The circuit can be assembled in a small package that does not require external connections except where desired. The light apertures may be adjusted to make the circuit operation relatively independent of the ambient light level, and, by connecting the switch S to the relay, the relay stays ON (as for an alarm) until the switch is opened. By replacing L_1 or L_2 by a resistor, the relay may be operated by interrupting a light beam. The simplicity, small package size, and long battery life make this relay control useful in many applications. With a 6.5-V mercury battery that has the volume of a single D cell the standby life is approximately the shelf life of the battery.

16.9 SWITCHING INVERTERS

The switching amplifier shown in Fig. 16.12 has two identical stages, and in each stage the input signal appears inverted at the output. Each stage is called an *inverter* because an ON input signal is converted to an OFF output signal. An inverter followed by a second inverter is called an *inverter-inverter*. Switching inverters are a basic component of pulse and digital circuits. The logic and counting circuits of a computer are usually a series of inverters. Flipflops, multivibrators, and triggers are usually inverter-inverters with positive feedback.

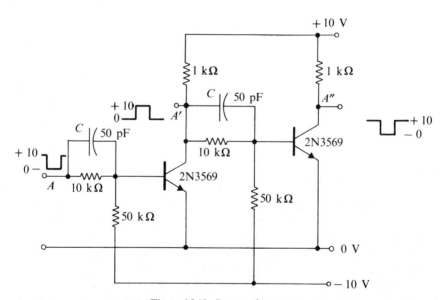

Figure 16.12. Inverter-inverter

Whenever a transistor is turned ON, electric charges are stored in the base-emitter capacitance, and the charge must be removed before the transistor will turn OFF. In a similar way charge must be stored in the capacitance before a transistor turns ON. The sum of the turn-ON and the turn-OFF times is the total switching time of the transistor, a figure of merit. A high-speed computer may require a total switching time of less than 0.1 μs (microsecond).

The switching time may be reduced by using small high-frequency transistors that have a low base-emitter capacitance. The turn-ON switching time is reduced by shunting the series resistor with a small speed-up capacitor, shown as C in Fig. 16.12. The turn-OFF time is reduced by applying an excess of reverse-biasing voltage to accelerate the removal of stored charge. The reverse bias is also needed to prevent turn-ON at high operating temperatures. For both reasons switching circuits usually have positive and negative supply voltages, as shown in Fig. 16.12.

A switching inverter is useful as a source of both equal and out-of-phase switching voltages. Perhaps the most common use of a two-stage inverter is in flipflops that are constructed by connecting the output terminal A'' back to the input A. The resulting feedback is a positive feedback that enhances the input signal. A flipflop locks itself in one state and can be switched to the other by coupling a pulse of the correct polarity into either collector. By coupling the pulse into both collectors via steering diodes, a flipflop switches alternately for each input pulse; in this way two positive input pulses produce one positive output pulse, and we have a scale-of-two divider or counter.

Flipflops and inverters form an important group of switching circuits in which transistors are triggered by a pulse to switch from ON to OFF and vice versa. Some of these circuits, known as **monostable triggers**, can be used to form a pulse of predetermined amplitude and time duration when triggered by a lower amplitude pulse. Other circuits use inverters as switching amplifiers in logic circuits that respond to a required combination of inputs. These, for example, may turn ON when any one of a group of inputs is turned OFF. Flipflops, triggers, and logic circuits are the basic building blocks of computers, scale circuits, and switching devices.

16.10 FLIPFLOPS (REF. 7)

A flipflop is a positive feedback amplifier in which one transistor is OFF when the other is ON until an overriding pulse forces the transistors to exchange roles. The exchange of states produces an output pulse having a fixed magnitude and a steep wave front. Because successive output pulses have alternate polarities, a succeeding flipflop can be made to respond only to every other pulse as a binary counter.

The flipflop in Fig. 16.13 is clearly very similar to the two-stage inverter (Fig. 16.12) except that an emitter resistor is shared by the emitters and the base resistors are lowered so that they can be returned to ground. Triggering is initiated by applying a steep-sided pulse or square wave that moves the Q-points of both stages into the active region. Because the positive feedback makes the system unstable, each Q-point moves the other until one is driven full ON and the other full OFF. The states exchange because the capacitor on the OFF-side initially has several times the charge of the other capacitor, and the ON transistor is turned OFF by the predominating charge. Although both emitters are subjected to the same pulse, the initial charge on the feedback

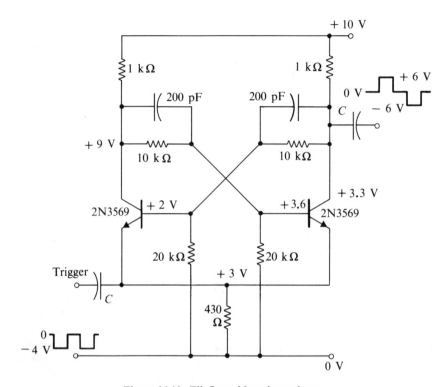

Figure 16.13. Flipflop with emitter trigger

capacitors forces the system to return to the opposite state, provided that the trigger pulse has sufficient energy to move the Q-points into the active (amplifier) region and is short compared with the decay time of the coupling capacitors.

The Q-point conditions of a saturated flipflop can be approximated by assuming that one transistor is removed from the circuit and that the other has the base, emitter, and collector shorted together. The base current may be neglected. The design of a good flipflop is complicated by the need to ensure that the transistors can be turned OFF at high temperatures when the collector leakage current is high, and by the need to ensure that the ON transistor is in saturation at low temperatures when the current gain is the lowest. The size of the capacitors is a compromise between being large to ensure reliable switching and small to permit fast switching.

Flipflops may be triggered by applying a pulse to both emitters, both bases, or both collectors. The trigger pulse may be coupled through a resistor, capacitor, or diodes. Depending on its polarity, the trigger may turn both transistors ON or OFF, and the collector-to-base capacitors in discharging make the transistors exchange states.

16.11 MULTIVIBRATORS (REF. 7)

A multivibrator is a 2-stage amplifier that is connected back to itself so that there is an excess of positive ac feedback. A multivibrator, as shown in Fig. 16.14, appears on the surface to be the same as a flipflop. However, the dc feedback in the multivibrator is reduced by separating the emitters so that the static Q-points are in the active region and are stable. The capacitors that cross-couple the collectors to the bases increase the ac loop gain and force the transistors to switch alternately ON and OFF between temporarily stable states. The astable, or free-running multivibrator, generates a square wave that is useful as a clock timing frequency, and the square wave is easily converted to a sawtooth by driving an RC integrator or converted into a pulse by an RC differentiator (see Chap. 14).

The switching period of a multivibrator is determined by the coupling capacitors and the resistors of the circuit. In turn, each capacitor is charged through the forward-biased base-emitter diode in less time than is required to discharge the other, so that the half period of a cycle is determined only by the discharge times. Since the transistor is reverse biased, the discharge time is determined mainly by the base resistor. Because switching is a nonlinear process and a calculation of the multivibrator frequency is only an approximation, the frequency shown in Fig. 16.14 is experimentally measured and changes with the supply voltage and the circuit loads. For a constant frequency a multivibrator is usually synchronized to a pulse obtained from a crystal oscillator.

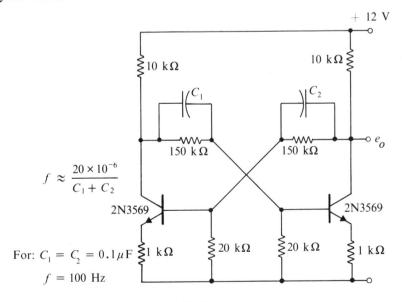

Figure 16.14. Multivibrator

16.12 MONOSTABLE MULTIVIBRATORS

A monostable multivibrator, or one-shot, is a flipflop that has one base ac-coupled and the other dc-coupled, as shown in Fig. 16.15. This coupling arrangement produces a stable and an unstable state. When triggered by a pulse, the mono switches to the unstable state and is held there until the coupling capacitor discharges. The circuit returns to the initial, or relaxed, conditions that are determined by the resistors of the circuit. The duration of the pulse output is determined by the capacitor C and the resistors through which it discharges, so that the duration of the output pulse is always the same. In the circuit shown in Fig. 16.15 the capacitor discharges through both bias resistors as if they were in parallel (by the Thevenin equivalent). The voltages indicated in the figure are those existing under relaxed conditions with the input transistor normally ON.

A mono is used to produce a pulse of predetermined amplitude and shape each time the mono is triggered. Monos are called *pulse shapers* and *pulse restorers*. They are used to produce a measured time delay after a triggering pulse and sometimes are used as frequency dividers.

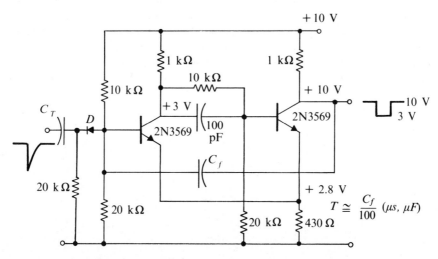

Figure 16.15. Monostable flipflop or frequency divider

16.13 SCHMITT TRIGGERS

A Schmitt trigger is a flipflop that has one RC feedback connection removed, as shown in Fig. 16.16. Because the emitters share a common resistor, the amplifier has a large positive dc feedback that drives the transistors into a stable ON or OFF state. When the input terminal is grounded, the input stage

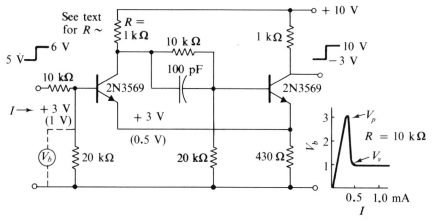

Figure 16.16. Schmitt trigger

is OFF and the output stage is ON. As the input voltage is increased, there is no change until the voltage reaches a threshold level—about 6 V—when both Q-points move into the active region. Because the emitters are regeneratively coupled, the gain is high: a small increase of the input voltage drives the output transistor rapidly from ON to OFF, and the input transistor is turned ON by the positive feedback. If the input voltage is lowered, nothing happens until the voltage drops to a reset level—about 5 V—where the transistors switch back to their initial states. The difference in voltage between which the switching is initiated by increasing and decreasing signals is called the *hysteresis voltage*, or *backlash*. By making the collector resistor 1 kΩ in the first stage and 10 kΩ in the second stage, we reduce the hysteresis voltage to 0.1 V. The small coupling capacitor steepens the wave front and speeds the switching.

A Schmitt trigger is useful for detecting a particular voltage level, for restoring waveforms, or for converting sinusoidal and nonrectangular signals to square waves—a process called *squaring*. Schmitt triggers are easily constructed to operate at frequencies above 1 MHz.

If the collector resistor in the first stage of a Schmitt trigger is increased to about 10 times the second stage collector resistor, switching produces a change in the common emitter voltage and a corresponding change in the input base voltage. The effect of this voltage change, shown in Fig. 16.16, is to reduce the input base voltage about 2 V at the trigger point as the input current is increased. A drop in voltage with an increase in current means that there is a negative resistance, which in this case is produced by positive feedback. Between the peak voltage V_p and the valley voltage V_v, the input resistance is a high negative resistance. Negative resistances are sometimes used to offset a positive resistance, as in an active filter, or to reduce the loss in a long transmission line.

16.14 TRANSISTOR AND FET GATES

Transistor gates may be used to switch low-frequency signals from one channel to another or to turn signals ON and OFF. If the gating signal is a steady high-frequency square wave and the channel frequencies are low or dc, the transistor is called a *chopper*. Choppers are used to convert dc and low-frequency signals to a high-frequency signal that can be amplified in an ordinary ac amplifier. After amplification the chopped signal may be reconverted by rectification or synchronous chopping to the original low-frequency signal.

The transistor gate shown in Fig. 16.17(a) is an effective switch for signals between 3 mV and 10 V, with the lower signal limit set by the fact that the ON voltage of the transistor is 1 mV. The transistors designed for chopper applications are used with the emitter and collector interchanged or inverted, and a diode is connected at the base to limit the gating signal when the transistor is reverse-biased. Because the gating signal is coupled into the signal channel by capacitance, transistor choppers are limited to operating frequencies below 1 kHz. Double-emitter transistors may be used to obtain a factor of 10 lower ON voltage, but the circuits are more complicated and require a transformer-coupled chopper signal.

The JFET circuit shown in Fig. 16.17(b) may be used either as a gate or

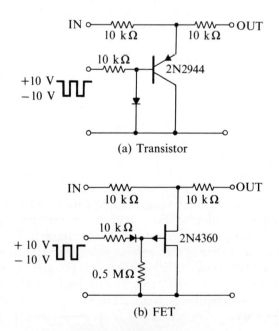

Figure 16.17. Gate or chopper

Sec. 16.15   MOS-FET Choppers   **319**

chopper. For chopper service the FET is driven by a square wave with a peak voltage that is about twice the pinchoff voltage of the FET. The chopper may be used at frequencies up to 10 kHz, and is satisfactory with input signals from 0.5 mV to 5 V, a dynamic range of 80 dB.

16.15 MOS-FET CHOPPERS

Insulated-gate FETs offer advantages in chopper applications when the chopping frequency exceeds 1 kHz. A MOS-FET chopper may be operated with a symmetrical, $+10$ V and -10 V, square-wave drive without the gate conduction problems of a JFET. Devices with a feedback capacitance less than 0.5 pF may be used in simple circuits at chopping frequencies as high as 100 kHz, or up to 10 MHz in carefully designed circuits.

The circuit in Fig. 16.18(a) illustrates the use of a single-gate MOS-FET as a series chopper where the mean of the source and load impedances is less than 1 kΩ. The circuit in Fig. 16.18(b) is for a shunt chopper that is used when the source and load impedances exceed about 1 kΩ. In both circuits the gate is shunted by a 510-Ω resistor and a capacitor C. For more exacting requirements a complementary MOS-FET pair may be used as a series-shunt

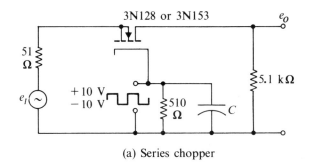

(a) Series chopper

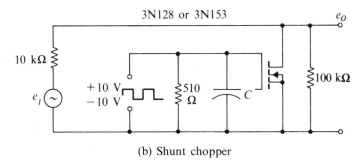

(b) Shunt chopper

Figure 16.18. MOS-FET chopper circuits

chopper with the gates driven in parallel, instead of using a phase-inverting drive as with transistors.

The capacitor C serves to reduce the slope of the gate signal and to reduce capacitance feed-over into the signal channel. In some circuits a sine-wave drive may give less cross-over than a square wave. The optimum value for C is best determined by trial while the output waveforms are observed with a low input signal. Generally, the optimum value for C makes the reactance of C at the chopping frequency approximately equal to the shunt resistance in ohms. The choppers shown in Fig. 16.18 may be used at frequencies up to 100 kHz. The minimum useful signal level is approximately 1 mV, and the signal is clipped on one side when the peak input voltage exceeds 0.5 V.

17 Laboratory Instruments and Methods

Most troubleshooting and circuit problems may be resolved quickly and easily with inexpensive and simple instruments. An ohmmeter, oscilloscope, and an oscillator require only a small investment and are the instruments used almost exclusively in a laboratory. Through analysis and study of working circuits in a home laboratory many persons have acquired the confidence and understanding needed to enjoy the challenges of practical electronic problems. The skills acquired in locating and correcting application difficulties are a valuable asset that may be developed into a lifetime of enjoyment and profit.

This chapter describes the instruments needed for electronic work in a small laboratory. Decribed are an arrangement of equipment and a circuit that simplify experimental studies and adjustments. Later in the chapter several instruments and power supplies are described that may be built as the needs develop for additional laboratory equipment.

17.1 LABORATORY INSTRUMENTS

Most application difficulties are resolved by a careful visual examination of the physical circuit and a few volt-ohmmeter (VOM) measurements. Most circuit problems may be found by searching for defective components, poorly soldered connections, and errors in assembling the circuit. These errors are sometimes hard to find and may be located only after a repeated search.

Sometimes leaving the problem and returning later for a fresh look at the circuit may resolve the difficulty.

Without doubt the ohmmeter is the most useful of all laboratory instruments, and an inexpensive model is the beginner's best purchase. With experience we acquire the habit of returning the selector switch of a meter to OFF or to a voltmeter position so that the meter is not carelessly damaged by connecting a sensitive current range to a power source. Because of the instrument's broad usefulness, two ohmmeters are desirable, and a better meter may be a good purchase after one has had the experience of damaging a first meter.

An inexpensive oscilloscope and an oscillator are needed for the development and adjustment of new circuits and for performance tests. An oscilloscope saves time by facilitating the location of many circuit problems, but except for waveform studies, most gain and performance studies may be executed almost as well with an ac voltmeter as with an oscilloscope. Only the most expensive oscilloscopes can be used for studies at high frequencies, and many radio technicians use only an RF oscillator, a diode ac voltmeter, and an ohmmeter. The diode voltmeter may be constructed as a probe to be used with your ohmmeter.

Several instruments and tools may be constructed with a few inexpensive components that may be at hand already. These include laboratory power supplies, a transistor gain tester, an ac voltage divider, a calibrated voltage reference, and simple amplifiers for increasing the sensitivity of your oscilloscope.

A pair of good long-nose and diagonal pliers that are made for electronic work are a good investment and a pleasure to use. A small 25-W soldering iron, an assortment of 20- or 22-gauge insulated wire, a simple wire stripper, and an assortment of small tools are needed.

Much time may be saved in experimental work by using resistors with a miniature clip soldered at each end. A supply of two resistors in each of the most frequently used values is adequate, and values that are multiples of 1, 2, and 4.7 make a good beginning supply. Two diodes and a few capacitors that may be clipped into circuits are frequently needed.

17.2 LABORATORY TECHNIQUES (REF. 3)

Whenever possible, a permanent but small laboratory should be set up where the equipment can be left in place and used when time permits. For occasional use, a laboratory can be set up on a kitchen table, but most development work and tests require more time than anticipated, and the experimenter needs a relaxed and uncluttered working area. The sketch in Fig. 17.1 suggests a convenient arrangement of instruments and circuit grounds.

A table with a Formica surface is about the most satisfactory working

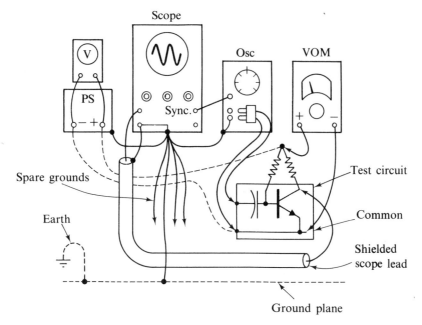

Figure 17.1. Laboratory set-up

surface for electronic work. A metal table provides a good ground plane to reduce 60-Hz pickup, and a wooden table may be given a ground plane by placing a sheet of aluminum foil between the table and a sheet of Formica. The table or the foil may be connected to a water pipe, if convenient, but for most purposes the foil reduces the pickup as long as the foil is connected to the ground side of your circuits and the circuits are close to the table and at least a foot from ac leads and lighting fixtures. The metal cabinets enclosing the oscilloscope, oscillator, main power supply, and the ground side of the oscilloscope should be connected to each other and the ground plane.

A 3-inch oscilloscope is large enough for home use and is not too long for use on a shelf, nor too large for a small table. A small scope has a fine trace that is as accurate at close range as a larger scope. A scope with an ac sensitivity of 4 mV/cm is adequate for most purposes, and a dc input is rarely needed. The oscilloscope is usually placed at one side of the work area. The oscilloscope input lead should be shielded for low-frequency work, but above 10 kHz a short unshielded wire is used to minimize feedback and to prevent capacitance loading of high-impedance circuits. The shield may be grounded at the oscilloscope and left without a connection at the probe end.

Several inexpensive kits for building your own audio-frequency oscillator are readily available, and an RF oscillator may be constructed if needed. Circuit studies require a way of reducing the oscillator output, and an oscillator equipped with a calibrated attenuator is called a *signal generator*. Audio-frequency oscillators generally supply frequencies from 20 Hz to 1-or-2 MHz, although a 100-kHz upper frequency is usually adequate. Audio oscillators supply at least 5 V with a 500-Ω load and 10 V open-circuited. When properly adjusted, the output signal is constant with frequency to within 1 or 2 dB for all but the top end of the frequency band.

For many purposes a *grid dipper* makes a very useful and satisfactory substitute for an RF signal generator. A grid dipper is an oscillator with a meter that indicates the oscillator's signal amplitude, and plug-in coils permit tuning the oscillator from 1 MHz to several hundred MHz. A circuit coupled to the dipper may be tuned by observing a decrease in the signal amplitude caused by energy transferred to the tuned circuit at resonance. The oscillator may be used to couple a test signal into RF amplifiers and to measure the frequency of another oscillator. A grid dipper may be purchased assembled or in kit form.

Several different power supplies may be needed in a small laboratory. Digital circuits need a regulated $+5$-V supply; portable instruments and radios need $+9$ V; automotive equipment operates on 12 to 16 V; aircraft and ships use 24 to 32 V; ac-operated audio equipment may use 30 to 34 V; and linear ICs and OP amps generally use equal $+$ and $-$ supplies in the 8-to-15 V range. Thus, the supply required in a small laboratory depends on the experimenter's field of interest and the amount of power needed. The experimenter may construct two or three of the power supplies described in this book, and the output voltage of these can be changed with an inexpensive IC voltage regulator for loads up to 1 or 2 A. For loads exceeding an ampere there is hardly any satisfactory substitute for a small storage battery.

A variable-voltage auto transformer has many uses around a laboratory. The continuously variable ac voltage is useful for examining the sensitivity of ac equipment to a voltage change, for applying ac power gradually to new equipment, for adjusting soldering-iron temperatures, and for reducing the output voltage of a rectified dc supply.

The ground side of the instruments and the test circuit should be connected as shown in Fig. 17.1. The ground side of the signal generator may be connected to the oscilloscope ground for low-gain measurements. For high-gain measurements the oscillator is best connected only to the input of the circuit under test, and the common side of the test circuit is connected to the oscilloscope ground. It is convenient to have several extra ground wires permanently connected to ground and supplied with clips at the free end to connect some of the many grounds needed in experimental work.

The dc power supply and the variable-voltage transformer that is used to change ac voltages may be located at arm's length from the work area, and all ac-power leads should be behind the instrument cases for shielding. Neither terminal of the power supply should be grounded because the supply may be connected to the circuit under test with the polarity changed from time to time. In a similar way, the oscillator should be grounded only by its connection with the test circuit. Grounds that are permanently connected should be made with care because a poor connection or an open ground may cause a considerable loss of time in resolving problems.

17.3 VOLT-OHMMETERS (REF. 3)

The volt-ohmmeter is used for measuring ac and dc voltages, dc currents, and dc resistances. An ohmmeter may use a sensitive microammeter with a switch that permits using the meter for different purposes and with different sensitivities. The most useful form of ohmmeter uses a sensitive meter that does not require an amplifier. These ohmmeters are built for rough service but may not remain accurately calibrated when used as an all-purpose tool. A VOM with an FET input stage offers a high input impedance but requires several batteries that may require replacement if the meter is left ON for long periods of time. The FET ohmmeter is usually more useful for laboratory measurements and may be kept accurately calibrated by using another VOM for rough service and everyday measurements.

For accurate measurements the internal resistance of a voltmeter should be at least 30 times the internal resistance of the voltage source being measured. The dc voltmeters of most ohmmeters used in electronic work have a sensitivity of 20,000 ohms per volt full scale. This statement means that a 3-volt scale has an internal resistance of 3(20,000), or 60 kΩ internal resistance. The 30-V scale has an internal resistance of 0.6 MΩ. These resistances are generally high enough so that the effect of the meter in reducing a voltage reading may be neglected. A more expensive ohmmeter with an FET input stage may have a 10-MΩ input impedance, but the simple ohmmeter has the advantage of not requiring an expensive battery supply and is usually more rugged.

The ac voltage ranges of an ohmmeter usually respond to either the average or the peak value of an ac signal, and the internal impedances may be much lower than the resistance of dc ranges. The ac ranges measure ac signals or voltages that are not superimposed on dc components, as in amplifiers. Ac superimposed on dc is measured by inserting a series capacitor, and most instruments offer this option when the signal is connected to an ***output terminal***. The lowest frequency at which the output reads the correct ac voltage should be determined by consulting the manufacturer's instructions or by testing the instrument with a variable frequency oscillator.

The dc current ranges of an ohmmeter must be used with care because the meter is easily damaged when a milliammeter is accidentally connected to a voltage supply. The meter is particularly vulnerable when used at the maximum sensitivity, and one should learn to avoid leaving a meter with the switch on a current range. In rare instances the voltage drop across the 1-mA, or more sensitive range, may affect a circuit and lead to an incorrect measurement and conclusion. The voltage drop may be 0.25 to 0.5 V on the sensitive range, and the internal resistance of the meter may exceed 3 kΩ.

Effective use of an ohmmeter requires at least a basic familiarity with the internal resistance of the lowest meter ranges. The voltmeter resistance is given on the face of the meter, and the ammeter resistance can be obtained by referring to the manufacturer's service manual. One should certainly know the ohmmeter circuit and memorize the open-circuit voltage and the short-circuit current of the low ranges. The short-circuit current should be known so that high-frequency transistors are checked using currents within the transistor rating. Similarly, the emitter diode of most transistors should not be subjected to reverse voltages as high as are produced on the high resistance range of some ohmmeters. The internal resistance of an ohmmeter is the resistance value read at the mid-scale corresponding to the range in use.

The ohmmeter is commonly used to make rough checks of the forward and reverse characteristics of diodes. Good low current silicon diodes usually show a reverse resistance of more than 20 MΩ. The forward resistance of diodes measured with an ohmmeter is variable and meaningless. Because the diode voltage drop tends to remain close to 0.6 V regardless of the current, an ohmmeter with a 1.5 V battery tends to read a little above mid-scale when the ohmmeter current is approximately the rated diode current. The current-voltage characteristics of a diode can be estimated by comparing the readings obtained on several ohmmeter ranges.

Zener diodes check as ordinary diodes when the ohmmeter voltage is below the breakdown voltage. The double anode Zener diodes are reverse-biased with either polarity and should show as open below the breakdown rating.

The polarity of a diode can be checked with an ohmmeter if we know which way the ohmmeter leads are polarized. Ohmmeters are made with either polarity, so the instrument should be checked, using a marked diode. The diode is an open circuit when the positive side of the ohmmeter is connected to the positive side of the diode. Remember that a diode can be connected to a voltage supply without a limiting resistor if the positive end is connected to the positive side of the circuit.

17.4 SIGNAL ATTENUATORS

The voltage gain of an amplifier is measured by means of a calibrated attenuator at the output of a signal generator. The measurement procedure begins with an adjustment of the attenuator to make the amplifier output signal 1 V when observed with an oscilloscope or an ac output meter. Then the oscilloscope input is moved to observe the amplifier input signal. The attenuator change required to make the input signal the same 1 V as observed at the output is a measure of the amplifier gain.

For high-gain amplifier studies a signal generator may need more attenuation in the output control than is provided by the manufacturer. Moreover, the usual 500-Ω output impedance is too high for many measurements because of errors produced when the circuit under study has an input impedance less than the attenuator impedance. These loading errors are reduced by constructing the 0-10-20 dB attenuator shown in Fig. 17.2(a). The resistors may be mounted on an isolated double-banana plug, and the output is easily switched with a miniature clip attached to the output cable. The attenuator terminates the signal generator in 510 Ω, as is sometimes required, and each step of the attenuator increases the signal loss by 10 dB, which is 5 per cent more than a factor of 3.

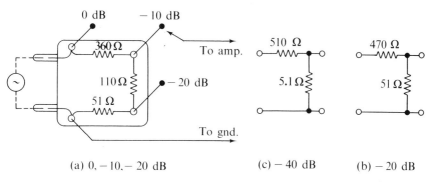

Figure 17.2. Attenuator plug-ins

The resistor values shown in Figs. 17.2(b) and 17.2(c) may be used to build 20- and 40-dB attenuators for additional loss and a low source impedance. However, to avoid loading errors, attenuators (b) and (c) must precede and not follow (a). For measurements requiring a 10-Ω source impedance the attenuator in Fig. 17.3 is designed to be used with an oscillator capable of driving a 500-Ω load with 5 V rms. The attenuator output is 1 V with a 100-Ω source impedance and is 100 mV and 10 mV with a 10-Ω source impedance. The attenuator may be preceded by the plug-in attenuators shown in Figs. 17.2(b) and 17.2(c). If a larger signal is required, a 500-Ω: 4-Ω transformer may be used to obtain 0.5 V with a 4-Ω source impedance.

For an oscillator lacking a satisfactory output control a 20-dB, 51-Ω, attenuator with 2-dB steps may be built, using the resistor values given in Fig. 17.4. This control offers the advantage of a 51-Ω maximum impedance that may be followed by the attenuators shown in Fig. 17.2. The 2-dB steps permit using an 11-step switch, and a 1-dB change at any position may be obtained with a push button.

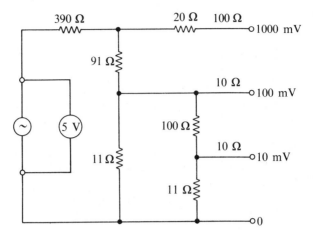

Figure 17.3. Signal source, low-impedance

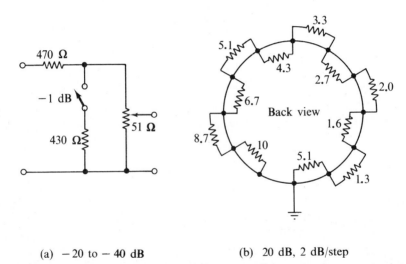

(a) −20 to −40 dB (b) 20 dB, 2 dB/step

Figure 17.4. Attenuator 2 dB/step with 1 dB push button

The connection between the output cable and the attenuator is a convenient place to insert a coupling capacitor that should be 100 μF, unless the amplifier input impedance is high compared with the attenuator resistance. For simulating a high-impedance source a resistor may be inserted at the amplifier end of the cable.

17.5 BREADBOARDS

A simple transistor stage may be assembled using clip-leads to connect a transistor and components, but the assembly tends to break apart and may be more time-consuming than a circuit soldered together on a breadboard. The breadboards used to construct experimental circuits have many forms that depend partly on the components and the form of the completed circuit. A simple terminal board with the components soldered in place has the advantage that the completed assembly is easily put away for later use or study. A perforated board, in which terminals are inserted where needed, offers convenient tie points and is better for complicated circuits. An inverted vacuum tube socket makes a convenient terminal arrangement for a round-can TO-99 IC, and the assembled amplifier and components may be easily connected into a larger breadboard circuit. Manufactured plug boards with sockets for inserting the wire leads of semiconductors, resistors, and flat-pack ICs are convenient for advanced experimenters, especially when using ICs, but these circuit boards are too expensive for occasional use. Moreover, they produce an assembly that is too cramped for low-frequency or power components, and capacitance feedback makes them impractical for high-gain audio and RF amplifiers.

Over the long run it is better to assemble circuits on terminal boards, using soldered connections. Place components that are less likely to be changed near the board and attach capacitors and transistors without cutting the leads. The collector load resistors should extend off the board, as shown in Fig. 17.5, and be brought together where they connect to the power supply or to a Zener regulator. The terminal boards may be supported by clamping the mounting studs between lengths of $\frac{1}{2}$ in $\times$ $\frac{1}{2}$ in (1 cm $\times$ 1 cm) iron bars. The iron bars give weight to the assembly and can be clamped together if they

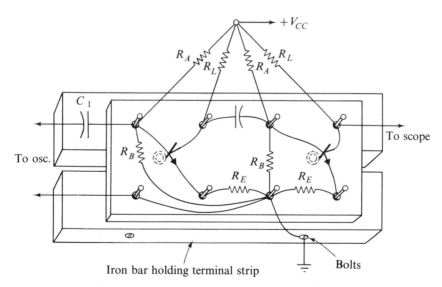

Figure 17.5. Terminal board with 2-stage amplifier

are drilled and tapped for two screws. The bars should be insulated by wrapping with tape, and sometimes they should be grounded.

17.6 TRANSISTOR TESTING

Well-designed circuits should operate satisfactorily if the transistors are neither open nor shorted and have a current gain of at least 50. With an ohmmeter, we can check most transistors easily and safely to establish that they pass these tests. The simplest test is to check that the transistor is not short-circuited. A transistor is a pair of back-to-back (coupled) diodes, having the base as the common side of both diodes. An ohmmeter that has one side connected to the base should show a low resistance when the other side is connected to either the emitter or the collector. With the ohmmeter connections interchanged, the meter should indicate the higher resistance of the reverse-biased diodes. Generally, the diodes do not check alike and the emitter diode tends to have the lowest forward and reverse resistance. Neither diode should be expected to show anything like the resistance ratio of a plain diode. When power transistors are checked, a suspected transistor should be compared with a new one of the same or similar type before a transistor is rejected as defective.

Punch-through causes a short-circuit from the collector to the emitter and is found by measuring the collector-to-emitter resistance. Good power transistors may show a reverse-biased collector-to-emitter resistance as low as 1 kΩ, and the resistance usually changes by a factor of at least 5 when the ohmmeter is reversed. With punch-through, the collector-to-emitter resistance is only a few ohms and does not change with the polarity. Transistors that behave as good back-to-back diodes and do not have punch-through are usually operative.

The emitter current and voltage ratings of high-frequency transistors should be considered before transistors or transistor circuits are checked by using an ohmmeter.

The current gain of a transistor can be measured with acceptable accuracy by using an ordinary ohmmeter and a resistor. This method uses the ohmmeter circuit to supply current; but, because β is proportional to the current, it is read on a convenient current or voltage scale. The meter is calibrated to read the correct current gain by selecting the resistor that supplies the transistor base current.

For power transistors the ohmmeter should be set on the lowest resistance range, and for low-power transistors the ohmmeter should be set on a range that reads 50 to 500 Ω at mid-scale. The high resistance ranges are unsatisfactory and may damage small, high-frequency transistors. Now, disregarding the fact that the meter is set on ohms, as shown in Fig. 17.6, a convenient meter scale, either for voltage or current, is chosen to represent the current gain and a number is selected to represent a current gain of 100. As shown in Fig. 17.6, the meter must be read at the low-current end of the scale where the voltage across the ohmmeter terminals is about 90 per cent of the open circuit voltage.

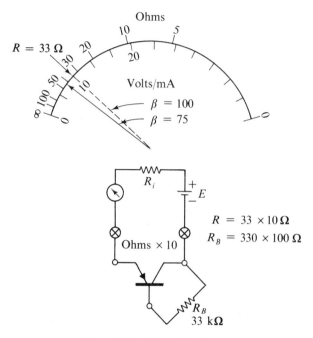

Figure 17.6. Ohmmeter β test

We calibrate the meter to read current gain by selecting the base resistor, as shown in Fig. 17.6. For example, the calibrating base resistor is 100 times the resistance at the scale number representing a current gain of 100. If the transistor has a current gain of 100, the meter will respond as if the collector circuit resistance is $\frac{1}{100}$ the base resistor. If the transistor has a current gain of only 50, the meter current will be one-half and read 50. *The transistor effectively reduces a fixed base resistance by the current gain, but the current is read as current gain rather than as a resistance.*

It is meaningless for practical purposes to require even a 10 per cent accuracy in a current gain measurement because current gain varies so much with the emitter current. This variation with the current can be observed by measuring the current gain on two different ohmmeter ranges. A measurement of β should always be made either to know that β is high compared with the amplifier S-factor, or to ensure that the transistor is reasonably good. There is no value in saying a transistor has a particular current gain unless the emitter current is given also.

Power transistors can be tested at higher currents than are attainable with an ohmmeter by using the circuit shown in Fig. 17.7. This circuit requires only a 10-V or 12-V power supply, a 100-mA meter, and a 10-kΩ resistor. Only the 10-kΩ resistor is needed if a suitable power supply is available. Many laboratory power supplies have a meter that will read 100 mA, and the 10 Ω that protects the meter can be omitted if the power supply is current-limited. A transistor is tested by connecting the power supply from collector to emitter, using the correct polarity. The collector current is read when the 10-kΩ base resistor is connected to the collector. (Actually, the base resistor can be connected to whichever side gives a reading.) Because the base current is 1 mA, the current gain is numerically the value of the collector current in mA. The transistor has to dissipate about 1 W during this test, and a heat sink is not usually required. Most small transistors that are rated to carry 100 mA will withstand 1 W power long enough to permit a meter reading, and the transistor will not be damaged if it can be held between the fingers during the test.

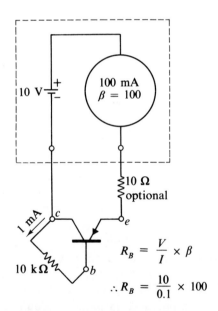

Figure 17.7. Power transistor β tests

The high range of a sensitive ohmmeter will show the equivalent collector leakage resistance of any good transistor. In a β measurement this resistance is negligible if the meter reading is large compared with the reading shown with the base resistor removed. The leakage current is most noticeable in testing power transistors.

17.7 TRANSISTOR CURRENT-GAIN METERS

A meter for measuring transistor current gains is easily built with a 10- or 100-mA full-scale meter. The circuit for the meter shown in Fig. 17.8 uses a 10-mA meter. The series resistor R_1 is 1500 divided by the full-scale meter

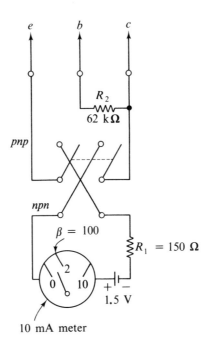

Figure 17.8. β test with 10 mA meter

current in mA, or $1500/10 = 150\ \Omega$. The base resistor R_2 makes the meter read one-fifth of full scale when $\beta = 100$. Hence, R_2 is 600,000 divided by the full-scale meter current, or $600,000/10 = 60\ k\Omega$ for a 10-mA meter. The scale may be marked with a line at one-fifth of full scale to designate $\beta = 100$, but the scale need not be calibrated because β values above and below the 100 mark can be interpolated visually with enough accuracy for practical purposes.

An ohmmeter may be equipped for β measurements by soldering the base resistor to an alligator clip. An *npn* transistor is tested by slipping the clip on the $+$ lead of the ohmmeter and connecting the clip to the transistor collector. Connect the opposite lead of the ohmmeter to the emitter and touch the free end of the resistor to the base. A *pnp* transistor is tested by interchanging the clips or the ohmmeter leads. Internal changes of the ohmmeter are not required.

17.8 LABORATORY POWER SUPPLIES

The simplest power supply for occasional or one-time use is a 9-V transistor battery or a lantern battery purchased at a hardware store. A battery has the advantage of complete freedom from ac hum and a low internal resistance. A lantern battery can supply as much as an ampere intermittently for several hours of tests. A battery made by series-connecting D-cells is alway useful in the laboratory, and a convenient power source is made by connecting the cells with 7-cm leads, soldered for a noise-free connection. The connections should be soldered with care to avoid excess heat, and the cells may be taped together in a convenient package. If the leads are stripped longer at one end, the bare wires provide convenient connections for intermediate voltage taps. A mercury cell cannot be soldered without shortcircuiting the cell.

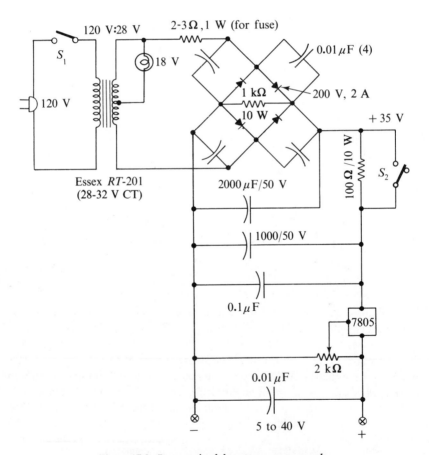

Figure 17.9. Inexpensive laboratory power supply

A multi-purpose, adjustable power supply for transistor amplifier studies may be constructed according to the circuit shown in Fig. 17.9. The power supply uses a rectifier-type ac transformer, a bridge rectifier, and a voltage-variable regulator. With the switch S_2 closed, the voltage may be varied from 5 V to 30 V with loads up to 1 A. With the switch S_2 open, the ac hum is reduced 30 to 40 dB, and 30-V output may be obtained with loads up to nearly 50 mA. An inexpensive voltmeter is convenient, but the supply voltage may be read with a second ohmmeter that has many other uses and is only a little more expensive than a panel meter. The multipurpose power supply has a broad range of uses in applications requiring a single-sided supply.

The 0.01 μF capacitors across the rectifiers in Fig. 17.9 are to prevent rectifier damage by line transients, and the capacitors tend to reduce audible transformer hum. The high-current supply has from 1 to 10 mV, peak-to-peak 120 Hz ripple, depending on the load current. With switch S_2 open, the ripple and noise are less than 1 mV p-p at 50 mA. The power supply may be used without the packaged regulator, and the output voltage may be adjusted with a variable-voltage transformer to change the line voltage. However, the packaged regulator reduces drift and the low-frequency noise caused by line transients and is well worth the additional expense.

17.9 AUXILIARY POWER SUPPLY

The laboratory power supply shown in Fig. 17.10 is useful as a source of equal + and − dc voltages or as a source of ac power. The supply is especially useful when driven by a variable-voltage auto transformer to obtain any ac or dc voltage between zero and the maximum. A transformer with a 24-V, 1-A secondary is readily available, and 30-V dc at 0.5 A may be obtained with a full-wave rectifier and a capacitor input filter.

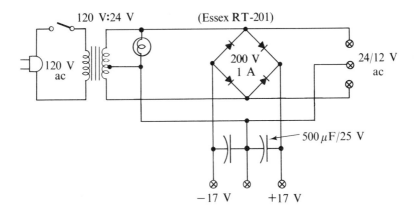

Figure 17.10. Auxiliary laboratory power supply

The circuit shown in Fig. 17.10 has the center-tap of the transformer connected to the midpoint of the filter capacitor. With the midpoint grounded the supply furnishes equal + and −17-V dc outputs with 120-V ac input Thus, with a variable ac input the dc supply may be used to power 0.5 A loads with 0 to ±15 V. The ripple voltage of a power supply increases with the load current and varies inversely with the capacitance of the smoothing filter. With a 100-mA load the 120-Hz ripple is approximately 5 per cent at the 17-V terminals. With loads less than 100 mA the ripple is proportionally lower. If the load is 0.5 A, the input capacitors should be increased to at least 1000 μF at 25 V.

17.10 VOLTAGE REGULATORS AND CURRENT LIMITERS

The auxiliary power supply described in Sec. 17.9 may be used with the voltage regulators or current limiters shown in Fig. 17.11. The current limiters in Fig. 17.11(a) may be set to limit the maximum load current to values between 2 mA and 100 mA. In this service the series transistors should have a small clip-on heat sink.

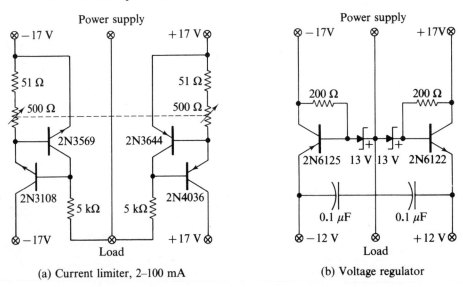

Figure 17.11. Limiter and regulator for power supply

The voltage regulators shown in Fig. 17.11(b) may be used to reduce the dc output to either 12 V or 9 V, depending on the Zener diodes. For 12-V output the regulators supply up to 1 A when mounted on a 10° C/W heat sink. Without heat sinks the regulators may be used with load currents up to

0.3 A. The 10° C/W heat sink may be a 7 cm × 6 cm flat sheet or an equivalent area, radiating on both sides in a 30° C ambient temperature.

With 8.2-V Zener diodes the output voltages are + and −9 V for low-voltage ICs or for battery-operated radios and instruments. With the 9-V output the load currents should be reduced to 0.5 A with heat sinks and to 150 mA without heat sinks. If desired, the current limiters may be used between the power supply and the regulators.

Transformers are available with a 17-V, 2-A, CT secondary for a + and −12-V dc nonregulated supply. A 12-V CT transformer may be used to obtain + and −8-V dc. Similarly, the transformer used with a low-voltage soldering iron may be used to supply power for an auxiliary dc supply.

17.11 INEXPENSIVE SHOP METER

A volt-ohmmeter useful for appliance and automobile servicing may be easily assembled from components available in almost any box of used electronic components. The most expensive component is a 1-mA meter that may be purchased used, or a more sensitive meter may be shunted to obtain the 1-mA, full-scale sensitivity.

With the circuit shown in Fig. 17.12 the meter may be used as an ohmmeter, as a 120-V ac voltmeter, or for trouble-shooting automobile 12-V dc circuits. As an extra bonus the 120-V ac meter may be used as a 60-V, full-scale dc meter. With a 620-kΩ external resistor the shop meter may be used to measure transistor current gains.

The circuit of the shop meter, Fig. 17.12, illustrates a suggested terminal arrangement viewed from the back side of the panel. As a way of avoiding

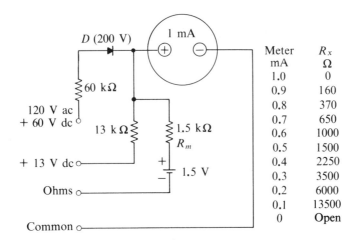

Figure 17.12. Inexpensive shop volt-ohmmeter

accidental short circuits in using the meter, a plastic box and panel are better than a metal enclosure. The resistor values given in the circuit diagram are suggested values, but it may be found necessary to trim the values by shunting with a 10-times larger resistor. The ac voltmeter is accurate enough for trouble-shooting if the meter reads full scale when connected to a wall outlet. The 12-V dc circuit is best adjusted to read about 90 per cent of full scale when connected to the 12-V battery of your auto. When adjusted to read 12 V at 90 per cent of full scale, the meter reads 13 V at full scale and may be used to indicate a fully charged battery when the engine is running and the alternator is operating correctly.

Trouble-shooting, for which this meter is intended, does not require a meter calibrated to read either volts or ohms. When the full-scale value is known, an estimate of a value that is not at full scale is usually adequate for the trouble shooting. The ohmmeter may be calibrated with a scale made from the scale values given in Fig. 17.12. If your ohmmeter does not read 1500 ohms at mid-scale, a correct scale may be made by changing all values by the percentage change needed to correct the mid-scale value. For a meter that does not read 1 mA full scale the external ohms R_x corresponding to any meter reading I may be found by using the formula

$$R_x = \frac{1.5}{I} - R_m \qquad (17.1)$$

where R_m, the total equivalent series resistance in the meter, is the external resistance needed to make the ohmmeter read at mid-scale. When $R_x = 0$, $I = 1.5/R_m$ is the full-scale meter current. Because the ohmmeter does not have an adjustable resistor, the meter is best made to read slightly more than full scale when the battery is new, and a D-size cell is recommended.

17.12 HIGH-FREQUENCY RF VOLTMETERS

The ac voltmeter of most VOMs may be used for measurements at audio frequencies. For some instruments an RF probe may be purchased to extend voltage measurements to 250 MHz. However, when a probe is not available, an RF probe is easily assembled for an occasional measurement and may be built into the housing of a felt-tipped pen. An RF voltmeter-probe that responds to the positive-voltage peaks of an ac signal may have either of the circuits shown in Fig. 17.13. In both circuits the rectifier is a high-frequency germanium diode, and the resistor in series with the meter makes the dc meter read rms volts. The meter reads rms volts when the resistor is 40 per cent of the meter resistance.

The circuit in Fig. 17.13(a) is used if the meter has a low dc input resistance, or is less than 100 kΩ. The 22-kΩ resistor shown in the circuit is

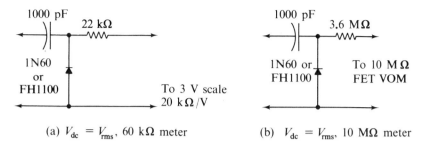

Figure 17.13. RF voltmeter circuits

correct for a 20-kΩ/V meter when used on a 3-V scale. On a 12-V scale the meter reads 25 per cent high, 2 dB, which is acceptable for many practical applications. Since the diode efficiency falls off below 1 V, a meter with a calibrated 3-V ac scale may be used for more accurate measurements at the low end of the scale.

The circuit shown in Fig. 17.13(b) is used with an FET VOM that has a 10-MΩ input resistance. As with the first circuit, the series-calibrating resistor is 40 per cent of the meter resistance, or is lower to improve small-signal accuracy. The ac voltmeter may be calibrated by increasing the capacitor in the diode circuit by a factor of 10 while comparing the RF probe at 10 kHz with the ac voltmeter of a VOM. Removing the extra bypass capacitor makes the RF probe inaccurate below 30 kHz but does not affect the high frequency calibration.

The impedance of a diode in a detector circuit is approximately one-half the dc load resistance. The ac input impedance of the diode voltmeter is approximately one-third the dc meter resistance up to a frequency at which the diode capacitance, approximately 2 pF, becomes the input impedance. The reactance of 2 pF is 1 kΩ at 100 MHz. Thus, if the voltmeter has a low capacitance and is carefully constructed to minimize shunt capacitance and series inductance (by using short leads), the meter should be accurate within 3 dB to 10 MHz in a 10-kΩ circuit and to 100 MHz in a 1-kΩ circuit. With a 100-Ω circuit the meter may be accurate up to 250 MHz. Above 100 MHz the physical dimensions of the circuit elements must be carefully decreased with increasing frequency and the RF voltmeter is best used only as an indicator of relative signal strengths. The RF voltmeter circuits have important applications when built into high-frequency circuits for ac signal monitoring.

The 1N60 germanium diodes have a 50-V breakdown rating that suggests keeping the voltage measurements below 10 or 20 V rms. Higher voltages may be measured by using a resistance voltage divider with 1 kΩ across the voltmeter input and a larger series resistor as needed to reduce the input voltage.

17.13 LINEAR AC VOLTMETER

A linear-scale ac voltmeter may be easily constructed with a single-stage transistor amplifier, a bridge rectifier, and a sensitive dc meter. The ac voltmeter shown in Fig. 17.14 has a sensitivity of 2 kΩ/volt and may be

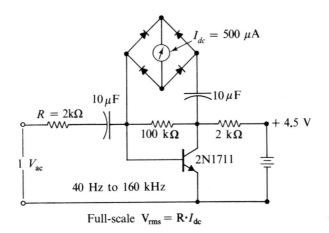

Figure 17.14. AC voltmeter

adjusted to give full-scale sensitivities upward from 100 mV rms. The meter may be operated on a 4.5 V or higher supply voltage, and the calibration is nearly independent of the supply voltage. The meter used in the voltmeter is a 500-μA dc meter and, as an ac meter, reads between 1-dB cutoff frequencies of 40 Hz and 160 MHz. The sensitivity is increased a factor of 5 by using a 50-μA meter. A higher sensitivity may be obtained by capacitor-coupling almost any low-power amplifier to the voltmeter input. The transistor operates as a high-gain amplifier, and there is little advantage in using an IC amplifier except to obtain increased linearity and a lower standby power. The voltmeter requires approximately 0.5 ma/volt of the supply voltage.

17.14 LABORATORY VOLTAGE STANDARDS

The calibration of meters and the maintenance of many instruments make a calibrated voltage standard a most useful accessory in an electronics laboratory. A high-quality voltage standard is often too expensive to own and has more accuracy than is needed in practical applications. However, with the circuit in Fig. 17.15 an inexpensive standard may be constructed for calibrating dc meters and for checking the incoming line voltage. The circuit is comprised of an isolation transformer for supplying the 120-V ac line to a

Sec. 17.14   Laboratory Voltage Standards

capacitor input rectifier and four Zener diodes that give convenient voltage values accurate to within one per cent of their initially measured values.

Voltage standards for calibrating the most common ranges of dc meters—namely, 3, 12, 30, and 150 V—are provided by a pair of series-connected Zener diodes. Because the voltage of the Zener diodes may vary as much as ± 5 per cent and vary with the supply current, the nominal diode voltages are selected to give a reference voltage that is less than the full-scale meter reading instead of a value that may exceed full scale. A calibration of the diode voltages may be obtained in almost any electronics laboratory or by using a meter new enough to be within the manufacturer's specifications. Perhaps the reference standard may be calibrated by a student in a technical school or by someone who works in an electronics laboratory.

The transformer is of an inexpensive low-power type, although almost any transformer with identical windings may be used as long as the winding resistances are between 1 and 2 kΩ. The capacitor is preferably a mylar or paper-type that does not load the supply with a leakage current, as with an electrolytic capacitor. The Zener diodes may be any one-half W, or preferably 1-W type, that is conveniently available, and a 5 per cent tolerance is adequate.

With the components shown, the reference voltages vary less than 2 per cent for a line-voltage change from 115 to 130 V. Throughout the United States the ac voltage at most customer locations is between 118 and 125 V. Thus, if the line voltage is between 110 and 125 V, the voltage values may be expected to be within 1 per cent of the values recorded when the voltages were initially measured.

The ΔV terminal shown in Fig. 17.15 is used to determine that the input voltage is adequate to ensure reasonably stable output voltages. The ΔV voltage should be read by connecting a 20,000 Ω/V meter between the ΔV

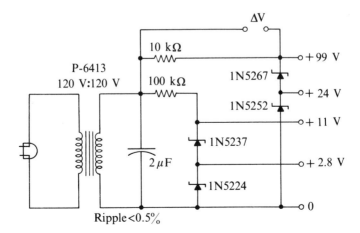

Figure 17.15. Practical Voltage Calibrator (voltages are nominal values)

terminal and the $+99$-V terminal. Before calibration of the reference voltages, a voltage-variable transformer should be used to ensure that ΔV is at least 5 V dc at the lowest expected line voltage. The ΔV values may be calibrated as a sensitive means of measuring the ac line voltage. As a rule, the ΔV voltage does not change more than 5 V from one place to another, and, as long as ΔV exceeds 5 V, the reference voltages may be assumed accurate within 1 per cent.

In addition to calibrating meters the reference voltages may be used to set the dc gain or calibrate an oscilloscope. With the oscillocope carefully calibrated ac meters may be checked by observing the meter reading obtained when a sine-wave signal source has a peak-to-peak deflection equal to a known dc deflection of the oscilloscope. A peak-to-peak ac meter is directly calibrated by this method, and an rms meter should read 0.354 times the peak-to-peak voltage.

The voltage calibrator may be used also as a voltage source for light loads. In particular, it is useful as a means of offsetting a part of a voltage so that a low-voltage meter may be used to observe small changes, as at the output of a voltage regulator. The 11-V reference may be used to observe the terminal voltage of an automobile battery with a meter on a 3-V range, and the condition of the battery or of the generator may be easily observed without inserting an ammeter. Surprisingly often a meter with a suppressed scale makes a measurement possible that cannot be accurately read on an available meter scale. In a similar manner the high-voltage standard may be used to increase the sensitivity of an ohmmeter for measuring high resistances. For example, an ohmmeter calibrated to read ohms with a 3-V internal battery may be connected in series with the reference 100-V supply and increase the ohms range by a factor of 103/3, or approximately 30. However, to avoid damaging the meter, large capacitors should be charged with the high-voltage supply before connecting the ohmmeter.

In the metropolitan areas of the United States the ac line voltage is maintained during times of light loads to within 1 percent of 120 V ac. Thus, the ac line voltage may be used to check the calibration of one or two scales of an ac voltmeter. Connecting a 400-V silicon diode in series with a dc meter and line voltage gives a dc check voltage for one or two scales of the dc voltmeter. With a half-wave diode rectifier and no filter the meter should read 54 V dc. With a full-wave rectifier the meter should read 108 V. However, the dc check cannot be made with an FET meter that has an input filter to remove 60 Hz.

Annotated Bibliography

References are selected for their value in supplementing the material of this book and for their availability. Detailed information concerning many circuits may be found in application notes that are distributed by semiconductor manufacturers. Detailed descriptions of electronic devices and circuit theory may be obtained from almost any electronics text the reader may have in his library.

1. Cowles, L. G., *Transistor Circuits and Applications*, 2nd ed. Englewood Cliffs, New Jersey: Prentice-Hall, Inc., 1974. A pratically-oriented analysis of junction and field-effect transistor circuits that uses simplified circuit calculations. Presents simplified explanations of biasing, feedback, and amplifier stability control. Describes amplifier gain calculations with many worked examples and includes laboratory experiments, electronic formulas, charts, and practical problems.
2. Cowles, L. G., *Transistor Circuit Design*. Englewood Cliffs, New Jersey: Prentice-Hall, Inc., 1972. A design manual that emphasizes feedback as a design tool with simplified methods for predicting amplifier stability. Covers circuits from dc to microwaves, audio and RF power amplifiers, and integrated circuits. Includes new material on transistor and FET circuit design, UHF and video amplifiers, and microwave devices. Describes microwave diodes as oscillators, mixers, detectors, varactors, and frequency multipliers. The Appendix includes design data, charts, and tables.
3. *The Radio Amateur's Handbook*, 52nd ed. Newington, Connecticut: The American Radio Relay League, 1975. A well-known handbook that contains a wealth of practical radio circuits and how-to-build information. Material of general interest includes electrical laws and formulas, vacuum tubes, semiconductors, radio communication circuits, measurements, and miscellaneous data. The Handbook is readily available and inexpensive.
4. *RCA Transistor, Thyristor and Diode Manual*. Somerville, New Jersey: Radio Corporation of America, 1971. A collection of data sheets describing the most

popular RCA semiconductor devices. Contains an outline description of transistor circuits, practical applications, and recommended circuits.

5. *RCA Power Transistors and Power Hybrid Circuits.* Somerville, New Jersey: Radio Corporation of America, 1972. Data sheets describing RCA power transistors and selection guides. Includes a few amplifier circuits and application notes on inverters, regulated power supplies, and hybrid linear power amplifiers. See also RCA Audio Power Amplifiers, APA-550, 1973 (40¢).

6. *RCA Linear Integrated Circuits, Application Notes Only.* Somerville, New Jersey: Radio Corporation of America, 1975. A collection of papers describing applications of integrated circuits for linear, IF, video, VHF, AM, FM, and TV applications. Includes MOS-FET applications as choppers, and RF, VHF, and UHF amplifiers.

7. *Transistor Manual*, 7th ed. Electronics Park, Syracuse, New York: General Electric Co., 1964. The GE Transistor Manual has many practical applications of transistors, silicon-controlled switches, unijunction transistors, and tunnel diodes. The circuits include audio amplifiers, radio receivers, digital computer elements, servos, and experimenter projects. The volume includes condensed data describing GE semiconductors recommended in 1964.

8. *The Semiconductor Data Library, Reference Volume*, 4th ed. Phoenix, Arizona: Motorola Semiconductor Products, Inc., 1973. A useful cross reference of registered and Motorola nonregistered semiconductors. The volume includes semiconductor selection guides, outline dimensions, and abstracts of Motorola Application Notes. (Some references to this volume refer to applicable Application Notes.)

9. TERMAN, F. E., *Electronic and Radio Engineering*, 4th ed. New York, New York: McGraw-Hill Book Company, 1955. An excellent radio engineering text, primarily concerned with vacuum-tube circuits. The book is a valuable reference source of information about components, circuits, amplifiers, modulators, noise, feedback, and electronic devices.

Appendix

TRANSISTORS

JEDEC No.	Type	β Min. Max.	V_{CEO} V	Power W	f_T MHz	Use	Source
2N1073A	Ge-pnp	20–60	−80	85	0.5	5 A AF power	M
2N1163	Ge-pnp	15–65	−35	90		25 A switch	M
2N1177	Ge-pnp	100–	−30	0.08		IF and RF amp.	R
2N1304	Ge-npn	40–200	25	0.15		Amp. or switch	R, M
2N1305	Ge-pnp	40–200	−30	0.15		Amp. or switch	R, M
2N1540	Ge-pnp	50–100	−45	90		5 A power	M
2N1711	Si-npn	100–300	50	0.8	70	Misc. AF or RF	F
2N1893	Si-npn	40–120	80	0.8	50	High volt. video	F, M
2N2219	Si-npn	100–300	30	0.8	250	UHF amp.	F, M
2N2484	Si-npn	100–500	60	0.36	15	Low noise, 10 μA	F, M
2N2869	Ge-pnp	50–165	−50	30	0.2	3 A AF power	R
2N2905	Si-pnp	100–300	−40	0.6	200	VHF amp. or osc.	F, M
2N2944	Si-pnp	80	−10	0.4	1	Inverted chopper	M
2N3053	Si-npn	50–250	40	0.1	100	Video amp.	F, M
2N3055	Si-npn	20–70	60	117	0.8	5 A med. volt.	F, M
2N3108	Si-npn	40–120	60	0.8	60	60 V, 150 mA	F
2N3440	Si-npn	40–160	250	10	15	Line volt., 20 mA	F
2N3568	Si-npn	40–120	60	0.3	60	Misc., 60 V AF	F
2N3569	Si-npn	100–300	40	0.3	60	Misc., 150 mA	F
2N3638	Si-pnp	30–	−25	0.3	100	Low β, 50 mA	F
2N3638A	Si-pnp	100–	−25	0.3	150	High β misc.	F
2N3644	Si-pnp	115–300	−45	0.3	200	High β RF, VHF	F
2N3694	Si-npn	50–150	40	0.3	200	10 mA RF, VHF	F
2N3716	Si-npn	50–150	80	150	2.5	High volt., 3 A	F, M
2N3739	Si-npn	40–200	300	20	15	Line volt., .25 A	M
2N3784	Ge-pnp	20–200	−30	0.15	800	UHF low noise	M
2N4036	Si-pnp	20–200	−65	5	60	Med. volt., 150 mA	F
2N4235	Si-pnp	30–150	−60	6	3	0.5 A med. power	F, M
2N4238	Si-npn	30–150	60	6	3	0.5 A med. power	F, M
2N4250	Si-pnp	250–	−40	0.2	40	Low noise	F
2N4258	Si-pnp	30–120	−12	0.2	700	UHF amp. or osc.	F

TRANSISTORS (cont.)

JEDEC No.	Type	β Min. Max.	V_{CEO} V	Power W	f_T MHz	Use	Source
2N4355	Si-pnp	100–400	−60	0.35	100	Misc. high β	F
2N4398	Si-pnp	15–60	40	200	4	15 A power	F, M
2N5191	Si-npn	25–100	60	40	2	4 A plastic pwr.	M
2N5194	Si-pnp	20–80	−60	40	2	4 A plastic pwr.	M
2N5876	Si-pnp	20–100	−80	150	4	6 A high volt.	M
2N5878	Si-npn	20–100	80	150	4	6 A high volt.	M
2N6122	Si-npn	25–100	60	40	2.5	2 A AF power	F
2N6125	Si-pnp	25–100	−60	40	2.5	2 A AF power	F
MJE-105	Si-pnp	25–100	−50	65		2 A plastic AF	M
MJE-205	Si-npn	25–100	50	65		2 A plastic AF	M
MJE2801	Si-npn	25–100	60	90		3 A plastic AF	M
MJE2901	Si-pnp	25–100	−60	90		3 A plastic AF	M

Note: F—Fairchild, M—Motorola, R—RCA

JUNCTION FIELD-EFFECT TRANSISTORS

JEDEC No.	Type and Basing	I_{DSS}, mA Min. Max.	g_m μmho	$V_{(BR)}$ volts	Use	Source*
2N3086	n-DSG	0.8– 3.0	400–1200	40	Low-gain amp.	C
2N3089	n-DSG	0.5– 2.0	300–2000	15	Low-noise amp.	C, N
2N3460	n-SDG	0.2– 1.0	800–2000	50	Amp., 5:1 I_{DSS}	S, N
2N3687	n-SDGC	0.1– 0.5	300–1200	50	Amp. 5:1 I_{DSS}	S, N
2N4220	n-DSGC	0.5– 3.0	1000–4000	30	VHF amp./Switch	S, N, M
2N4339	n-SDG	0.5– 1.5	800–2400	50	Low-noise, 3:1 I_{DSS}	S, N
2N4360	p-SDG	3.0–30.0	2000–8000	20	Amp., Gen. Pur.	F, N, M
2N4868	n-SDGC	1.0– 3.0	1000–3000	40	Low 1/f noise	S, N
2N5265	p-SGDC	0.5– 1.0	900–2700	60	Amp., 2:1 I_{DSS}	M
2N5358	n-SDGC	0.5– 1.0	1000–3000	40	Amp., 2:1 I_{DSS}	S, N, M
2N5360	n-DSGC	1.5– 3.0	1400–4000	40	Amp., 2:1 I_{DSS}	S, N, M
2N5484	n-DSG	1.0– 5.0	3000–6000	25	VHF-UHF amp.	F, N, M
AD-832	n-SDG	0.4– 1.0	70–500	40	Dual FET, $I_G < 0.1$ pA	Analog
U-310	n-SDG	20.0–60.0	10000–20000	25	VHF-UHF amp.	S

*Note: C—Crystalonics, F—Fairchild, N—National, M—Motorola, S—Siliconix

INSULATED-GATE FIELD-EFFECT TRANSISTORS

JEDEC No.	Type* and Basing	I_D or I_{DSS} mA	g_m μmho	$V_{(BR)}$	Use	Source**
2N4351	n-E, SDGC	3–	1000–	25	600 switch	M
2N4352	p-E, SDGC	3–	1000–	25	600 switch	M
3N128	n-D, DSGSb	5–25	5000–12,000	20	VHF, 1 gate	R
3N138	n-D, DSGSb	15	6000	35	RF, 1 gate	R
3N187	n-D, DGGS	5–30	7000–18,000	20	VHF, 2 gates	R

*Type: E—Enhancement, D—Depletion.   **Mfgr.: M—Motorola, R—RCA

MISCELLANEOUS SEMICONDUCTOR DEVICES

No.	Type	Use	Source
2N1671B	Unijunction	Oscillator, 25 μA	G.E., Mot.
2N6027	Prog. Unijct.	Oscillator, 2–5 μA	G.E., Mot.
UA-740	IC OP amp.	FET input, 10^6 MΩ	Fair., Nat.
UA-741	IC OP amp.	General purpose	Fair., Nat.
LM-372	IC AGC amp.	AM, 0.05–2 MHz	Nat.
UGH7805	IC Regulator	+5 V, 0–1 A	Fair.

ZENER DIODES—400 mW and 1 W

V_Z	JEDEC Nos.		V_Z	JEDEC Nos.	
2.4	1N4370,	1N5221	11.	1N962,	1N4741
3.0	1N4372,	1N5225	13.	1N964,	1N4743
3.6	1N747,	1N4729	16.	1N966,	1N4747
4.3	1N749,	1N4731	20.	1N968,	1N4747
5.1	1N751,	1N4733	24.	1N970,	1N4749
6.2	1N753,	1N4735	30.	1N972,	1N4751
7.5	1N755,	1N4737	36.	1N974,	1N4753
9.1	1N757,	1N4739	43.	1N976,	1N4755

Standard Tolerance $\pm 5\%$, $\pm 10\%$, $\pm 20\%$

SIGNAL AND RECTIFIER DIODES

JEDEC	Type	Use	Source
1N60	Ge-30 V	UHF signal	Sylvania
1N64	Ge-20 V	Gen. Purpose	Sylvania
1N82	Si-25 V	UHF, 50 mA	Int. Rect.
1N91	Ge-100 V	Rect., 1 A	Motorola
1N459	Si-175 V	Signal, 3–50 mA	Sylvania
1N4002	Si-200 V	Rect., 1 A	Motorola
1N4004	Si-400 V	Rect., 1 A	Motorola
1N4006	Si-800 V	Rect., 1 A	Motorola
1N4114	Si-19 V	Low-ohms Zener	Motorola
1N4720	Si-100 V	Rect., 3 A	Motorola
1N5245	Si-15 V	Zener, 0.5 W	Motorola
FH1100	Schottky	UHF signal	Hew. Pac.
RO-700	Schottky	UHF signal	Motorola
MBD101	Schottky	UHF signal	Motorola
MVS460	Tuning	FM/TV tuners	Motorola

SEMICONDUCTOR INTERCHANGEABILITY

Most low-power transistor circuits operate satisfactorily when similar devices with equivalent current gains are substituted, provided the bias is adjusted. Circuits operating at high frequencies or switching at high speeds require transistors that meet the frequency and speed requirements indicated in the data sheets by the cutoff frequency f_T or the switching times. The demands on power transistors are more exacting, and many devices are inferior even though more expensive. The substitution of power transistors is not recommended because the recommended types have been carefully selected. For

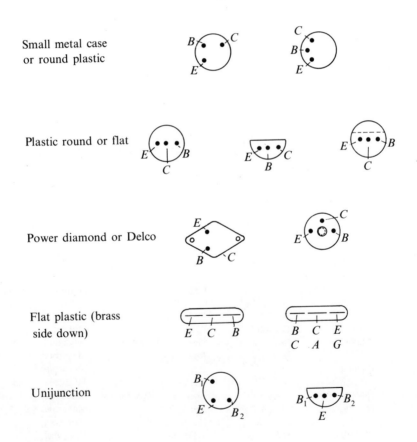

Transistor Basing Diagrams

FETs: No standard. Find gate with ohmmeter

Diodes: The cathode, or plus, is the marked end

equipment repairs, a better device should be substituted, if possible, and the bias should always be adjusted.

Field-effect transistors are improving rapidly, and substitutions can be made providing the g_m and the I_{DSS} ratings are similar. New types are priced reasonably, while the prices of the early types remain high. Amplifier applications require high g_m at low I_{DSS} values and the most inexpensive devices generally prove unsatisfactory.

Silicon diode types are not specified in the circuits because almost any 400 V, or 200 V, 1 A diode should be satisfactory. For experimenter purposes the voltage rating of diodes should be 3 to 4 times the rms voltage input to the diode. The current rating should be 3 to 5 times the diode dc current. Excellent high-current high-voltage diodes may be obtained at reasonable prices. Diodes produced in small quantities may be surprisingly expensive.

COMPOSITION RESISTOR VALUES

Standard EIA resistance values from 1.0 to 10 ohms are listed in the table below. Resistors are available in decade multiples of the listed values up to 22 MΩ. All values are available in 5 per cent tolerance, and those in bold type are available in 10 per cent tolerance.

STANDARD RESISTANCE MULTIPLES

1.0	1.6	**2.7**	4.3	**6.8**
1.1	**1.8**	3.0	**4.7**	7.5
1.2	2.0	3.3	5.1	**8.2**
1.3	**2.2**	3.6	**5.6**	9.1
1.5	2.4	3.9	6.2	**10.**

RMA RESISTOR COLOR CODE

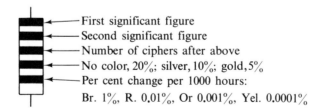

— First significant figure
— Second significant figure
— Number of ciphers after above
— No color, 20%; silver, 10%; gold, 5%
— Per cent change per 1000 hours:
Br. 1%, R. 0.01%, Or 0.001%, Yel. 0.0001%

Black	0	Green	5
Brown	1	Blue	6
Red	2	Purple	7
Orange	3	Gray	8
Yellow	4	White	9

The first four colors are used most frequently. The cipher color is useful for evaluating the order of magnitude of a resistor. A gold band in the cipher color means × 1/10 and silver means × 1/100.

COLOR CODE EXAMPLES

Colors	Ohms	Tolerance
Brown, black, red	1000	20%
Red, red, red	2200	20%
Red, red, black, silver	22	10%
Red, red, gold	2.2	20%
Green, brown, green, gold	5.1 Meg	5%

DECIBEL FORMULAS AND TABLE

By definition the decibel is a logarithmic measure of power ratios:

$$dB \equiv 10 \log_{10} \frac{P_2}{P_1}$$

Ratios greater than 1 represent a power gain, +(plus) dB. For ratios less than 1, representing a loss, the dB value is found using the reciprocal of the ratio and the dB value is negative, −(minus) dB.

For convenience, the dB is used as a measure of voltage ratios, where:

$$dB = 20 \log_{10} \frac{E_2}{E_1}$$

However, voltage ratios expressed in dB units do not represent power ratios unless the voltages are measured across equal resistances.

Also, 0 dBm is used as a 1-mW power reference level.

A PRACTICAL DECIBEL TABLE

dB	Voltage Ratio	Power Ratio	dB	Voltage Ratio	Power Ratio
0	1.00	1.00	14	5.0	25
1	1.12	1.26	16	6.3	40
2	1.26	1.59	17	7.1	50
3	1.41	2.00	20	10.0	100
4	1.58	2.51	25	17.8	316
5	1.78	3.16	30	31.6	10^3
6	2.00	4.00	40	10^2	10^4
8	2.51	6.31	60	10^3	10^6
10	3.16	10.0	80	10^4	10^8
12	4.0	15.9	100	10^5	10^{10}

Appendix

FREQUENCY CLASSIFICATIONS

Abbreviation	Frequency	Wavelength	Classification
VLF	–30 kHz	–10 km	Very-low frequencies
LF	30–300 kHz	10–1 km	Low frequencies
MF	300–3,000 kHz	1,000–100 m	Medium frequencies
HF	3–30 MHz	100–10 m	High frequencies
VHF	30–300 MHz	10–1 m	Very-high frequencies
UHF	300–3,000 MHz	100–10 cm	Ultrahigh frequencies
SHF	3–30 GHz	10–1 cm	Superhigh frequencies
EHF	30–300 GHz	10–1 mm	Extra-high frequencies

Amateur Radio Bands		*Television Bands*	
		Channels	Frequencies
80 m	3500–4000 kHz		
40 m	7000–7300 kHz		
20 m	14000–14350 kHz	2–4	54–72 MHz
15 m	21.00–21.45 MHz	5–6	76–88 MHz
10 m	28.00–29.7 MHz	(FM)	(88–108 MHz)
6 m	50.0–54.0 MHz	7–13	174–216 MHz
2 m	144.–148. MHz	14–83	470–890 MHz

THERMAL RESISTANCES (RESISTANCE GIVEN IN °C/W)

Transistor without Heat Sink		*Insulating Washers*	
Epoxy types	200–500	None	0.2–0.5
TO-18	200 up	Beryllia	0.3–0.5
TO-5	150 up	Aluminum	0.5–1.0
Diamond power	25–50	Mica	0.5–1.0

Heat Sinks		*Flat Sheet (One Side)*	
Clip-on for TO-5	25–50	3 cm × 3 cm	50
Small-fin types	10–40	10 cm × 10 cm	8
Large in still air	1–5	30 cm × 30 cm	2

MAXIMUM RECOMMENDED TEMPERATURES (°C)

Germanium junction	100–125	Thyristors	100–125
Silicon junction	150–200	Unijunction	100–125
JFETs	150–175	Silicon diodes	150–200
MOS-FETs	175–200	Zener diodes	150–175

REACTANCE-FREQUENCY CHART

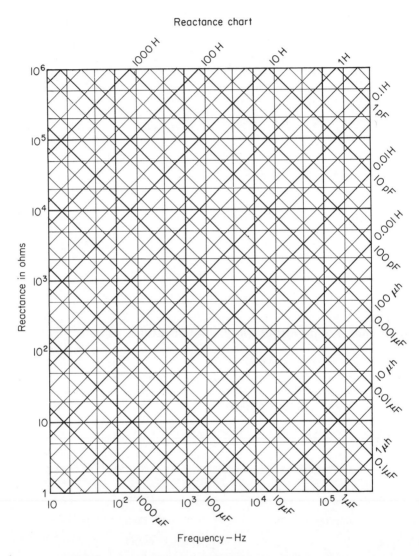

Note: All intermediate lines are at 2 and 5. For frequencies multiplied by 10, 100 or 1000, use μF and H reduced by 10, 100, or 1000.

Useful values:

$1\ \mu\text{F} \approx 1\ \text{H} \approx 10^3\ \Omega$ at   160 Hz
$1\ \text{pF} \approx 1\ \mu\text{H} \approx 10^3\ \Omega$ at   160 MHz
$1\ \text{nF} \approx 1\ \text{mH} \approx 10^3\ \Omega$ at   160 kHz

Index

Index

A

Alpha, 15
Amplifier:
  audio-frequency, 46, 114-43, 197, 280-91, 296
  automatic gain control, 206, 292-301
  bridge, 156-58
  broad band (see Amplifier, wideband)
  cascade, 51, 206, 216, 219
  class A power, 125-35
  class B power, 125-29
  collector feedback, 21, 27, 55, 122
  combined feedback, 29-32, 48
  common base, 11, 16, 68, 210, 219
  common collector, 11, 16, 33-37, 51-55, 212, 219
  common emitter, 17-27, 32, 114-17, 217
    biasing, 28
    bypassed, 31, 44, 46, 203
    feedback, 135-43
  complementary emitter follower, 37

Amplifier (cont.)
  complementary symmetry, 138-43, 280, 288
  compounds (see Pairs)
  Darlington (see Darlington compound)
  dc, 68, 74-76, 81
  differential, 146
  direct-coupled, 39-58, 65-76, 121, 216
  driver, 134-43
  electrometer, 82
  emitter coupled, 49
  emitter feedback, 22-27, 47
  emitter follower (see Amplifier, common collector)
  feedback, 39-58, 158
  field-effect, 59-86, 117, 206
  frequency response, 3-5 (see also RC coupling)
  general purpose, 31, 43-46, 69-74, 280
  hearing aid, 280
  high-fidelity, 288
  high-gain, 43, 46-48, 64, 80-83

Amplifier (cont.)
  instrument, 148, 297
  IC (Integrated circuit), 140, 144-59, 185, 290, 301
  intercom, 280
  intermediate-frequency, 201-5
  iterated, 2
  line, 45
  low-impedance, 68, 116, 200
  low-noise, 81, 112-24
  low-power, 68, 127-33, 280
  medium-frequency, 201-5, 209, 216-18
  microphone, 28
  millivoltmeter, 75
  MOS FET, 59, 77-85, 221
  multi-stage, 38-58
  narrow-band RC, 60-62, 199 (see also Amplifier, tuned)
  nonblocking, 119
  operational, 124, 140, 151, 184
  oscilloscope, 32, 66
  phase-inverting, 56, 305 (see also Inverter)
  phono preamplifier, 121, 282
  power, 125-43, 288-91
  preamplifier, 32, 66-68, 112-24
  push-pull, 134
  radio-frequency, 83, 203-7, 299
  RC-coupled, 39, 60-64, 78
  relay, 19, 36, 53
  remote, 66-68, 75
  self-biased, 249
  single-ended, 137
  single-sided power, 129-34
  single-stage, 16-37, 63-66, 77, 129
  solenoid, 53
  summing, 147, 286
  switching, 158, 304-6
  tape recorder, 123
  temperature compensated, 65, 71
  threshold, 19
  tone-control, 285
  tuned, 83, 172, 195-208, 299

Amplifier (cont.):
  two-stage, 38-58, 65-75, 114
  UHF, 84-86, 205-8
  unity gain, 146
  universal, 31, 43-46
  variable Q, 154, 198-201
  video, 209-224
  wide-band, 41, 72, 79, 84-86, 146, 205
  zero-biased, 63
Analog computer, 151
Attenuators, 232, 293, 326
Automatic gain control, 292-301

B

Battery power, 120
Bessel filters, 181-84
Beta, 8, 15, 330
Beta cutoff, 212
Bias:
  adjustment, 4
  CE amplifier, 23-29
  FET amplifier, 77
  power amplifier, 126, 142
  self, 249
  zero, 63
Bibliography, 343
Breadboards, 328
Bridge:
  rectifier, 90
  steering, 235
  Wien, 263
Burglar alarm, 309
Butterworth filters, 173-78, 185-87

C

Capacitance multiplier, 108
Capacitor:
  coupling, 39, 60-64, 154
  emitter bypass (see References 1 and 2)
  rectifier smoothing, 88

Index

Chopper, 319
Chebyshev filters, 178-81, 188
Clamp, 248
Clipper, 243-48
Compensation, 22-24
Computer, analog, 151
Converters, 267-69, 273-75
Crossover distortion, 126, 136, 290
Crossover network, 291
Crowbar, 106
Current gain (see Alpha, Beta, S-factor)
  definition, 6, 8, 15
  measurement, 330
  single-stage, 6, 10, 23
  transistor, 345
Current limiter, 105, 336

D

Darlington compound, 51-55
DC restorer, 248
Decibel, 350
Demodulation, 237
Detector:
  diode, 239
  square-law, 239
  synchronous, 237
Differential amplifier, 146
Diode, 8
  breakdown, 307
  characteristics, 347
  hot carrier, 239
  radio-frequency, 239, 347
  rectifier, 89
  Schottky, 239
  Shockley, 307
  varactor, 242
  Zener (see Zener diode)
Diode circuits, 225-249 (see also Rectifier)
  attenuator, 232, 294
  chopper, 236
  clamp, 248

Diode circuits (cont.):
  clipping, 243
  crowbar, 228
  cutout, 226
  detector, 239
  gate, 249
  limiter, 246
  meter protector, 230
  mixer, 241
  modulator, 236-40
  noise suppressor, 229, 246
  polarity protection, 227
  power reducer, 227
  steering, 235
  switch, 226-36
  transistor protector, 231
  tuning, 242
  voltmeter, 235, 240
Distortion, 126, 135, 138, 290
Dual gate FET, 84

E

Electrometer, 82
Emitter capacitor (see References 1 and 2)
Emitter feedback, 22-27
Emitter follower (see Amplifier, common collector)

F

Feedback (see also References 1 and 2)
  collector, 21
  combined, 29, 48
  emitter, 22
  instability, 278
  Miller, 6, 164, 216
  neutralization, 201
  series, 6, 39-43
  shunt, 6, 43-48
  significant, 5, 21

FET, 8, 59-86 (*see also* Amplifier, field effect)
  chopper, 319
  figure of merit, 59
  insulated gate, 77, 83-87, 346
  junction-type, 59, 346
  resistor, 252, 293, 298
  switch, 318
  transconductance, 346
Figure of merit (*see* FET)
Filter, 160-94
  active, 164-94
  band-pass, 189
  band-rejection, 190
  bridge-T, 192
  high-pass, 187
  interstage, 160-65
  parallel-T, 190-94
  RC, 60-64, 160-64
  smoothing, 88
  types, 168
Flipflop, 313
Frequency divider, 316
Frequency response, 3-5, 60-64, 162

G

Gain:
  control, 293
  definitions, 6
Gain-impedance relation, 9-12
Gate, 249, 318

H

Heat sink, 13

I

Impedance (*see also* Miller effect)
  CB input, 68
  CC input, 11, 34
  CE input, 6, 17
Insulated-gate FET (*see* MOS FET)
Integrated circuits, 144-59

Integrator, 149-53
Interstage coupling, 60, 162
Inverter, 271-76, 312
Iterated circuits, 2

L

Laboratory:
  experience, 320
  instruments, 321-42
  meters, 337-40
  troubleshooting, 321

M

Meter:
  current gain, 333
  electrometer, 82
  linear ac, 340
  millivoltmeter, 75
  voltmeter, 235, 240, 337-40
  volt-ohmmeter, 325
Miller effect, 7, 164, 216, 220
Mixer, 241, 286-
Modulators:
  amplitude, 238
  diode, 236
  FET, 237
MOS FET, 77-85, 319 (*see also* Amplifier, MOS FET)
Multivibrator, 265, 315

N

Negative resistance, 201
Neutralization, 201
Noise, 112-18
  suppressor, 229, 245-48

O

Ohmmeter, 325, 337
Ohm's law, 12

Index

Oscillator, 250-79
 blocking, 268
 Clapp, 254
 Colpitts, 253
 crystal, 257
 flyback, 267
 Franklin, 250
 Miller, 258
 phase-shift, 259-62
 Pierce, 257
 pulse, 270
 radio-frequency, 252-58
 relaxation, 265
 reasonant circuit, 252
 ringing, 267
 tuned, 255-57
 twin-T, 262
 voltage controlled UHF, 276
 Wien bridge, 263

P

Pairs:
 CE-CE, 50
 collector feedback, 54
 Darlington, 51-55
 feedback, 57
 FET-transistor, 65-76
 power, 134
Parallel-T, 190-98
Parasitics, 278
Peaking (*see* Compensation)
Phase inverter, 56, 304
Pinchoff voltage, 59, 64
Power amplifier, 125-43
Power dissipation, 13, 27, 143
Power supplies, 87-111, 324, 334-37
 regulated, 92-109, 336
Pulse:
 response, 223
 shaper, 316
Punch through, 330 (*see also* Second breakdown)

Q

Q-factor, 195
Q-point, 4

R

RC coupling, 60-64, 162-64
Reactance chart, 352
Rectifier, 87-91
 voltage doubler, 91
References, 343
Regulator, 92-109
 transistor, 99-107
 Zener diode, 94-98
Resistor, 349
Ripple, 88

S

Schmitt trigger, 316
Second breakdown, 138, 231
S-factor, 6
Silicon controlled rectifier, 307
Silicon controlled switch, 306
Squaring, 317
Switch:
 latching, 306
 light operated, 308, 311
 transistor, 302
Switching, 302-20
 amplifier, 304
 inverter-inverter, 312

T

Tape preamplifier, 123
Thermal resistance, 351
Tone control, 282-86
Transistor, 8
 breakdown, 138, 231
 breakdown pair, 307
 characteristics, 345

Transistor (*cont.*):
　chopper, 318
　gain control, 295
　gate, 318
　interchangeability, 348
　switching circuits, 302-20
　temperatures, 351
　testing, 330
Transistor gain-impedance relation (TG-IR), 9-12
Trigger, 313, 316
Tuned amplifiers, 195-208
Tuned circuit, 195
Tuning interaction, 202
Twin-T, 190-98

U

Ultra-high frequency (UHF)
　amplifiers, 84, 205-8
　mixer, 241
　oscillator, 276
　tuner, 241
　voltmeter, 240, 338

V

Varactor, 242
Video amplifiers, 209-24
Voltage controlled oscillator, 276
Voltage controlled resistor, 232, 252, 293, 298
Voltage gain, 5, 10, 15, 16
Voltage multiplier, 91
Voltage regulator, 92-109, 336
Voltage standards, 340
Voltmeter, 235, 240, 337-40
Volt-ohmmeter, 325

W

Wave shaping, 243-46, 316 (*see also* Filter)
Wien bridge, 263

Z

Zener diode, 94-98, 246, 347